TIN IN ANTIQUITY

For My Parents

TIN IN ANTIQUITY

its mining and trade throughout
the ancient world with particular
reference to Cornwall

R. D. Penhallurick

THE INSTITUTE OF METALS
LONDON

Book 325
published by
The Institute of Metals
1 Carlton House Terrace
London SW1Y 5DB
© 1986 R. D. Penhallurick
ISBN 0 904357 81 3

British Library Cataloguing in Publication Data
Penhallurick, R. D.
 Tin in antiquity: its mining and trade throughout the
 ancient world with particular reference to Cornwall
 1. Tin industry—History
 I. Title
 338.4′76696′093   HD9539. T5

 ISBN 0–904357–81–3

Library of Congress Cataloging in Publication Data

Penhallurick, R. D. (Roger David), 1940–
 Tin in Antiquity

 Bibliography: p. Includes index
 1. Tin Industry—England—Cornwall—History.   2.
 Tin Industry—History.
I. Title. HD9539. T6G865   1986.   338.2′7453′094237   86–7302

Typeset by
Fakenham Photosetting Ltd
Fakenham, Norfolk

Printed and made in England by
Adlard & Son, Dorking, Surrey

Contents

List of Illustrations

List of Maps

Foreword

The Institute of Metals is publishing a series of books on the history of metallurgy, so it was with great interest that I, as historical advisor to the Book Committee, heard of this book by Roger Penhallurick.

There are two areas of the world noted for their tin deposits: Malaysia and Cornwall. Since the early history of the first has not yet been studied it was clear that the first history of tin would have to be about Cornwall, and it was appropriate that it should be written by a Cornishman.

As we see from his previous works, the author has wide interests. As Assistant Curator of the Royal Institution of Cornwall, County Museum at Truro, he has had excellent opportunities for meeting those involved in the Cornish tin industry and for discussing wider aspects with mining engineers, many of them of Cornish origin, returning from foreign parts to their native land. So while the emphasis of this book is on tin in Cornwall it discusses the history of tin in the world and thus becomes a more important work. Such a treatment not only allows us to learn more of the world scene but to fill some of the gaps left by lack of evidence in the Cornish scene itself.

This book is much needed. It is the first attempt at a comprehensive history of the early metallurgy of tin and follows that small but delightful book by Ernest Hedges on 'Tin in Economic and Social History.'

October 1985
R. F. Tylecote

Preface

The principal aim of this book is to describe the tin occurrences of Africa, Asia, and Europe: those deposits well known in the mining world and those apparently 'conjured up' by some archaeologists as an explanation of where Near Eastern civilizations in particular obtained their precious supplies. The book divides itself quite conveniently into three parts in which the emphasis of discussion differs.

Part 1 examines the minor tin deposits of Egypt and the much larger ore-bodies found south of the Sahara. It deduces from the archaeological evidence whether any of these sources were available to the Mediterranean and Near Eastern civilizations in antiquity.

Part 2 examines the claims put forward for tin occurrences in various Near Eastern countries — usually, it would appear, by people with little or no understanding of tin and the environment in which it occurs. These are compared with what the geological and mineralogical literature describes as proven localities for tin to decide (really by little more than intelligent guesswork) where the early civilizations obtained their tin before the discovery of cassiterite in Europe. The early working of tin deposits in eastern Asia is also examined.

In Europe the position is different. There is no doubt about where tin occurs. Only Ireland appears as a country some have claimed was an early tin producer, even though economic geologists would today give it little serious thought as a potential tin-field. To 'prove' mining in prehistoric Europe means relying largely on such archaeological evidence as the distribution of artifacts made of tin bronze, augmented by such items as pottery repaired or decorated with tin foil, and the recovery of datable tin ingots from shipwrecks. Most of the European tin deposits lack any inherent proof of early exploitation. There is no central European tin equivalent of the early copper mines found in the Balkans at Rudna Glava, Aibunar, and elsewhere. Only in France is there unequivocal evidence from the tin deposits themselves that early man worked the alluvial tin and tin-rich lodes by open-cast methods.

The final and longest part of the book deals with Cornwall, where the emphasis is again different. Here it is not a question of debating whether or not tin was exploited from the early bronze age onwards, from c. 2000 BC, or even whether it lapsed into oblivion at some subsequent period between then and the historical period of Cornish mining in the 12th century AD. The evidence is overwhelming that tin has been won in Cornwall continuously over the past 4 000 years. There is more evidence from Cornwall than from the rest of the world's tin-fields put together. Such is the richness of Cornish finds from alluvial deposits that it is possible to examine in turn all the tin deposits of the country, catalogue the finds from each tin-stream while carefully noting the find circumstances and the physical nature of the deposits, to prove beyond reasonable doubt that the finds are the relics of early tinners. This, too, can be backed up by finds from several archaeological sites. These demonstrate that the inhabitants not only knew how to smelt tin but were familiar with the local stream-tin, some of which appears on sites of different periods.

Gold is frequently mentioned in the book; this is far from irrelevant. In alluvial deposits tin and gold are common associations and the only minerals normally occurring together that were used in antiquity. It may well be that cassiterite was first discovered by prospectors after gold. However, this cannot be proved until such time (if it ever comes) as it is known precisely where tin was first discovered and the geology of the deposits. The occurrence of gold in Cornwall is stressed in order to emphasize its widespread occurrence, albeit in small quantities. This fact has been continually overlooked by many archaeologists who have credited Ireland with an importance as a gold producer that it does not appear to merit on close examination.

The choice of the word 'Antiquity' rather than 'Prehistory' in the title of the book is deliberate, for while the earliest evidence is important and given weight, the use of tin in literate societies is examined. In Cornwall the period of interest extends up to the 12th century when the first records of tin production appear. The book's final chapter on 'Tin Production in Dark Age Cornwall' thus forms a background for those interested in the mediaeval and later aspects of Cornish mining history. This subject has been, and continues to be, painstakingly documented by numerous people since the time of Richard Carew with whose delightful and apt prose the introduction is headed.

Roger D. Penhallurick
Truro, 1985

Acknowledgments

Writing this book has been a spare-time occupation since 1978, the fruition of a student ambition of 15 years earlier to achieve for Cornish antiquity what Oliver Davies had accomplished for 'Roman Mines in Europe'. Being employed by the Royal Institution of Cornwall at the County Museum in Truro considerably lightened that part of the task, with so much information readily available. The real burden came in gathering data from ever widening geographical frontiers. If the end-product is deemed satisfactory it will be due, in no small degree, to the large number of people willing to offer their help and advice.

Firstly I must thank my colleague H. L. Douch, Curator of Truro Museum, whose knowledge of Cornish history is encyclopaedic, and secondly the staff of Truro City Library who rarely failed to obtain for me copies of all the publications I asked for. Dr A. V. Bromley and Dr Leslie Atkinson of Camborne School of Mines were always ready to answer my enquiries. I am particularly grateful to Ivor Moyle, Hon. Curator, and C. V. Smale, President of the Royal Geological Society of Cornwall, and to Colin Edwards of the County Record Office, for much valuable help. Prof. K. F. G. Hosking, an acknowledged world-expert on the geology and mineralogy of tin, carefully read the whole of the MS and proposed necessary amendments. Dr Noel Barnard similarly vetted the section dealing with China, and Prof. I. R. Selimkhanov read the section on the USSR. Peter Embrey of the Dept. of Mineralogy at the British Museum (Natural History) not only provided useful information but also arranged for the analysis of Cornish gold by Dr C. T. Williams. Prof. R. F. Tylecote gave me much encouragement throughout and was instrumental in arranging publication by The Institute of Metals. Prof. George Rapp read the sections dealing with the Near East and kindly obtained ^{14}C dates of wood from Cornish tin streams. Thanks also to Lyn Greenwood for producing the index. Special mention must be made of A. La Spada at the International Tin Council, London, for much detailed information on lesser-known tin deposits. I must also thank A. R. Mowbray & Co. Ltd for permission to reproduce the story of St John the Almsgiver from *Three Byzantine Saints* translated by E. Dawes and N. H. Baynes.

The other friends and colleagues who helped in various ways are, with apologies for any omissions: Barbara Adams, Dr D. P. Agrawal, Carol Andrews, Nancy Aronson, Donald Bailey, Justine Bayley, Neil Beagrie, Stephen Bird, George Boon, Prof. V. V. Breskovska, Colin Bristow, Dr Jill Burke, Charles Burney, Simon Camm, Clare Conybeare, Paul Craddock, Prof. U. Dann, Tanya Dickenson, Dr Vit Dohnal, Paul-Marie Duval, Alan R. Eager, George Engel, P. J. Felder, Colin French, R. I. Gait, Sandy Gerrard, Prof. P. R. Giot, K. R. Greenleaves, Dr R. J. Harrison, Dr Axel Hartmann, F. Herbert, Prof. Charles Higham, Catherine Johns, Jennifer Kennedy, G. de Leeuw, Ted Long, Dr I. H. Longworth, Dr Ellen F. Macnamara, Dr Jiří Majer, P. R. S. Moorey, Elisabeth Munksgaard, Stuart Needham, J. P. Northover, Nessa O'Connor, Oliver Padel, Dr Susan Pearce, the late J. Penderill-Church, Henrietta Quinnell, Dr C. A. Ralegh Radford, Dr T. A. Reilly, Beth Richardson, Edgar Samuel, Dr Fulvio Lo Schiavo, Dr Colin C. Shell, Andrew Sherratt, Charles Smith, Miss J. Stancliffe, Dr C. J. Stanley, John W. Taylor, John Thackray, Nicholas Thomas, Prof. A. C. Thomas, F. H. Thompson, J. H. Trounson, John Watton, Dr G. Weisgerber, Robert Wilkins, Dr René Wyss.

I must also thank the owners of several artifacts from Cornish tin streams who wish to remain anonymous.

I especially thank the administrators of 'the Sir Arthur Quiller-Couch fund', Truro, for a grant towards the cost of research.

Finally, at a more homely level, I must thank my parents with whom I began my labour of love and my wife with whom I completed it, for their forbearance while I was endlessly cloistered in my study. My wife also read the whole of the draft copy and pointed out unclear passages and other shortcomings. With such worthy helpers errors of fact can be laid at the author's feet and nowhere else.

Note on dating

In this book many dates appear in conventional calendar form: 1000 BC or AD 1000. An explanation is needed for radiocarbon (^{14}C) dates. These are issued by the laboratory in years bp (before present). 'Present' is taken as being AD 1950. Thus the University of California, Riverside, date of worked oak from the Pentewan tin streamworks appears as UCR-1828 (the sample number), 4140 ± 100 bp. Deducting 1950 years gives 2190 ± 100 bc. 100 is the standard deviation which means that there is a 68% probability of the date falling between 100 years either side of the central date, and a 95% probability of it falling between 200 years (two standard deviations) of the central date: 2390–1990 bc. ^{14}C dates are not as precise as popularly supposed and they require calibration against calendar dates obtained from tree-ring studies principally on the long-lived bristle-cone pine. The most recent published calibration is by J. Klein *et al.* (*Radiocarbon*, 1982, **24**, (2), 13–150). Using their tables with calibrated ranges of 95% confidence, 4140 ± 100 bp appears as 3015–2415 BC. Many published dates will appear as ± 150 or more, while many recently published dates will have standard deviations of ± 50 years or less.

In compiling this book it has not always been possible to know whether published dates in calendar years have been calibrated or not, so that many of the dates must be regarded as tentative.

Introduction

In travelling abroad, in tarrying at home, in eating and drinking, in doing aught of pleasure, or necessity, tin, either in its own shape, or transformed into other fashions, is always requisite, always ready for our service.

Richard Carew, 1602, *The Survey of Cornwall*

TIN: THE GEOLOGICAL BACKGROUND

Tin is an element of such importance that metal-using civilizations could not have developed in the way they have without it. Ernest S. Hedges, former Director of the International Tin Research Council, examined in an admirable way the scientific and artistic functions fulfilled by tin throughout history, with chapters on modern uses that will be a revelation to some readers.[1]

Of all the metals used in antiquity, tin is one of the rarest. Its average crustal concentration is only 2 parts per million (ppm) compared with 55 for copper, 13 for lead, and a massive 50 000 for iron. Among the scarcer metals silver averages at 0·07 ppm and gold at 0·004 ppm.

Tin occurs widely in over 50 minerals.[2] Most are extremely rare, of recent discovery, incapable of exploitation, and frequently occurring in regions totally devoid of any recoverable tin minerals. It is possible that the rare calcium–tin–silicate malayaite contributed to some tin recovered chemically in modern workings at Pinyok in Thailand, while a few tin-bearing sulphide ores have been mined to a limited extent in Bolivia and north-east Siberia.[3] Best known is the tin–copper–iron sulphide stannite, popularly called 'bell-metal ore', first discovered at Wheal Rock, St Agnes, in the late 18th century, but hardly more than a curiosity in Cornwall. There is some thought that in antiquity stannite might have been smelted to produce a natural tin–copper alloy, but this has not been satisfactorily demonstrated. Tin-sulphides are often discovered in mines already being worked for the tin-oxide cassiterite (as happened at Wheal Rock), and the sulphide may not itself be recoverable. However, stannite may be found without associated cassiterite as in Langkawi Island, Malaysia (K. F. G. Hosking pers. comm.).

Only one mineral, cassiterite (SnO_2), accounts for virtually all the tin that has ever been recovered. Its great advantage is that it is a stable oxide that remains unaltered when weathered out of lodes to form concentrations of alluvial or placer deposits commonly, if sometimes misleadingly, known as 'stream tin'. Within the lodes cassiterite can be altered by ascending solutions to form stannite, as must have happened at Wheal Rock. In Cornwall, alluvial pebbles were often the size of 'split peas' or eggs, but could be much smaller or weigh several pounds. In colour, stream tin varies from pure white – rare in Cornwall but not uncommon in Malaysia – through various shades of red and brown to jet black. The intensity of colour depends upon the amount of included iron.

A particularly beautiful banded variety, for long known only from Cornwall, is 'wood tin', appropriately so-called from the concentric rings of varying shades (*see* Fig. 1). Its origin is problematic. It has been assumed that wood tin is the oxide formed through the decomposition of stannite, which may be partly true. However, in Cornwall primary wood tin has been found in several lodes, notably in a section of Wheal Vor at a depth of 365 m. It 'appeared here as a cementing material binding together fragments of killas [clay-slates], and some of it was sprinkled over with brilliant crystals of tin oxide...'[4]

Tin is a problem metal. Not only is the archaeologist uncertain about where prehistoric man obtained his ores before the discovery of European deposits, but the mineralogist is equally puzzled by the metal's genesis. The ultimate origin of metallic elements is the Earth's mantle. The basic igneous rocks derived from it at mid-ocean ridges, or extruded through volcanoes on the continents, contain copper, lead, zinc, manganese, and other metals in plenty. But not so tin. Tin has been detected among the fumarolic products of Mount Etna, at the mid-Atlantic and Indian Ocean ridges, and in the basic rocks of, for example, the Merensky horizon of the Bushveld Complex, South Africa. However, such environments are hostile to the formation of cassiterite, which needs an 'acid' or granitic hothouse in which to develop. It has been

1

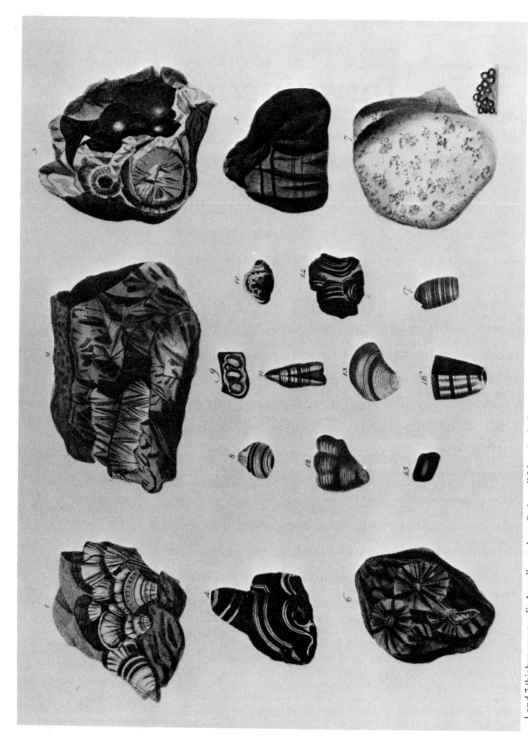

1 and 7 'higher quarter, St Austell – perhaps St Austell Moor in the Pentewan valley; weight 141·8 g
2 Goss Moor: weight 312·5 g 3 Gaverigan, just east of Indian Queens on the edge of Goss Moor; weight 226·8 g
1 **Pebbles of stream tin of the wood-tin variety from various Cornish localities in the area of the St Austell granite**
(Rashleigh, 1797)

suggested that fluorine was a prime 'carrier' transporting the element from depth up to the granite as a stannic fluoride. Hosking has tentatively suggested that some, perhaps all, of the tin in Cornish lodes derived from muscovite altered from biotite within the granite.[5] However, he was forced to admit at the Fourth World Conference on Tin in 1974:[6]

> We do not know the source of the tin: we do not know how it travels to the site of deposition: we do not know the chemical reactions leading to the deposition of cassiterite or of any other primary tin species: we are not certain if the economically important primary tin deposits are always, sometimes, or never genetically related to the 'granites' (or their volcanic counterparts) with which they are spatially related.

Such problems need not concern the archaeologist, who need only remember the simple facts that there are no recoverable concentrates of cassiterite without granitic rocks somewhere in the vicinity, and that comparatively few of the world's granites are noticeably stanniferous. One can sympathize with the Cornish mining engineer who is said to have summed up the problem of the occurrence of tin with the enlightened words: Where it is, there it is.

THE USE OF PURE TIN

Cassiterite is not a difficult mineral to smelt, and the resulting tin is often of very high purity. Cornish ingots are frequently over 99% tin metal. Such high purity was almost obligatory, for customers could easily test the quality by simply bending a thin bar of the metal. When cold, tin emits a creaking noise referred to in Cornwall as 'the cry of tin', a property that disappears if lead or some other impurity is introduced in minute amounts. In the pure state tin is of limited value. Its softness and low melting point (232°C) are features that make it eminently suitable as the major constituent of modern solders and printers' type, but in antiquity unalloyed tin was rarely used. Because it could be beaten into thin sheets like gold leaf, tin was sometimes used for decoration.

Examples detailed later include the decoration of iron age saddles from Pazyryk in Siberia (Fig. 11) and, more spectacularly, some Hallstatt pottery from lakeside dwellings in Switzerland (Fig. 25). Vessels made of pure tin are inferior to those of pewter; some formerly assumed to be tin are now known to be pewter with a low lead content. An early example is the Egyptian 'pilgrim' bottle from Abydos (Fig. 2), dated to *c.* 1500 BC, which serves to demonstrate that the alloying properties of tin were fully appreciated by metal smiths in the 2nd millennium BC.

In Cornwall, spindle-whorls of pure tin are known from Trevelgue and Bussow; the former is in good condition but the latter is crumbling to dust, a fate common to many objects of pure tin. A few Romano-British vessels of pure tin have also been found in Cornwall and are discussed in the section dealing with finds from the Carnon Valley tin stream in Chapter 25 (*see* Fig. 100).

To appreciate the real importance of tin in antiquity it is necessary to examine the development of copper and bronze metallurgy.

THE DEVELOPMENT OF COPPER METALLURGY AND THE USE OF TIN BRONZE

The earliest metals used by man were gold and copper. The former generally occurs as the pure or 'native' metal, though often with a silver content which may be high enough to produce a pale yellow alloy called electrum. Copper also occurs native, but the earliest use of both these metals cannot be accepted as the beginning of metallurgy, since they would have been regarded in the same manner as any non-metalliferous stones.

The softness of native copper, as of native gold, would be an advantage in the manufacture of sheet-metal utensils, but when a harder tool, specifically a cutting instrument was required, the copper had to be hardened. Experience soon demonstrated that cold-hammering (forging) hardens native copper and makes it less malleable. This can only be done to a limited degree, for as Coghlan[7] points out, the cold-working of native copper is not practical without annealing, i.e. softening the metal to 400–500°C for a short period followed by cold-hammering. Annealing was possible in an open fire, which easily attains temperatures of around 700°C. Fully work-hardened in this way, the hardness of a 99% pure copper can be increased from 45 HV (i.e. the hardness as tested by a Vickers diamond pyramid) to 115 HV. Melting native copper at around 1 083°C is not effective as the result is far too soft.

The great metallurgical advance came with the realization that a variety of ores of copper, usually attractively coloured, contained metal which could be extracted. This was not possible in an open fire, but the required temperatures in excess of 1 000°C were attainable in many pottery kilns known from pre-Dynastic Egypt and elsewhere. The potter, used to ores for paints, may well have been responsible for the advance in metallurgy. Certainly in China it is the potters who are credited with the rapid discovery of tin bronze *c.* 2000 BC.

The type of copper ores selected partly depends on the nature of copper deposits. In copper lodes it is usual to find a distinct zoning of different minerals with depth. The upper zone occurs above the water-table, or, as in the case of Cornish lodes, above the depth to which the water-table had descended in the desert conditions of Permian and Triassic times. At present in Cornwall the water-table is rarely below 12 m, yet the minerals in question have been found as low as 360 m at Phoenix United.

The upper zone is known as the oxidizing zone, characterized by the presence of native copper, copper oxide (cuprite), copper carbonates (malachite and, less frequently, azurite), and a host of others. There are, in fact, more minerals of copper than of any other metal because copper is so susceptible to alteration, both by hot, late-stage emanations from depth soon after the filling of the lode, and by percolating ground waters at any subsequent period. Copper is very reactive to other elements within the wall-rock of a lode. The weak acids, such as carbonic acid, that circulate in ground waters may be intensified by the liberation of arsenic from a variety of sulpharsenides, notably arsenopyrite, which is very common in the Cornish lodes. For that reason secondary copper arsenates were a speciality of many Cornish copper mines, especially those in the Gwennap area.

Not only is a vast array of secondary copper formed in the upper zone, but a good deal of copper may be leached out as soluble sulphates which percolate downwards to be redeposited below the water-table in an economically invaluable zone of 'secondary enrichment'. This, too, contains many mineral species of which chalcocite is one of the most important. This particularly rich, grey copper sulphide, popularly known as redruthite in Cornwall, contains over 70% copper compared with only 33% for chalcopyrite, the 'primary' copper sulphide initially intruded into the lodes. Other useful enriched ores include bornite (the 'horse-flesh' ore of the Cornish miner), and the rarer tennantite and tetrahedrite. Below the zone of secondary enrichment, which is of limited thickness in Cornwall but greatest in St Just-in-Penwith, the lodes pass into an unaltered zone of primary mineralization or 'protore'. Because of its depth and lower grade of recoverable ore, this may not be economic to work. The classic locality for the study of hydrothermal lodes is Cornwall, and those wishing more detailed information could do no better than to read some of the invaluable papers by Hosking.[8,9]

In recent years it has become clear that arsenical coppers were more important and used more extensively in antiquity than formerly realized. This does not imply that arsenic was refined specially from arsenopyrite or some other mineral: rather it implies that an arsenic-rich copper ore such as tennantite was smelted to produce a metal that would be a natural arsenic–copper alloy.

The addition of only a few per cent arsenic to copper greatly increases the tensile strength of cold-hammered tools. Experiments referred to by Coghlan[10] have achieved a hardness of up to 207 HV on the cutting edge of a tool, while the part not cold-worked had a hardness of only 60 HV. Maréchal, using a copper alloy with 8% arsenic, attained a hardness of

252 HV with a metal cold-rolled to the extent that its thickness was reduced by 75%.

Using arsenical coppers was undoubtedly a health hazard. It may not be too fanciful to see the description of the smith Hephaestos as a lame god as an echo of the debilitating effects of arsenic fumes, notwithstanding that Hephaestos was supposedly born deformed. Metallic arsenic is volatile, and although little arsenic would be lost in the smelting of oxidized coppers rich in arsenic, a considerable loss could be expected in the treatment of arsenic-bearing sulphide ores.[11] Probably for these reasons tin was more highly prized than arsenic.

The copper–arsenic alloy is sometimes called arsenic–bronze, but some metallurgists prefer to reserve the term 'bronze' for the deliberate alloying of copper with tin. Both metals occur in many of the Cornish lodes, even at the same depth, but this is not universal. Even in Brittany and the Erzgebirge, in many respects geologically similar to Cornwall, the tin-bearing lodes rarely contain copper. Some copper deposits, for example in Afghanistan (to choose a region where very early low-tin bronze has been found), could have produced a natural tin bronze with a few per cent tin. Ordinarily, however, tin had to be traded from the few localities where it occurred in workable quantities to be added deliberately to the copper melt.

A mixture of 90% copper and 10% tin was aimed at in antiquity. Maréchal experimented with various percentages and found that with 8% tin the bronze, when rolled to reduce its thickness by half, produced a hardness of 229 HV. Tools can be work-hardened to a considerable extent, so that the cutting-edge of a flat axe with 10·1% tin, and a palstave with only 7·7% tin, both gave readings of 239 HV.[12] Therefore a precise percentage of tin was not critical as the final hardness was achieved by work-hardening.

The properties of alloys of various compositions were well understood in the early bronze age, and their quality 'was not surpassed to any marked extent in the later periods'.[13] Allen, Britton, and Coghlan found that a bronze with over 13% tin was difficult to work. Yet in south-west England this amount was commonly exceeded. Of 29 Cornish bronzes in Truro Museum analysed by P. J. Northover (*in litt.* 4 Oct. 1980), eight yielded between 12·96% for the Carnpessack sword and *c.* 18% for a haft-flanged axe found near Blue Anchor. Several bronzes in Exeter and Taunton Museums contain even larger amounts of tin, especially those from the Taunton hoard with one high-flanged palstave containing 22·9% and a socketed hammer with 23% tin.

The addition of a second metal to copper, either arsenic or tin, not only increases the hardness of the product. Pure copper is not easy to cast successfully as it develops gas bubbles that seriously weaken the end-product. However, the addition of arsenic or tin somewhat lowers the temperature at which melting occurs and improves the casting by producing a more fluid melt that cools to a denser, less spongy metal. This was particularly important after the early bronze age, when more complex shapes were cast in closed moulds.

Because of the scarcity of tin in the Near East, arsenical coppers continued in use long after the advantages of tin bronze were understood. It has been reasonably claimed that the slow and painstaking development of an iron technology began here because tin was so rare, and the rising population of urban civilizations created a demand for metal goods. Iron is by far the most abundant metal, though until the discovery of the carburization of iron (the addition of carbon to form steel) in the early 1st millennium BC, good quality tin bronze was far superior to iron. This was a fact recognized throughout the ancient world and well summed up in a Chinese text credited to Duke Han,[14] a ruler in 685–643 BC:

> The lovely metal [bronze] is used for casting swords and pikes; it is used in company of dogs and horses. The ugly metal [iron] is used for making hoes which flatten weeds and axes which fell trees; it is used upon the fruitful earth.

An abundance of raw materials for the manufacture of bronze in China was responsible for the slow take-over

by iron. In Greece, by contrast, the scarcity of tin encouraged the swift adoption of iron working in the 11th and 10th centuries BC.[15]

The development of iron technology and its bronze age background have received full coverage in a recent publication by Wertime and Muhly.[16] It is beyond the scope of the present work to delve deeply into the subject of early metallurgy, and those seeking more information should consult the appropriate authorities listed in the References to this Introduction.

NOTES AND REFERENCES

1 HEDGES, 1964
2 TAYLOR, 1979; EVANS, 1980
3 HOSKING, 1974
4 COLLINS, 1912, 117
5 HOSKING, 1964, 239
6 HOSKING, 1974
7 COGHLAN, 1975
8 HOSKING, 1964
9 HOSKING, 1950
10 COGHLAN, 1975
11 TYLECOTE, 1976, 7–8
12 COGHLAN, 1975
13 ALLEN *et al.*, 1970
14 COTTERELL, 1981, 89
15 SNODGRASS, 1980, 49–51
16 WERTIME & MUHLY, 1980

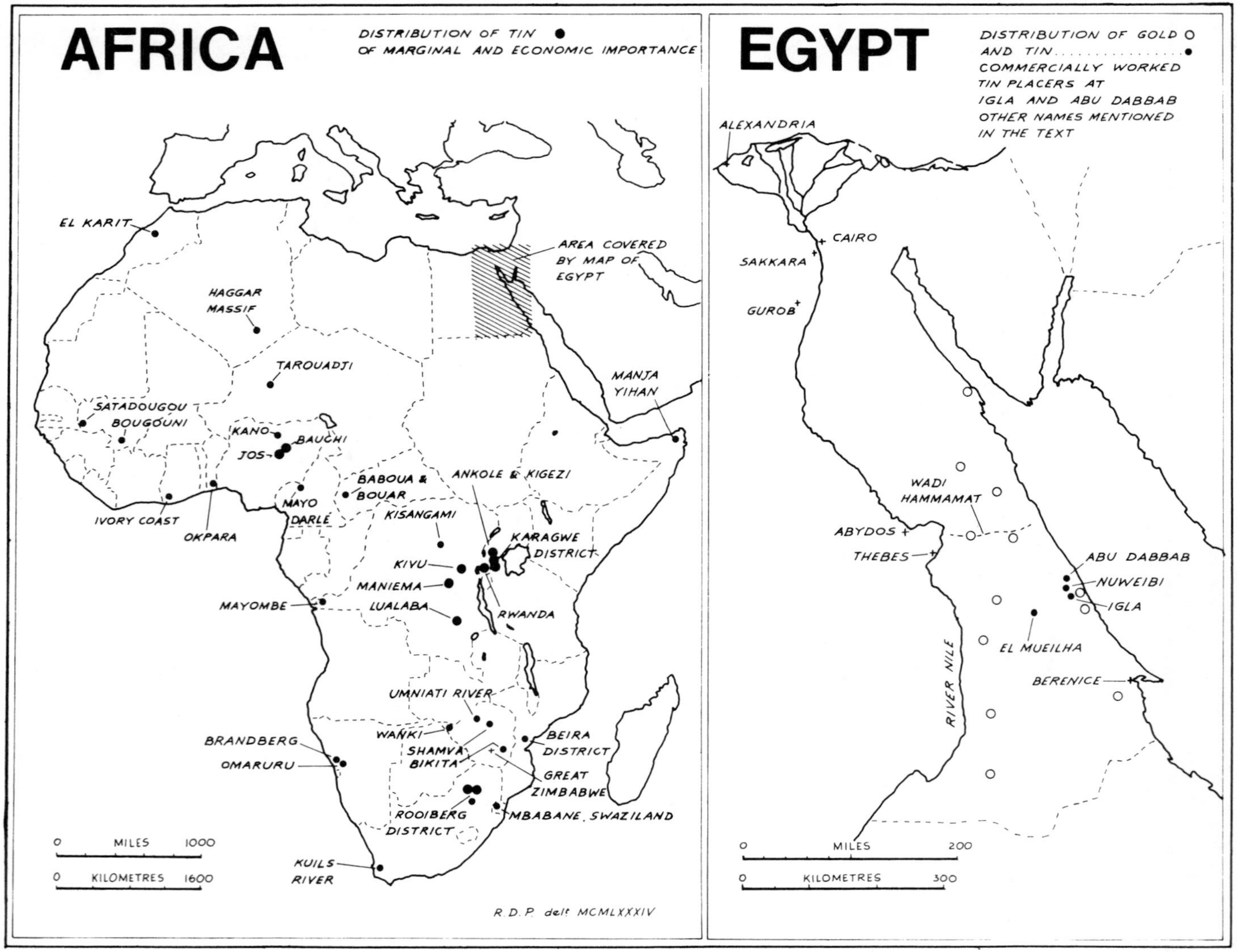
AFRICA
DISTRIBUTION OF TIN ● OF MARGINAL AND ECONOMIC IMPORTANCE
EL KARIT
HAGGAR MASSIF
TAROUADJI
AREA COVERED BY MAP OF EGYPT
MANJA YIHAN
SATADOUGOU
BOUGOUNI
KANO
BAUCHI
JOS
BABOUA & BOUAR
ANKOLE & KIGEZI
MAYO DARLE
IVORY COAST
OKPARA
KISANGAMI
KARAGWE DISTRICT
KIVU
MANIEMA
MAYOMBE
LUALABA
RWANDA
UMNIATI RIVER
WANKI
BRANDBERG
SHAMVA
BIKITA
BEIRA DISTRICT
OMARURU
GREAT ZIMBABWE
ROOIBERG DISTRICT
MBABANE, SWAZILAND
KUILS RIVER
0 MILES 1000
0 KILOMETRES 1600
R.D.P. delt MCMLXXXIV
EGYPT
DISTRIBUTION OF GOLD ○
AND TIN ●
COMMERCIALLY WORKED TIN PLACERS AT IGLA AND ABU DABBAB
OTHER NAMES MENTIONED IN THE TEXT
ALEXANDRIA
CAIRO
SAKKARA
GUROB
WADI HAMMAMAT
ABYDOS
THEBES
ABU DABBAB
NUWEIBI
IGLA
EL MUEILHA
RIVER NILE
BERENICE
0 MILES 200
0 KILOMETRES 300

Map 1

PART 1

AFRICA AND ASIA

⚜⚜⚜⚜⚜⚜⚜⚜⚜⚜⚜⚜⚜⚜⚜⚜⚜⚜⚜⚜⚜⚜⚜⚜⚜⚜⚜⚜⚜⚜⚜⚜⚜

1   Tin in Africa

The northern hinterland of the Sahara does not immediately spring to mind as a tin-field, yet small deposits have been found here. Since World War II it has been discovered in the Hoggar massif of southern Algeria, notably in the El Milla–Coolo area. There are several minor occurrences in the Moroccan Atlas, and at El Karit near Oulmès a few tonnes are produced by open-cast mining.[1]

None of these deposits were known in antiquity, and the only north African deposits which have so far attracted archaeologists are those in Egypt. Much of the Eastern Desert has proved to be a rare-metal province containing, among other things, tin (cassiterite), tungsten (wolfram), and tantalum associated with granites dated between 500 and 700 m.y. (million years).[2] Cassiterite was found at El Mueilha early in 1934.[3] Cassiterite found at Igla in 1940 was smelted on the spot the following year, while in 1942 more was discovered at Abu Dabbab. In 1960 tin was found at Sabaloka, the sixth Nile cataract north of Khartoum, 'near major archaeological sources of bronze in Nubia'.[4] More recently traces have been found in granite on the Arabian side of the Red Sea.[5] The cassiterite in the Eastern Desert is found associated with wolfram in quartz veins. In 1971 Soviet and Egyptian geologists made a detailed study of the tin-bearing quartz veins and stockworks (networks of fine veins) at Igla, Abbu Dabbab, and Nuweibi. All proved to be economically unworkable, though placer deposits at the first two sites contained recoverable reserves of at least 1 200 tons of cassiterite in ground containing an average content of about $2 \cdot 5$ kg m^{-3}. Placer deposits are also known at El Mueilha and at Sabaloka. In wadis at all these sites the torrential rains that fall once or twice a decade have left 'a sparkling trail [of cassiterite] in every riffle capable of catching the heavier grains'.[6]

While this is of great interest to the economic geologist, it is a different matter to make out a case for Egyptian tin production in the days of the Pharaohs. In antiquity the Eastern Desert was famous for its gold, much coveted by foreign rulers, as revealed in the Amarna letters dated to the closing years of the reign of Amenophis III (1386–1349 BC).[7] The persistent demand of Tushratta, King of the Mitanni, whose daughter was married to the Pharaoh, was:[8]

> send gold quickly, in very great quantities, so that I may finish a work I am undertaking, for gold is as dust in the land of my brother.

Egyptian exploitation of auriferous quartz veins was described in the 2nd century BC by Agatharchides, whose account survives in the work of Diodorus Siculus (iii, 12–14). The Egyptians also worked the more easily developed placer deposits as far back as Dynasties I and II (3050–2705 BC), for nuggets have

2   A 'pilgrim bottle' of low-lead (6%) pewter from Abydos (photo Ashmolean Museum, Oxford)

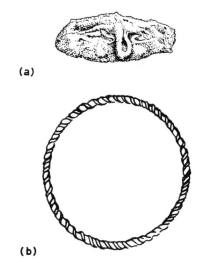

(a)

(b)

3a **Oval bezel of a New Kingdom ring made of tin, from Gurob, inscribed with a *nfr* 'good luck' sign flanked on each side with a wedjet or Horus 'hawk' eye, also for good luck; length 19 mm (re-drawn from a photo in Thomas, 1981)**

 b **Bangle made of one strand of pure tin wrapped around another, from Town IV at Thermi, Lesbos; dia. 42 mm (re-drawn from Winifred Lamb, 1936)**

been found in a tomb of this period. Gold was plentiful, and an early Dynasty I tomb at Sakkara contained wooden pilasters decorated from floor to ceiling with strips of embossed gold.[9] One of the oldest surviving maps, now in the Turin Museum, is a papyrus of Dynasty XX (1185–1070 BC) depicting workings in the gold-bearing Wadi Hammamet east of Thebes.[10]

While gold and tin are not uncommon associates, as will be described in detail in the section on Cornwall, they are not often found together in Egypt. There the gold is mainly in a narrow belt of Miocene and younger sediments bordering the Red Sea,[11] though it is said to be associated with the tin at El Mueilha.[12] Near the El Mueilha deposits were found eight inscriptions identifying this as the area to which Pepi II (2355–2261 BC) sent men for 'stone'.[13] This cannot be construed to mean prospecting for tin, for if the granite here is 'second class', as Wertime reported, it is more likely that the stone masons (or whoever were sent there) departed with a bad report. This would explain the apparent lack of subsequent interest in the area.

If it is assumed that the Egyptians were aware of their tin reserves, shortage of water would have hindered treatment. However, the occasional rains that turn the wadis into natural dressing-floors leave trails of cassiterite caught in the riffles on the surface. Returns would have been small and perhaps of little interest to the conservative Egyptians; they would have done better to concentrate slaves and scarce

water on treating the gold placers, which could be exchanged for imported tin. The archaeological evidence does not indicate local exploitation of tin, while surviving texts favour the view that it was imported. Harris[14] refers to expressions that connect gold with the rock or the ore from which it was obtained, as well as to the washing of gold. If the gold had been accompanied by tin, and the tin was recognized as such, then some reference to it would be expected. There is none. Silver was extracted from gold and recognized as a white colour variety of gold. They would surely have been astute enough to identify cassiterite had the brown or black grains been associated with the gold.

Yet objects of pure tin, or nearly pure tin, are known from Egypt.[15] The most impressive is a 'pilgrim bottle' (Fig. 2) originally described as pure tin but now known to contain 6% lead. It was one of a number of rich articles in an early XVIII Dynasty (*c.* 1570–1400 BC) tomb inserted into a XII Dynasty brick shaft at Abydos. Such vessels were popular at Mycenae and throughout the Near East, though not made of tin. The use of pure tin in Mycenaean Greece is discussed in Chapter 22, but is relevant here because of the discovery of another object of pure tin in Egypt, the bezel of a ring from Gurob (Fig. 3), a town of Middle Egypt on the edge of the desert about 25 km west of the Nile. It flourished during the XVIII Dynasty (1570–1293 BC) but declined during the reign of Ramesses V (i145–1141 BC) and was abandoned soon after. Metal working must have been practised in the town, as the finds include part of a crucible containing copper. More importantly, the town imported a good deal of Mycenaean pottery.[16] It is possible, therefore, that at this late date tin was reaching Egypt from a European source through Mycenaean traders. The question is, where was tin obtained before the discovery of European sources? A tin bracelet is known from Thermi on the island of Lesbos dated to *c.* 3000 BC (Fig. 3b), while the Egyptians are said (with reservations) to have used tin bronze as early as the IV Dynasty (2630–2524 BC).[17]

Around 2000 BC tin was lost to the Egyptians. James Muhly[18] fitted the deprivation into the First Intermediate Period (*c.* 2181–2140 BC) when relations with Syria were broken off, implying that the Egyptians then depended on an eastern source for their tin. The diplomatic break may be referred to in 'The Admonition of Ipuwer', though this is sometimes taken to be contemporary with the Second Intermediate Period (1784–1570 BC). Renewal of contact with Byblos and Syria brought a return of tin bronze to Egypt, though it was never abundant. As late as 1800 BC an axe and a knife contained only 0·2 and 0·3% tin, probably as a natural trace element.[19] The Egyptians were notoriously conservative and made little use of tin bronze until its advantages were

4   Syrians carrying ingots of copper and tin (photo Metropolitan Museum of Art, New York)

promoted in the 17th century BC by the Hyksos rulers, who introduced the horse, the chariot, and other such novelties from Asia.[20]

Tin and lead are frequently confused in written texts. In modern Arabic *rasàs* means both. Egyptian inscriptions are not always clear about which metal is meant. Flakes of limestone (*ostrakka*) from Deir el-Medina, dated to around 1150 BC, are inscribed with *djehty*. This generally refers to lead, but could indicate tin, which was specifically called *djehty hedj*, 'white lead', comparable to the Latin *plumbum album* for tin. Janssen quotes prices for vessels of lead in the reigns of Ramesses II and IX.[21] The annals of the all-conquering Thutmosis III inscribed at Karnak after his Asiatic campaigns (*c.* 1475–1465 BC) list ingots of lead from Syria and Cyprus, while tin, *djehty* and variants, also came from Syria. This is the earliest appearance in Egypt of a word for tin. In the tomb of Rekh-mi-Re' at Thebes (*c.* 1445 BC), Syrians are depicted bringing slate-grey ingots of tin, each with a hole for ease of handling at one end.[22] Similar ingots are drawn in the tomb of Amen-em-opet (993–984 BC) at Thebes (Fig. 4). The first two Syrians are carrying ox-hide ingots of copper, and the back two carry what were originally thought to be blue-grey ingots of lead, but which are now considered to be tin.[23] Whether the Syrians were carrying European tin being traded in the Eastern Mediterranean, or tin from an eastern source, is an open question.

In later history texts refer to the importation of tin from the west. The evidence of the *Periplus of the Erythraean Sea* (*c.* AD 60) is dealt with in a later section. Here may be noted the short Greek treatise on Alchemy by Stephanos of Alexandria (*fl.* AD 610–641), a professor of Herakleios.[24] In an obscure piece in his second lecture 'on the making of gold with the help of God', Stephanos writes:

> The vapour is the unfolding of the work, the level manifestation, the thread bought with silver, the air-displaying voyage, the Celtic nard, the Atlantic Sea, the Brittanic metal. . . .

ή βρεττανικη μεταλλος can only refer to tin. Stephanos' account is a reminder of the much quoted story that appears in the Life of his near contemporary John the Almsgiver, Patriarch of Alexandria, who died in AD 614 (*see* Appendix).[25] Disregarding the miraculous element, the story is important in confirming that there was nothing unusual in the idea of Cornish tin reaching Egypt in the post-Roman period.

Before the opening up of the European tin deposits, the only likely source for Egyptian tin supplies lay in Asia. The notion that the Egyptians, the Phoenicians, and even the Mycenaean Greeks had access to tin in Nigeria or southern Africa cannot be taken seriously. Africa south of the Sahara is well endowed with tin, but any hopes that the history of mining in Zimbabwe might be pushed back to the 1st millennium BC and

5 **'Hand paddocking': digging out the tin-bearing gravels by hand at an unnamed tin working in northern Nigeria in the early 20th century**

attributed to Egyptian or Phoenician contacts have been well and truly quashed. 'A torrent of undigested Phoenician trivia' is Fagan's opinion of one book he reviewed.[26] Whatever the white man chose to believe for political purposes, the remarkable site of 'Great Zimbabwe' is a monument to Bantu culture, which flourished between the 11th and 15th centuries AD as a result of a lucrative trade with the Arabs of the Mozambique coast.[27] Gold was the main attraction, and copper was mined at Messina on the south bank of the Limpopo, and at several of the Rooiberg mines 400 km to the south in the Transvaal. Tin also came from Rooiberg. The mine may be somewhat older than the ^{14}C date of c. AD 1450 obtained from some of the mine timber. Only 5 km east of the mine was found an impure bronze ingot with a 7% tin content. The extensive Rooiberg workings are believed to have yielded at least 2 000 t of tin metal. Shafts were sunk to around 20 m, the suspected limit of their hoisting system, using stone hammers, steel gads, and the age-old technique of fire-setting, which will be detailed in Chapter 12. Similar stone hammers of dolerite, with shallow hand-holds on the sides, were used up to the turn of the 20th century at Loolekup copper mine in Bechuanaland, and cannot have differed greatly from those used when mining began there in about AD 770.[28]

Tin bronze was scarce and largely confined to Zimbabwe and the Transvaal. The metal most commonly used by the early inhabitants of South Africa was iron.[29] The earliest evidence for its use is AD 270±55 at Silver Leaves in Pretoria, and AD 460±50 for iron smelting at Broederstroom in the Transvaal.[30] Recently a ^{14}C date of AD 96±220 has been published for an iron-using site at Machili near the Zambezi River.[31] Iron-working people moved into South Africa in the early years of the 1st millennium AD from Nigeria, the cradle of metal working south of the Sahara.

The accompanying table shows the importance of Nigeria as a tin producer. Before its discovery by Europeans in 1901 in the Bauchi plateau, tin was being smelted by native families at Lerui in Kano province from locally won, alluvial stream tin. A good deal of the alluvial tin is buried under basalt lava flows and basaltic clay up to 36·5 m deep, but much occurs in river deposits no more than 6 m deep. Southwood[32] described how early in the 20th century the natives would dig up the shallow deposits using hand labour ('calabash it'), presumably packing it in gourds, and would take it for sale once a week or so (Fig. 5). The quality of the tin was quickly judged by the effective rule-of-thumb method of 'taking a cigarette tin full and striking it off'; if it weighed 793 g it would yield 70%

tin metal. The rich placers are derived from tin stockworks in the Jos Plateau granites. Pebbles of 'native tin' coated with cassiterite have been described, though as in Cornwall, the native tin must be attributed to the residue of abandoned smelting sites. Cassiterite from the Kogin Delume ('River of Tin' in Hausa) was cast into '12 inch long strings of about ⅛ inch diameter, which are produced by pouring the molten metal on to an 18 inch high semi-circular bank of clay which is perforated by dry guinea-halms [straws]'.[33]

Tin-in-concentrates: annual average production 1964–74 (ITC statistics)

	Tonnes
Burundi	100*
Cameroun	36
Morocco	13
Niger	66
Nigeria	8 144
Rhodesia	600*
Rwanda	1 338
South Africa	1 983
South-west Africa	771
Swaziland	12*
Tanzania	173
Uganda	140
Zaire	6 204

*=estimated production

A tradition of Nigerian tin streaming has led to speculation of a possible prehistoric link with Egypt. Allowing for the fact that a species of monkey now found in the Lake Chad region and in Ethiopia may be depicted on a fresco at Knossos, Crete, *c.* 1500 BC,[34] the likelihood of such early contact with Nigeria is remote in the extreme. Only the problem of finding tin sources for the Ancient World allows archaeologists to keep an eye (if not a mind) open on the subject. Not only is evidence of such an early Nigerian trade lacking, but the sequence of adoption of a metal-using culture south of the Sahara is the reverse of that north of the desert. In equatorial Africa iron came before bronze, introduced, it is believed, through the Saharan trade into west Africa in the closing years of the first millennium BC.[35] Trade may have been conducted by the Carthaginians through Berber tribesmen. The earliest Nigerian iron-smelting furnaces at Taruga near Nok, on the edge of the central plateau, yielded ^{14}C dates centred around 440, 300, and 280 BC.

Many of the justly prized Benin bronzes are, in fact, made of brass, though a group of 49 manillas (arm or leg rings) and bracelets belonging to the 13th century AD are of tin bronze. The advent of a bronze and brass technology may not be much earlier than AD 1000. Arabic records testify to the export of copper 'to the land of the negroes' from the 11th century onwards, a trade supported by the discovery, *c.* AD 1100, of a wrecked southward-bound caravan in the Mauritanian desert.[36] The cargo consisted of 2 000 brass rods; such rods may well have been the prototypes of the Nigerian tin strings already referred to, which were bundled up like faggots for sale as far away as the coastal towns of Nigeria.[37]

The wild improbability of an early Egyptian tin quest in equatorial Africa is neatly summed up by J. V. Luce in a discussion about equally fanciful trips to the New World:[38]

> Crete for the Egyptians lay at the western limit of the World. There is no evidence that the ancient Egyptians ever looked, much less went, any further west. Speculation about the Old Kingdom 'explorers' and 'colonists' diffusing Egyptian culture far and wide have no basis in fact and are implausible, given that the Egyptians never even explored their own river to its upper reaches.

NOTES AND REFERENCES

1 ITC, 1964; SCHUILING, 1967
2 SABET *et al.*, 1973
3 ALFY, 1946
4 WERTIME, 1978
5 GARSON, 1977, 161
6 WERTIME, 1978
7 AMENOPHIS III (or AMENHOTEP III), 1386–1349 BC. Egyptian chronology is controversial, but of little importance in this book. The dates are those compiled by Claus Baer, 1976 (University of Chicago, duplicated typescript). Alternative dates are given, for example, by Ruffle, 1977, 62
8 ALDRED, 1972, 152–3
9 EMERY, 1961, 228
10 SCAMUZZI, 1965
11 FORBES, 1964, vol. VIII, Ch. 5
12 POSS, 1975 (the archaeological sections of this book are unreliable)
13 WERTIME, 1978
14 HARRIS, 1961
15 LUCAS, 1948, 286
16 THOMAS, A., 1981, item 405
17 TYLECOTE, 1976
18 MUHLY, 1973, 245, 332
19 TYLECOTE, 1976
20 ALDRED, 1971, 126
21 JANSSEN, 1975, 442
22 DAVIES, 1943
23 *Bull. Metropolitan Mus. Art.*, New York, 1932, 191, Fig. 13
24 TAYLOR, 1937, 130
25 DAWES & BAYNES, 1948, 216–18
26 FAGAN, B. M., 1970, in a book review in *Antiquity*, **41**, 320–2
27 PHIMSTER, 1974; HUFFMAN, 1974

28 MORE, 1974
29 KÜSEL, 1974
30 MASON, 1974
31 FAGAN, 1981
32 SOUTHWOOD, 1946
33 NICHOLAS (1904) in Fawns, 1905, 139–41

34 TYLECOTE, 1976
35 VAN DER MERWA, 1980
36 SHAW, 1965; SHAW, 1978
37 NICOLAUS (1904) in Fawns, 1905
38 LUCE, J. V., quoted by Glyn Daniel in 'Editorial', *Antiquity* (1972), **46,** 262

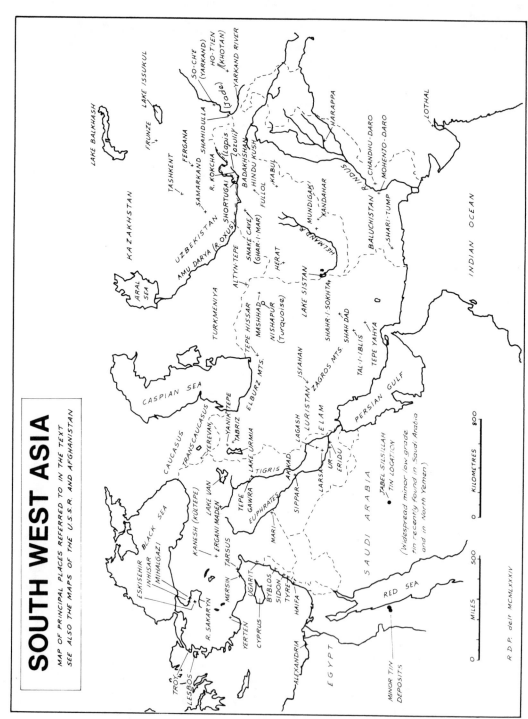

SOUTH WEST ASIA

MAP OF PRINCIPAL PLACES REFERRED TO IN THE TEXT
SEE ALSO THE MAPS OF THE U.S.S.R. AND AFGHANISTAN

LAKE BALKHASH

FRUNZE   LAKE ISSUKUL

SO-CHE (YARKAND)
HO-TIEN (KHOTAN)
(Jade)
YARKAND RIVER

TASHKENT
FERGANA
SAMARKAND   SHAHIDULLA
KAZAKHSTAN

SHORTUGAI   Lapis   (Jade)
BADAKHSHAN   HARAPPA
R. KOKCHA   HINDU KUSH
UZBEKISTAN   FULLOL   KABUL
AMU DARYA (OXUS)   CHANDHU-DARO
MUNDIGAK   MOHENJO-DARO
KANDAHAR   SHARI-TUMP
ARAL SEA   BALUCHISTAN
INDUS R.

LOTHAL

INDIAN OCEAN

TURKMENIYA   ALTYN-TEPE
SNAKE CAVE (GHAR-I-MAR)
HERAT
HELMAND
CASPIAN SEA   TEPE HISSAR   MASHHAD
NISHAPUR (Turquoise)   LAKE SISTAN
SHAHR-I-SOKHTA   SHAH DAD

YANIK TEPE   ELBURZ MTS.
TABRIZ   TAL-I-IBLIS
CAUCASUS   LAKE URMIA   ZAGROS MTS.   TEPE YAHYA
TRANSCAUCASUS   YEREVAN
LAKE VAN   ISFAHAN   PERSIAN GULF
KANESH (KULTEPE)   LURISTAN
ERGANI MADEN   LAGASH
ESKISEHIR   TARSUS   AKKAD   ELAM
INHISAR   MERSIN   TIGRIS   LARSA   UR   ERIDU
MIHALGAZI   TEPE GAWRA   SIPPAR
BLACK SEA   EUPHRATES
R. SAKARYA   MARI
YERTEN   UGARIT   JABEL SILSILAH
CYPRUS   BYBLOS   TIN LOCATION
SIDON   SAUDI ARABIA
TYRE   (Widespread minor low grade
HAIFA   tin recently found in Saudi Arabia
TROY   and in North Yemen)
LESBOS   ALEXANDRIA
EGYPT   RED SEA
MINOR TIN
DEPOSITS

0   500   MILES

0   800   KILOMETRES

R.D.P. delt MCMLXXIV

Map 2

2   *Tin in the Near East*

Rekh-mi-Ré' ('Wise of God'), vizier of Egypt *c.* 1470–1445 BC, built his own magnificent tomb. Among its many wall paintings depicting the work involved in his exalted office is a scene showing porters carrying metal ingots (Fig. 6). One man carries on his shoulder an ox-hide ingot of copper; following are two men, each with a basketful of small oblong ingots. The accompanying text describes them as:[1]

> bringing Asiatic copper which his Majesty carried off from his [Syrian] victory in the land of Retenu in order to cast two doors of the temple of Amun.

According to Davies,[1] temple doors were made of copper in the mixture of six parts to one, the 'one' presumably being tin. The meaning of Asiatic copper is disputed. Lepsius described it as 'a variety to which was attached a great price'.[2] Harris gives reasons for supposing that it was a special copper alloy, and its notably light colour could well indicate a high tin content. Whatever its precise meaning, it cannot be doubted that tin as pure metal or in bronze ingots, before its importation from European sources, must have come from the east, in all probability through one of the Levantine ports. It might be assumed from this that it would be a simple matter to pick out the tin-rich areas on an economic map of the Near East and weigh the pros and cons of their ancient exploitation. Unfortunately it is not that easy. Archaeologists have peppered maps of the Near East with tin sources which investigation finds to be figments of the imagination or, at best, minor occurrences unlikely to have been known in antiquity.

Tin has been found recently in North Yemen and, to a greater extent, in the north-central and north-eastern parts of the Shield in Saudi Arabia. The present writer is indebted to K. R. Greenleaves of Riofinex Ltd for supplying relevant details in June 1984. The tin is related to a late-phase potassic granite intruded *c.* 560–530 million years ago and thus similar to the known minor occurrences in Egypt and Sudan. Tin content is low and subsidiary to wolfram, though nowadays it could prove profitable where bulk tin is of the order of 0·25%. Some samples yielded values as high as 6%, though the source grade and volume, geomorphology and climate, are not conducive to tin

placer possibilities. Certainly, no stream tin has been found so far. The most significant cassiterite yet found is at Jabel Silsillah (26°6′ N, 42°40′ E) associated with minor copper and zinc sulphides introduced during a late hydrothermal lateration phase of one of five successive intrusions of acidic granite in a ring complex. This is largely covered by young sediments, but tin-bearing greisens are visible through the cover in places suggesting recoverable reserves to depths of 50 m or more. None of the occurrences are likely to be more significant than those in Egypt. There is no evidence of ancient tin working in Saudi Arabia, though there are old workings that seem to have been either abortive or related to associated or spatially coincident copper/zinc or gold mineralization.

It is necessary, therefore, to look for a different source for Egyptian tin in antiquity. Trade between Egypt and the Levantine port of Byblos dates back to predynastic times. Wainwright believed he had found the source of the tin traded along this route when in 1934 he published an article inspired by the statement that 'Tin deposits in the Kesserwan district were examined and approved by Australian engineers.'[3] Further enquiries revealed that the Australians, Sams and Robinson, had prospected behind Byblos (modern Jubail) in 1910 and found silver and copper in commercial quantities. Wainwright's informant had been given to understand that 'the tin and copper are found together'. It was thought that both these ores had been washed down the rivers Nahr Ibrahim (River Adonis) and Nahr Feidar (River Phaedrus), which enter the Mediterranean only a few kilometres south of Byblos. Copper does not form alluvial deposits, only minor traces as in some Cornish tin streams. The Byblos tin is a fiction, either as alluvials or as lode deposits in the Kesserwan mountains east of Byblos. No tin has been reported reliably from here. The most likely explanation is that the Australians automatically filed a claim to prospect for tin along with other minerals they hoped they might be lucky enough to find. In any case, Wainwright's informant reported that 'there are no signs of mining operations by the ancients in the Kesserwan district'.

Tin ingots were certainly traded through the Levant in the 2nd millennium BC, for in the autumn of 1976

6   Drawing copied from the paintings on the east half of the south wall of the passage in the tomb of Rekh-mi-Ré' at Thebes (from plate LIII, Norman de Garis Davies, 1943)

two ingots were found 'in the sea near the Phoenician port of Dor, south of Haifa (Fig. 7)'.[4] Over a period of years local fishermen raised about 7 t of copper and tin ingots, which were sold to scrap dealers in Haifa for re-melting. Two surviving tin ingots are now in the Museum of Ancient Art at Haifa; their date is uncertain. A late date, after *c.* 800 BC, has been suggested on the assumption that the ingots are from a 'Tarshish ship' carrying tin from Tartessos in southern Spain. Alternatively they may be earlier as they bear stamps which resemble Cypro-Minoan symbols used in Cyprus and Ugarit *c.* 1500–1100 BC. It is not known whether the ingots were entering or leaving Haifa, or whether they had any connection with tinless Cyprus. It is tempting to see them at the western end of a lengthy overland caravan trail, perhaps even through Ugarit where they were weighed and officially stamped. Whatever their origin they are a spur to search for the tin-field which produced them.

Turkey has been claimed as a likely source of cassiterite. Bissing wrote that in pre-Atatürk days (before 1923) tin was mined at Eskisehir.[5] These may be the workings said to have existed 'in former times' at Mihalgazi in the Sakarya basin where 'the ore was richer in comparison with other occurrences'.[6] According to Ryan,[7] tin occurred at Inhisar, Mihalgazi, Akçasu, and Koyunli in the Sakaryn valley where there are extensive exposures of acid granite. In spite of this promising start, there is very little evidence for tin in Turkey, a view held by Wertime[8] who found no tin at Balekesir, another traditional site for it, reinforcing

the view of Turkish geologists that no workable deposits are in the country. Traces of the tin sulphide, stannite, may occur in some copper deposits, as claimed at Arakli, but at best there is little tin of more than mineralogical interest.

Such traces are unlikely to have been deliberately exploited in antiquity, and analyses of Anatolian metal work only confirm the scarcity of tin bronze in Turkey in the 3rd millennium. Branigan thought that in level II at Troy (*c.* 2500–2200 BC), some 500 years after the founding of the city, tin bronze was the dominant metal.[9] But this is not the case. Only two artifacts from level II contained tin: one with a little over 2%, the other with 10·4%. The rest are arsenic bronze – or arsenic–copper alloys, as some metallurgists prefer to call them. Childe[10] supposed that Troy obtained its tin from Bohemia, but even if it is accepted that the single artifact with 10·4% tin was imported, such an early central European source is out of the question. A trade route through the Balkans looks geographically attractive, especially as Bulgaria was favoured with precociously early copper mining dating back to the 5th and 4th millennia. Chernykh, in his detailed history of ancient Bulgarian metallurgy,[11] while suspecting Anatolian influence as the spur that initiated it, believes that the Balkan–Carpathian metallurgical 'province' developed independently. Moreover, tin bronze did not come into use in Bulgaria until the middle bronze age (about the beginning of the 2nd millennium), and remained scarce. Indeed, there are fewer awls and needles of tin bronze than in the

7   Two late bronze age tin ingots from the sea near Haifa, Israel; *left* 31·4 × 19 × 3·7 cm, weight 11·4 kg, 94–83%Sn; *right* 32·4 × 21·6 × 3·6 cm, weight 11·9 kg, 95·3%Sn (courtesy Museum of Ancient Art at Haifa)

'eneolithic' (*c.* 4800–3800 BC) when some pieces containing 6–10% tin did occur, though without adequate explanation why.

Low-tin bronzes (those with about 2% or less tin) are generally considered to be 'natural' bronzes as a result of smelting copper ores containing traces of tin, which need not be in the form of cassiterite. (There are well over 50 naturally occurring tin-bearing minerals, mostly of great rarity.) Such bronzes are known from Mersin (Yumuk Tepe) on the south-east coast of Turkey. A metal stamp from level XVI (*c.* 4300 BC), and pins from levels XIV and XIII, contained 2·6, 2·1, and 1·3% tin respectively. There was much eastern influence at Mersin which was also 'open to elements from the Anatolian plateau', so there is no need, as Charles Burney concluded, to evoke long-distance trade from outside Anatolia to account for very low tin content bronzes.[12] Recent analyses of some Anatolian artifacts[13] yielded a maximum of 0·8% tin and an average of less than 0·2% in objects of the mid-3rd millennium BC (early bronze age II). Only in the closing years of the early bronze age were there dagger blades with 6·5 and 7·1% tin at Yerten (modern Korkuteli) in south-west Turkey. This, and a similar rise in the use of tin bronzes in Troy III and IV (2200–1900 BC)[14] must be accepted on present evidence as evidence of trade with a tin-bearing area outside Turkey.

Links with Mesopotamia were strong, especially under the Akkadian empire founded by Sargon around 2370 BC. Barbarian influence put a temporary stop to this, but by 1900 BC Assyrian merchants were again well known in Anatolian towns. Nowhere is the evidence better than in the *karum*, or trading centre, at Kanesh (Kültepe) in Cappadocia. Business documents in the form of clay tablets tell of the Assyrian desire for Anatolian metals. Copper was obtainable from, among other places, Ergani Maden in eastern Turkey, where mining dates back perhaps as far as 7000 BC. The deposits are vast, yielding 20 000 t annually nowadays.[15] Gold could have come from the Troad, the extreme western part of Anatolia, thus drawing the whole country into the trading network. A copper trade down the Euphrates is extremely ancient; the river's original name was *Urudu* or 'copper river'.[16] Silver is widespread in Turkish mineral deposits, and as Mellaart states,[17] 'The whole purpose of sending Assyrian merchants to Anatolia was to ensure a steady supply of Anatolian silver and some gold.' In exchange they gave cloth and tin, 'transported by caravans of black donkeys bred in Assyria'. They made a profit on the cloth of 100%, and on the tin of 75–100%. The quantities traded could be considerable; a cargo of 410 talents of tin (more than 12 t) is once mentioned, though for some curious reason tin prices are never recorded. Trade with Kanesh continued until *c.* 1757 BC when Hammurabi of Babylon destroyed Mari (900

km up the Euphrates) and a period of wars followed which reduced 'central Anatolia, once rich, to a land of ruins'.

The Kanesh tablets give no indication of where Assyrian tin came from. Gurney favoured the Caucasus.[18] This mountain range is rich in copper, but claims that 'scattered finds of pieces of cassiterite have been reported since 1882, and the presence of tin . . . in several copper mines'[19] are denied by modern Russian scholars. It has to be accepted, on the authority of Selimkhanov, that 'deposits of tin, the existence and exploitation of which are mentioned since the 19th century . . . have never existed in the Caucasus'.[20] As occurs widely in copper deposits, natural traces of tin are widespread producing 0·01% tin or less in bronze age artifacts. More tin occasionally occurs, though never enough to suppose the existence of a cassiterite deposit that could be mined, or streamed, and traded. Chernykh[21] demonstrated the lack of tin in his description of the late bronze age hoard from Constanta in Romania. Two sickles of Kuban type, made of north Caucasian metal, contained only 0·01 and 0·025% tin. Korenevsky[22] describes metal work from the Tly burial ground south of Ossetia in the central Caucasus. The early artifacts are arsenical bronze, or alloys of arsenic, antimony, and copper. Tin bronzes appear only from the 12th century and soon predominate.

In the Transcaucasus, that is between the Caucasus mountain range and the Turkish–Iranian border, tin bronze was again little used. Even in the late bronze age arsenic alloys predominated in some areas such as western Georgia.[23] Of 54 objects of the 3rd and 2nd millennia BC analysed by Selimkhanov,[24] only three contained a sufficiently high tin content to indicate a deliberate alloy:

Dagger blade	Stepanovan	11·54% Sn
Awl	Stepanovan	12·60% Sn
Axe-adze	Eshektepe	10·08% Sn

Stepanovan and Eshektepe are near Erevan (or Yerevan) in Armenia. Here, and south of the Soviet border, in the early 2nd millennium was made a polychrome painted pottery called Urmia ware after the Iranian lake west of Tabriz where the pottery developed. It is found, for example, at Geoy Tepe to the west of the lake, where tin bronzes appeared around 2000 BC. A pin with 10% tin and a bangle with 5% belong to the period.[25] The pottery and the use of tin bronze appear to have spread north into the Transcaucasus, both becoming scarcer through Georgia.

Urmia ware was found at the Armenian site at Metsamor where, it is claimed, cassiterite was smelted; the process was assisted with the use of phosphorus-rich 'brickettes' made from animal bones ground up with clay. This flux supposedly led to a 'significant reduction in the temperature required for smelting'.[26] Such a method has not been reported anywhere else, and doubts have been expressed about the authenticity of the claim (R. F. Tylecote, personal communication). Even if it is true it does not mean that the tin ore had been obtained locally.

Mongait asserted that there was tin in Transcaucasia,[27] writing in the 1950s when Soviet geologists were prospecting for tin in such areas as the Urals and the Caucasus.[28] There are several claims for deposits of tin ore on the Soviet side of the border. Mellaart, writing of the bronze industry west of Yerevan, said that alluvial tin had been found at Paleoaraks at the foot of Mount Aragats.[29] Karajian mentions tin in the Kurbaba Mountains near Tillek, also between Sahend and the River Araxes, as well as near Migri on the Araxes, and in Hejenan.[30] Information is not easily obtained about Soviet tin deposits, but none of these supposed Transcaucasian deposits are mentioned in mining or mineralogical literature dealing with the Soviet Union. Muhly[31] writes of tin arriving at Sippar and Mari on the Euphrates being referred to once in the ancient texts as 'tin from the Nairi-lands, a name given generally to the area encircling Lake Van and Lake Urmia'. This tin source, he continues, 'is enshrined in the modern literature', even though 'no convincing evidence for it has yet been published'. Texts dealing with the Assyrian trade are dated to *c.* 1950–1850 BC, and the Akkadian texts from Mari and elsewhere cover the period 1800–1750 BC. Muhly has constructed a map of the trade routes,[32] one going as far north as Hasanlu, only 160 km south of Tabriz and less than 320 from the Soviet border on the Araxes river, all of which is rather academic if there is no convincing evidence that tin could have travelled south along it.

The texts from Mari show a way out of the difficulty by also recording tin being shipped up the Euphrates, presumably from the Persian Gulf, pointing to a distant origin involving maritime trade. But what of tin in the rest of Iran? It is a huge country, larger than the combined total areas of the UK, Eire, the Low Countries, France, and Iberia.

In 1912 William Gowland, Professor of Metallurgy at the Royal School of Mines, lectured on 'The Metals in Antiquity'.[33] 'In Persia', he said, 'we are on safer ground; in Khorosan, van Baer has discovered very ancient mines of tin ore.' Unhappily, van Baer's mines do not exist. There is no evidence of workable tin in Iran. If there really is tin near Mashed, or between Shārud and Astarabad in the Elburz Mountains,[34] it remains to be proved. Writers have often assumed the presence of tin without question. Ghirshman, for example, states that the country's wealth 'included gold, extracted in Medea, copper and tin'.[35] Wulff goes further: 'Tin ore is found in northern Persia at Mt Sahand near Tabriz, in the Qaradag ranges, both close

to the copper mines, and on the southern slopes of the Alburz near Astarabad and Sarud. Stream tin and gold are found near Kùh-e Zar (Damgan district), and Kūh-e Baban, between the copper mines of Anarak and Isfahan, and twenty-two miles west of Mashed at Robat-e Alokband again near copper mines.'[36] No references are given for this formidable catalogue, and there is nothing to suggest there could be more than a trace of tin of mineralogical interest.

In the mid-1960s, the Geological Survey of Iran claimed to have found tin in the black sands of the Caspian Sea near the mouth of the Safid Rud or 'White River', though an expedition by the University of Minnesota later found no sign of it there. Iranian geologists have reported traces of tin in borings near old copper workings at Chah Kalapie, north-west of Neh village in the eastern Dashte Lut, but no recognizable cassiterite deposits have been found anywhere in the region.[37] The most that could be expected from here is smelted copper yielding a tin bronze with no more than 1–2% tin.

The Arab geographer Muqadasi stated that tin occurred at Hamadan, 560 km south-west of Tehran. As Muhly wrote, 'a mineral zone running roughly from Hamadan to Tabriz seems to fit all the evidence for the Near Eastern tin trade'.[38] All that is lacking is any supporting geological evidence. It is not credible to assume that this region, over 320 km in length, which had yielded stream tin in prehistory, was so thoroughly worked out as to leave no trace. It also happened to have been exploited at just that period in geological time when the deepest roots of the tin lodes had been totally removed to supply the last parcel of stream tin that the granite was capable of yielding.

The tin content of much mediaeval Islamic metalwork is low, though mirrors contained 10–12% tin, and goblets known as *Haft Jush* goblets commonly contained 20% tin, as did contemporary Iranian coinage.[39] Wherever Muqadasi thought tin came from, according to Paul Craddock a text dated AD 1301 by Abu Qasim on glazing gives only foreign sources for tin: 'Frankenland', which could well mean Saxony and England rather than France; China, which might also include Malaysia; and the region of the Bulgars (i.e. the middle Volga), itself a tinless area but through which, it may be assumed, Siberian tin was being traded. Whatever the value of each of these sources was to mediaeval Iran, tin from within the country's boundaries is not mentioned.

By no means all prehistoric Iranian bronzes were of tin bronze when the potential of that alloy was fully understood, a feature equally apparent from the analyses of bronzes from Iraq, Syria, and Palestine.[40] The most famous of all Iranian metalwork are the Luristan bronzes, first reported from shallow graves plundered in vast numbers by Lur tribesmen in the Zagros region. The lack of controlled excavation and the dispersal of the bronzes to museums and art collections throughout the world make dating and provenance difficult or impossible. However, it is now realized that the bronzes cover a wide range in time, from *c.* 2500 BC to *c.* 500 BC. In analyses of 32 examples, one contained no tin, four others less than 2%, with progressively more in others rising to between 11 and 12% in three bronzes, and 12·7% tin in a horse's cheek-piece shaped like a griffin. What the analyses of such bronzes indicate is that a consistent supply of tin was not available, a clear sign that it had to be imported from a long distance. Moorey[41] did favour 'local Persian sources' for his Luristan bronzes, particularly in the area around Tabriz and towards the Caucasus, which has now been discounted as improbable.

In the year of Gowland's lecture mentioned above, the Americans F. and E. Hess published their exhaustive bibliography of tin.[42] There are only two references to Iran. The first, in a German work, merely states the presence of tin in the country; the other is a paper by J. Mactear based on his visit to Persia in 1893.[43] Mactear saw no tin himself; he quotes verbatim an amusing report which a Dr Riach sent to Lord Palmerston in 1837. Riach described a journey up the Anngert, a stream about 90 km north-east of Tabriz. After a march of about three-quarters of an hour: 'We came to perhaps the most wonderful mine of tin in the world' – an open-cast working in granite discovered by Mr Robertson the superintendent and Mr Rowe – 'an experienced miner brought up in the Cornish mines. . . . If this quarry be so rich as it is said to be it is quite evident that there is enough of this metal in Karadagh to supply the world.' Happily for the Cornish mines of the 1830s, this tin bonanza must have been a phantom. As J. H. Collins (a name familiar to anyone with the slightest interest in Cornish mining) said in the discussion following Mactear's paper, 'Dr Riach did not know much about tin and I doubt whether Mr Rowe knew much about it either, though he was a Cornishman. I think it would be very hard to get anybody to believe that such a deposit as this is described to be, and found as long ago as 1837, has remained untouched for more than half a century.'

NOTES AND REFERENCES

1   DAVIES, 1943, 53–4
2   HARRIS, 1961, 57
3   WAINWRIGHT, 1934
4   ANON., 1980; MADDIN *et al.*, 1977
5   VON BISSING, 1932
6   DE JESUS, 1978
7   RYAN, 1957, 62–3
8   WERTIME, 1978
9   BRANIGAN, 1974, 64
10   CHILDE, 1939, 43
11   CHERNYKH, 1978
12   BURNEY, 1977; MELLAART, 1978, 25 gives *c.* 5200 for the earliest appearance of 'a small amount of tin' in bronze tools from Mersin, Cilicia
13   MOOREY & SCHWEIZER, 1974
14   TYLECOTE, 1976, 21
15   RIDGE, 1976. Mine production of copper ores in Turkey in 1970 exceeded 27 000 tons and were less than 20 000 tons in 1971. Most came from Ergani Maden (Maden = mine in Turkish) where the copper belt extends *c.* 20 km east–west
16   HAWKES, 1977, 159, 167–8
17   MELLAART, 1978, 50–1
18   GURNEY, 1952, 86
19   FIELD & PROSTOV, 1938; GIMBUTAS 1965, 485, refers without reference to 'one insignificant place in the northern Caucasus' where tin occurred
20   SELIMKHANOV, 1962 and 1968
21   CHERNYKH, 1981
22   KORENEVSKY, 1981
23   GIMBUTAS, 1965, 485
24   SELIMKHANOV, 1962
25   TYLECOTE, 1976, 22
26   BURNEY, 1977, 130
27   MONGAIT, 1965, 128
28   ANON., 1956
29   MELLAART, J. 1968, *Anatolian Studies*, 187–202, quoted in Muhly 1973, 463 note 793
30   KARAJIAN, H. A., *Mineral Resources of Armenia Anatolia*, 1920, 186, quoted in Lucas 1928
31   MUHLY, 1973, 305–6
32   MUHLY, 1973*a*
33   GOWLAND, 1912
34   LUCAS, 1928
35   GIRSHMAN, 1978, 26, 58
36   WULFF, 1966, 3–4
37   WERTIME, 1978
38   MUHLY, 1973*a*, 409
39   CRADDOCK, 1979
40   MOOREY & SCHWEIZER, 1972
41   MOOREY, 1964; MOOREY, 1971, and the review by N. K. Sandars in *PPS* 1972, 438f
42   HESS & HESS, 1912
43   MACTEAR, 1895

3   Tin from 'Meluhha'

The royal inscriptions from Mari and Larsa tell of the greatest prosperity and trade before the destruction of those towns by Hammurabi in about 1757 and 1761 BC respectively – the 35th and 31st years of his reign, usually dated to 1792–1750. According to the Larsa texts, merchants went there to purchase copper and tin: the copper came from Magan in Oman, via Tilmun (Bahrain), but the origin of the tin is left in question. Tin mines in north-west Iran or the Transcaucasus are highly unlikely, as discussed in the last chapter. Fortunately, there is evidence for another tin source in texts from Lagash.

Lagash, about 50 km east of Larsa, was of minor importance except under the governorship of Gudea (c. 2143–2124 BC). His inscriptions indicate extensive trade: gold from Cilicia in Anatolia, marble from Amurra in Syria, and cedar wood from the Amanus Mountains between these two countries, while up through the Persian Gulf or 'Southern Sea' came more timber, porphyry (strictly a purplish rock), lapis lazuli and tin.[1] One inscription has been translated:

> Copper and tin, blocks of lapis lazuli and *ku ne* [meaning unknown], bright carnelian from Meluhha.

This is the only reference to tin from Meluhha. Unfortunately, the location and extent of Meluhha are uncertain. If it appears at all on maps it is usually in the Sind region of the lower Indus valley, often with a question mark. More will be said below on lapis lazuli, but it must mean one of two things: either Meluhha was a name vague enough to embrace Badakhshan (the northernmost province of Afghanistan) as well as some portion of the Indian subcontinent including the Indus valley, or 'tin from Meluhha' means that the metal came from some port in Meluhha – just as 'copper from Tilmun' means copper from elsewhere shipped through the island of Bahrain. Whichever interpretation is correct,[2] the result is the same. Tin must have come from somewhere in India, or from elsewhere along a trade route down the Indus valley.

India is not without its tin locations, rare though these are. The mineral has also been found as a constituent of curiosity value in the gem-bearing gravel (*iliam*) at Balangoda in Sri Lanka.[3] The mineralized zone of Malaysia extends into southern Burma and north-west Thailand. Tin from 'British India' was almost exclusively from Burma, where it occurs in Tenasserim State between the Tavoy River and the border to the south. It also occurs at Mauchi in southern Shan near the Thai border.[4]

The largest deposits in India proper are in the Hazaribagh district of Bihar. 'Old workings' are said to exist,[5] but are unlikely to predate British rule. Tin was found in 1849 at Nurungo (or Nurgo), apparently by natives smelting ores mistaken for iron; they then took the resulting shiny tin metal for silver.[6] Attempts at working the ore in 1864 and in the early 1890s were not successful;[7] production was insignificant. In the years 1912–17, the only output was 0·05 ton in 1914 and 0·35 ton in 1915. The Nurungo veins were followed to a depth of 18 m, then abandoned because of poor values and too much water.[8]

The belief that tin from this part of India, from Burma or Malaysia, could have supplied the Indus and other civilizations in prehistory does not stand up to investigation. The maritime trade between the Near East and Malaysia is discussed in Chapter 9. Here it is sufficient to point out that contemporary with the bronze-using Indus civilization, the inhabitants of the Ganges valley and the Hazaribagh district remained copper-using. Their spears, harpoons, and other artifacts, all of copper, are found with pottery which itself shows no influence from the Indus civilization. However, it overlaps it in time and continues in use after the waning of the Indus cities c. 1800 BC.[9]

Tin of 'scientific' interest is said to have been found at Bastar north of Madras.[10] It also occurs in a tourmaline–pegmatite at Hosainpura, Palampur.[11] There is nothing recoverable here, nor is it likely to have been known before its discovery in September 1903. Suggestions that the Indus civilization may have derived its tin from an Indian source can thus be ruled out.

Tin bronzes from Gujarat are at the southernmost limit of Indus influence. The copper could have come from Rajasthan, though copper ingots at the port of Lothal, at the head of the Gulf of Cambay, suggest imports from Oman or some other Near Eastern copper mining district. Tin supplying Harrapa and Mohenjo-Daro, most famous of the Indus cities, may

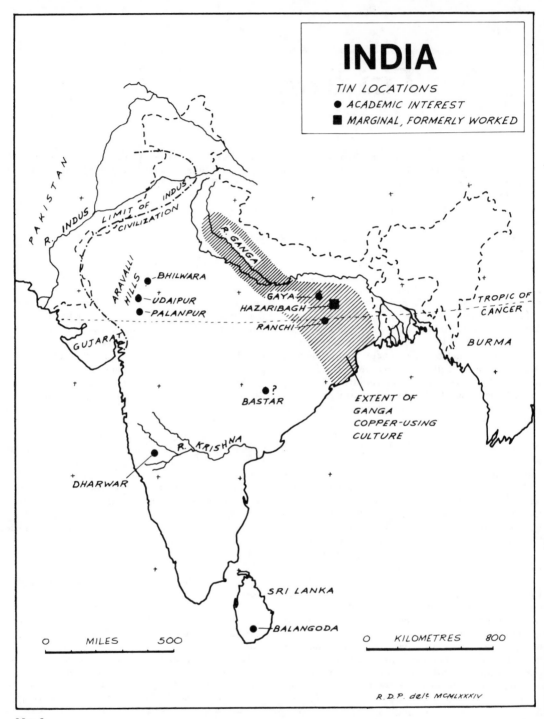

Map 3

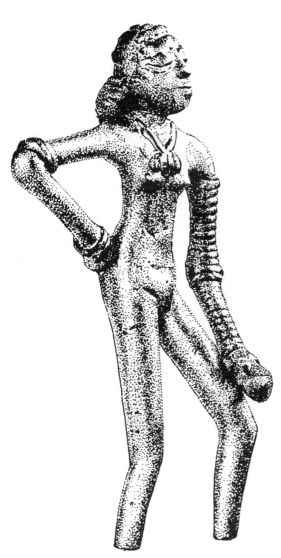

8   **Dancing girl from Mohenjo-daro; tin-in-bronze (11–13%); height 11 cm (re-drawn from a photograph supplied by D. P. Agrawal)**

have been sent overland to Lothal for export, though the scarcity of tin in the Indus cities makes this idea unconvincing.

At Harappa three copper alloys were used in the period 2500–2000 BC: copper and up to 2% nickel; copper and up to 5% nickel; copper with *c.* 10% tin and a trace of arsenic. Ingots of tin as well as of copper were found at Harappa.[12] The rarity of the metal is seen at Mohenjo-daro where, of 64 artifacts examined, only nine were of tin bronze (Fig. 8).[13] Ingots of tin bronze have also been found at Chanhu-daro. Yet in spite of its scarcity, tin bronze was widely used. Its occasional abundance and, in the case of the bronzes

from Luristan in southern Iran, the high quality of the tin bronzes produced, equally underline the fact that a rich source of tin existed somewhere. A tin deposit cannot have been worked out in antiquity leaving no tell-tale trace for the modern prospector to discover. Were every last gramme of economically recoverable tin extracted from Cornwall, the mineralogist would still be confronted with background readings of distinctly anomalous concentrations of tin in certain soils and stream sediments. A tin-field cannot disappear without trace.

The problem in antiquity in the Near East was to obtain a regular supply. In the 3rd and 2nd millennia BC there was 'always the possibility of a tin famine when political circumstances disrupted the lines of supply'.[14] That the lines were readily broken is a demonstration of their length in politically unstable areas. Furthermore, natural hazards might be equally disruptive. The logical conclusion is that the tin came from a remote area.

Tin was not the only prized mineral sought after in remote areas. It is instructive to look at the quest for two semi-precious stones, which admirably demonstrates the extraordinary journeys prehistoric men endured to obtain them, and leads us by degrees to identifying distant tin-fields.

Among the semi-precious material recovered from Harappa is jade,[15] or, more accurately, nephrite – a tough calcium–magnesium–aluminium–silicate of variable colour. It was especially valued in China from the 4th millennium onwards, though not until the Shang bronze age, after about 1600 BC, were techniques developed to carve it in intricate fashion. This relatively intractable substance has a hardness of 6·5 on Moh's scale, roughly equivalent to a steel knife. Nephrite is rare on archaeological sites in the Near East. Schliemann found it in level II at Troy (2500–2200 BC) – eight axes of green jade and two of white.[16] Green nephrite is comparatively widespread in hornblende-schists in Central Asia and eastern Siberia. Schliemann's suggested origin in the region of Lake Baikal – over 6 000 km from Troy – is needlessly far. The earliest Chinese jade probably did come from the Vostochnyy Sayan, west of Lake Baikal, where rolled masses are found in the rivers. However, for much of Chinese history, nephrite was obtained from equally distant Central Asia. If this latter region supplied Schliemann's jade, it is still over 4 500 km from Troy. The principal rivers regularly 'fished' by the Chinese for jade pebbles were in the mountains south of Khotan (Ho-t'ien) in Sinkiang province; the upper Yarkand Daria (or Yeh-erh Ch'iang Ho) and the Yorungkash (Yü-lung-k'a-shih Ho), which were famed for their white 'mutton fat' jade, and the Kara Kash where the prize was black jade.

Nephrite was also quarried in the Kara Kash valley, notably at Shahidulla (Sai-t'u la) in its upper reaches some 125 km south-west of the town of Yarkand (So-ch'e). Bushnell[17] quotes a Manchu author who wrote in 1777 that the Mohammedan natives 'ride up on yaks beyond the snow limit, light fires to loosen the jade, and dig out large lumps with their picks and rolled down the precipice into the valleys below'. This is a clear description of fire-setting, the age-old, universal mining technique also employed to extract lapis lazuli. From the Kara Kash trails lead over the mountains into Kashmir and Afghanistan. The Kara Kash is less than 160 km from the headwaters of the Oksu which becomes the Amu Daria or River Oxus, in part the border between the USSR and Afghanistan. Sir Mortimer Wheeler assumed that the Harappan nephrite beads came from Chinese Turkestan, and there is evidence for a considerable merchant class at Harappa which, with the attendant caravans of camels or pack-horses, is discussed by Stuart Piggott. Even goats may have been used: Piggott describes how the long-haired creatures may still be seen 'coming over the passes to India from Ladakh [in Kashmir], each with a little pack of rock-salt slung across its back'.[18]

The strikingly beautiful, semi-precious lapis lazuli is prized among artists as 'natural' ultramarine, an extremely costly pigment because of the remoteness of its parent mineral and because its hardness necessitates long and troublesome levigation in its preparation. It is a metamorphosed limestone coloured by lazurite, a sodium–aluminium–silicate. Like nephrite it was found at Harappa and in level II at Troy, but these occurrences are recent compared with its presence in the Gerzean neolithic of Egypt from about 4000 BC. It was not common; out of *c.* 2 000 graves in Egypt's predynastic cemetery at Nagada, only 11 contained beads of lapis lazuli.[19] At Tepe Gawra and elsewhere in Mesopotamia it probably occurred earlier than in Egypt. By 3500 BC it was common at Tepe Gawra, while the wealth of Shahr-i-Sokhta on the eastern border of Iran depended in part on the making of articles of alabaster and lapis lazuli. Thousands of finished and unworked articles have been excavated there. Its supposed scarcity in the Akkadian period, from around 2445 BC on the 'high' chronology, is illusory. Its disappearance from the graves at Ur is not surprising since Ur was in decline. Agade, the new capital, has yet to be discovered somewhere near Kish. *The Curse of Agade*, written *tempore* Naram-Sin (*c.* 2389–2352 BC) describes how the buildings were 'filled with gold, silver, copper, tin, and lapis lazuli' which were sent to the granaries 'as if it were only barley'.[20] In the Indus valley cities, lapis lazuli was a rarity: two beads and a 'gamesman' were found at Mohenjo-daro, three beads and a fragment of inlay at Harappa, and four beads at Chanhu-Daro.[21]

Whether lapis lazuli was scarce or abundant owing to varying fortunes of the trade or merely as a result of taste, it can be confidently claimed that all used in the

Near East came from the same area. Rumours of its occurrence in Iran can be traced back to Dionysius Periegetes, a Greek poet of the 1st century AD, who wrote in his *Descriptio Orbis Terrarum* that the land of Ariana (the eastern province of Persia) 'is unfruitful but yields many beautiful gems, and the country rock is traversed everywhere by veins holding lapis lazuli'.[22] Assyrian cuneiform inscriptions recorded lapis lazuli at Mount Damāvand, 70 km north-east of Tehran; Shah Abbas I unsuccessfully tried to revive mining there. Whatever Shah Abbas was after, it was not lapis lazuli. What he and Dionysius described was clearly plentiful and without doubt turquoise, itself used for ornamental purposes and found as beads at Mohenjo-daro and Tepe Gawra. Turquoise is a phosphate of copper and aluminium with a hardness of 6 on Moh's scale – about the same as lapis lazuli. It is not uncommon as a secondary mineral in copper deposits. The best-known locality for it today is Nishapur in north-east Iran, about 80 km west of Mashhad, where it encrusts and occupies 'cracks in a brecciated Tertiary lava'.[23] This admirably fits Dionysius' account. In any case, lapis lazuli is a rock, not a vein-filler.

This leaves only one locality as the principal source for lapis lazuli today and in antiquity. Of his journey through Afghanistan, Marco Polo wrote (AD 1271) that 'in another mountain are found the stones from which is made lapis lazuli of the finest quality in the world'. Darius the Great (522–486 BC) boasted that the lapis lazuli used in the construction of his palace at Susa came from Sogdia, ancient name for the region embracing parts of the USSR and Afghanistan including the province of Badakhshan. Four mines are known today 335 m above the Kerano-Munjan valley through which flows the Kokcha north from the Hindu Kush to the Amu Darya – the Oxus of the ancients. This is 'bare inhospitable country' with deep ravines and widely scattered settlements linked by rocky trails open for less than half the year.[24] Such extreme conditions were no deterrent to the determined trader. Marco Polo's epic journey was sufficiently long ago to give a fair idea of the time involved in travel in prehistory. From Badakhshan to Pashai (Kabul or somewhere near) took ten days; another seven took him into Kashmir.

Alexander the Great crossed this great range – the Caucasus as the Greeks confusingly called it – in pursuit of the pretender Bessus Artaxerxes in 328 BC, after wintering at the foot of the Hindu Kush and, naturally, founding another city of Alexandria. 'The crossing of the Caucasus, undertaken in the early spring, was an achievement which, for the difficulties overcome and the hardships of cold and want endured, seems to have fallen little short of Hannibal's passage of the Alps. The soldiers had to content themselves with raw meat and the herb of siliphon as a substitute for bread.'[25] Eventually reaching the Oxus 'after two or three days through the hot desert', he took his army across in the skin boats used by the natives then and now (or at least very recently) and advanced on Maracanda, the modern Samarkand. Alexander carved no new trail. He passed through established villages and along routes which must have been followed for several millennia by traders in lapis lazuli and, it will be argued, in tin; for whether Alexander knew it or not, when he reached the great valley of the Oxus he was in tin country. In 328 BC, Alexander founded here Alexandria Eschate (Alexandria the Ultimate), modern Kokand, to mark the limit of his empire on the bank of the Syr-Dar'ya. To the east lies Fergana, famous for its horses and its tin. What the above account shows is that immense distance and hardship was no barrier to trade, even before the 3rd millennium BC. What applied to a semi-precious stone such as lapis lazuli must also have applied to the trade in scarce metals, above all tin.

The mechanism of such trade can only be guessed at. Time may have been of small consequence so it is not necessary to envisage impatient traders sent by the lapis lazuli workers of Shahr-i-Sokhte – still less Egyptians – hanging about the foothills of the Hindu Kush waiting for the snows to clear before making the annual summer dash to Kerano-Munjan. Crawford found that among the nomads of Afghanistan it is common to wait a whole year for goods paid for in advance.[26] Trade in lapis lazuli may have been a little more efficient than that, but if it bears any resemblance to the metals trade it is no wonder that tin remained generally scarce in the Near East.

South of the Siberian Lake Teletskoye at Pazyrik were excavated the justly famous 'frozen tombs'.[27] Among the wealth of material preserved here, tin was found to have been used unalloyed as a foil for decorating several articles. Little is said of the location of the tin workings and nothing of how the ore was extracted. All that can be learned is that tin was obtainable 'on the south-western borders of the High Altai, where the copper also was probably worked'. An expedition in the 'Western Altai' in 1935–7 examined 'ancient tin workings dating from the fourteenth to the third centuries BC', and reported that the Andronovo tribes (named after the type-site north-west of Krasnoyarsk on the upper Yenisey) dug 'deep subterranean workings for obtaining copper, tin, and gold'.[28] Among the highly valued items found at Pazyrik were eight head-dresses belonging to eight splendid Fergana horses, animals known to have been prized even in China. Those at Pazyrik or their ancestors, originated in the Fergana basin nearly a thousand miles to the west. Apart from horses, knowledge of tin probably also spread to Pazyrik and the upper Irtysh tin-fields from Fergana sometime in the 2nd millennium BC.

The nature of the tin-bearing deposits of Fergana is not easily deduced from the available literature. Ore-bodies extend, probably discontinuously, for at least 1 000 km through the mountains from the Sarydzhaz region in the north-east to at least as far south-west as Maikhoura. Tin also occurs with wolfram in quartz–topaz greisens and related rocks in the Karaganda region of central Kazakhstan, though here the tin is 'poorly distributed',[29] so unlikely to have been known before its recent discovery. How much tin is mined now is not known, but the account given by Materikov illustrates the interest taken in the region by Soviet geologists. Mines were opened up about 1940 in the Kirgiziya Khrebet west of Lake Issyk-kul'.[30] Apart from tin mining in the highlands surrounding the Issyk-kul' basin, tin from the Salair range south-east of Novosibirsk, as well as concentrates from Kazakhstan, were smelted in Novosibirsk.[31] The Fergana region is not important now, but there has certainly been production at Sarybulak where 'the valuable components of its ores are tin, lead, zinc, copper, antimony and silver'. Placer tin deposits also exist (it would be curious if they did not), though Materikov implies that none are worked nowadays. They should be as widely scattered as the parent lodes, so could occur near Maikhoura within relatively easy reach of prehistoric traders from the Indus and Mesopotamian civilizations.

The antiquity of mining in this part of the Soviet Union is discussed by Litvinskii[32] who led an expedition in June 1946 to early tin workings in the Ziyaddinskii and Zerabulaskii hills of eastern Uzbekistan. Bronze age mining had already been suspected in the Fergana region because of the high tin content of bronze artifacts. Historical evidence for mining is fragmentary, but local production is suggested by Makhdisi (AD 985), who referred to the export of tin vessels from Robindzhan (or Arbindzhan) identified with the archaeological site of Ramadzhan-tepe just east of Zerabulak.

Only 23 km south of Zerabulak lies the 'kishlak' or village of Changalli, and 1·5 km north-east of Changalli is the kishlak of Kochkarli, at both of which are tin deposits. At Changalli a series of outcropping tin-bearing quartz lodes had been worked opencast to a depth of 10–12 m. Those at Kochkarli extended over about 150 m and, though partially filled through weathering, were probably 7 or 8 m deep and 4 m wide at the bottom. Perhaps as a result of weathering, no evidence for fire-setting or the use of wedges was found, but at Changalli there were many implements made of 'hornstone'. One piece of green-glazed pottery at Kochkarli, and two pieces of unglazed wheel-thrown pottery at Changalli are probably mediaeval. Pottery collected from the suspected mine settlement, a tepe 80 m west of Changalli, pointed to the 10th and 12th centuries as the latest period of occupation.

In the eastern Ziyaddinskii mountains, 30 km south

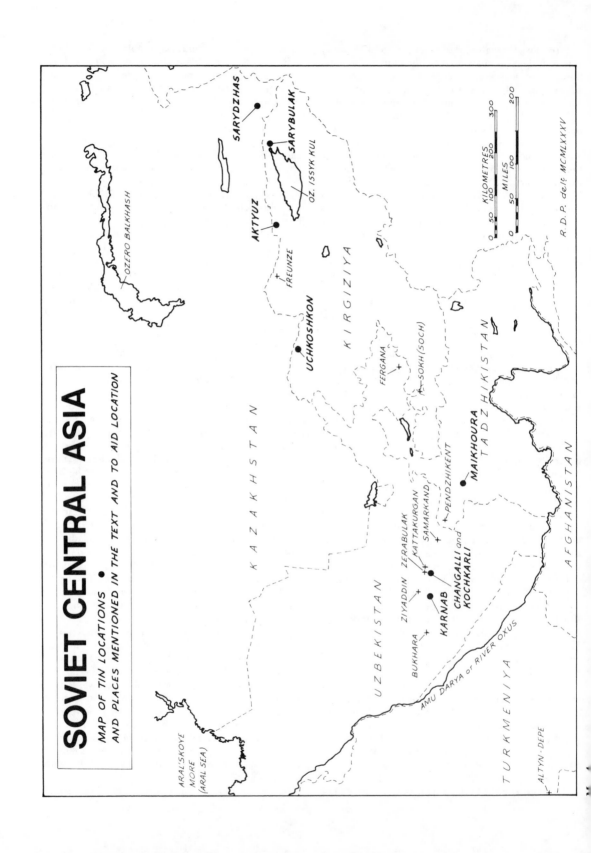

SOVIET CENTRAL ASIA

MAP OF TIN LOCATIONS ●
AND PLACES MENTIONED IN THE TEXT AND TO AID LOCATION

KAZAKHSTAN

KIRGIZIYA

UZBEKISTAN

TADZHIKISTAN

TURKMENIYA

AFGHANISTAN

ARAL'SKOYE MORE (ARAL SEA)

OZERO BALKHASH

SARYDZHAS

SARYBULAK

OZ. ISSYK KUL

AKTYUZ

FREUNZE

UCHKOSHKON

FERGANA

SOKH (SOCH)

MAIKHOURA

PENDZHIKENT

SAMARKAND

KATTAKURGAN

ZERABULAK

ZIYADDIN

CHANGALLI and KOCHKARLI

KARNAB

BUKHARA

AMU DARYA or RIVER OXUS

ALTYN-DEPE

R.D.P. del? MCMLXXXV

KILOMETRES
0 50 100 200 300

MILES
0 50 100 200

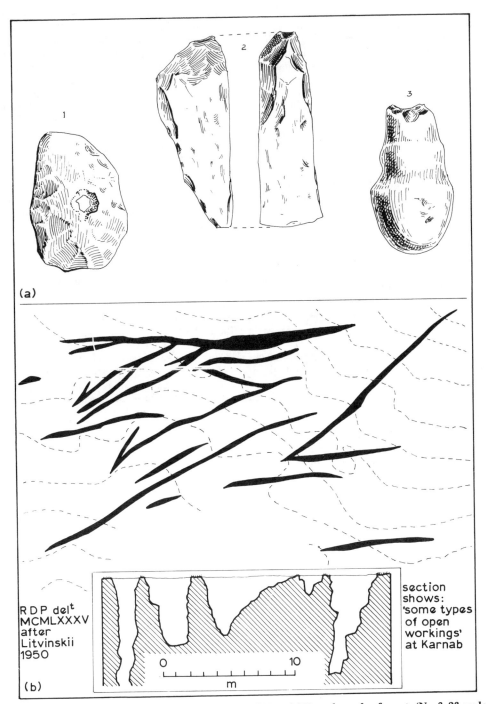

9*a* '**Mining implements of stone from Changalle and Karnab**' 1 and 3 Karnab, made of quartz (No. 3, 23 cm long)
   2 Changalle, made of 'hornstone' (no scale)
 b Plan showing 'group of ancient workings' on tin veins at Karnab (no scale or contour intervals)

of Ziyaddin, lies the kishlak of Karnab where more extensive mine workings covered 10 ha and 'a significant quantity of tin was obtained' (*see* Fig. 9). Open-cast pits varied between 10 and 100 m long following the lodes which sometimes joined. Most workings were 2–10 m deep and 1–8 m wide. In one case a working reached a depth of 18 m, stopping at the water-table below which the rich ore was left untouched. Some open-cast pits led into underground galleries up to 80 m long. Hundreds of stone hammers of several types were found; 'a whole heap lying in one corner of an underground chamber'. Curiously, no metal tools were found. There were many pottery sherds, and a skeleton buried in an already collapsed working was accompanied by a jug at the head and a bowl at the feet – a style reminiscent of pottery belonging to the 5th and 6th centuries AD, though dating was uncertain as a jug of a later mediaeval type was also found.

Litvinskii concludes that tin was probably mined here in the first centuries AD and ceased when the levels above the water-table had been fully exploited. In any case, during the Mongol invasions of the 12th century, many towns were destroyed (Robindzhan in 1158) and others 'went to the devil', effectively putting an end to mining. Litvinskii's paper does not prove tin mining in Uzbekistan in the bronze age, but it does prove early working in the region with the best documented occurrences of economically workable tin in central Asia north or south of the Soviet border.

South of the Amur Darya lies Afghanistan where in the 1970s Soviet geologists discovered 135 locations with 'indications of tin', though only two, 'Vostok' and Magn, were considered suitable for further investigation.[33] There may be a little tin in south-west Badakhshan, but to the south in Nuristan some 10 000 km^2 contain tin-bearing pegmatites. West of Kabul pegmatites 35 m wide and 300 m long occur north of Besud, although the tin is 'extremely sporadically distributed in the vicinity of the greisenized pegmatites' and does not exceed 3·5%. Wertime reported that traces of tin occurred in copper deposits at Mukur, 160 km south of Besud.[34] In the south-west, tin was found notably in the vicinity of the River Arghandab, which joins the Helmand flowing to Lake Sistan on the Iranian border. The cassiterite is in small quartz veins forming 'nest-like' concentrations of up to a few per cent, together with such typical associates as tourmaline, wolframite, molybdenite, and beryl. There is even a little gold in places. A Cornishman could imagine himself at home in such an environment. Amounts of tin may be small, but some alluvial deposits can be anticipated, and the Russian discoveries here are of particular interest to archaeologists who, until now, have not looked favourably on the assertion of Strabo (*c.* 54 BC–AD 24) that the Drangae, inhabitants of the region of Lake Sistan, 'who otherwise are imitators of the Persians in their mode of life, have only scanty supplies of wine, but they have tin in their country'. It

can no longer be claimed that Strabo was misled or misinformed by travellers' tall stories.

Over 300 km north of Sistan is the town of Herat. The plutonic rocks of the 'Herat zone' are also tin-bearing, as in the Bulghaja granite in a band 5–18 m wide and 600 m long. The cassiterite content is no more than 0·47%, with sulphides of copper up to 1%, and lead also 1%. In this complex are 'Vostok' and Šand; the latter deposit contains only about 1% tin, but the former contained up to 6%. In Cornwall at the present time, ores are worked which contain as little as about 0·6%.

It is too early yet to assess the importance of the Afghan discoveries. Tin is certainly widespread, and one wonders whether it really does peter out before reaching the Pakistan and Iranian borders. Kohl noted the discovery by Soviet geologists of apparently 'huge deposits' of tin along the Iranian border north of the Helmand basin 'in association with numerous archaeological sites'.[35] The results of excavations might be exciting in view of Strabo's words, but 'huge deposits' may be a too glowing report of recoverable ore, and it would be unwise to optimistically view Afghanistan as an eastern equivalent of Cornwall or Iberia. Until more data are available it seems safer to regard the tin-fields north of the Amur Darya as a more likely source of metal in prehistory.

The archaeological evidence from Afghanistan is not unequivocal. Analyses of metal finds from the Snake Cave at Ghar-i-Mar, near Mazar-i-Sharif, on a tributary flowing north to the Amu Darya, pose more questions than they answer. A few objects proved to be of smelted copper ores containing no tin. What is surprising is the discovery in 1962 of corroded pieces of sheet metal bearing traces of an embossed design and made of a low tin content bronze.

68·32%Cu; 5·15%Sn; 0·17%Fe; 0·01%Ni.

The remainder is siliceous matter. The uncorroded metal is thought to have contained nearer 7% tin.[36] These fragments came from the deepest level in the Snake Cave, contemporary with the earliest occupation dated by ^{14}C to around 5487 and 5291 bc.[37] (5487 bc is too early to be callibrate; 5291 ± 100 BC = 6585–5595 BC.) If this dating is acceptable, not only is this metal the earliest tin bronze known from anywhere, but it is also an isolated occurrence far older than its nearest rival and quite unrelated to the main development of bronze age metallurgy. The Snake Cave bronze can be justifiably regarded as a freak. Pieces of pins or rods found in 1966 in a 'goat culture' neolithic site at Darra-i-Kur in Badakhshan are also low-tin bronzes, but probably date between about 2190 and 1800 BC.

Cu	Sn	Fe	Ni	Pb	Zn	Ag	Non-metals
70·98	4·91	0·07	0·01	0·65	0·03	—	23·10
90·44	5·23	0·17	0·08	2·56	—	0·08	1·44

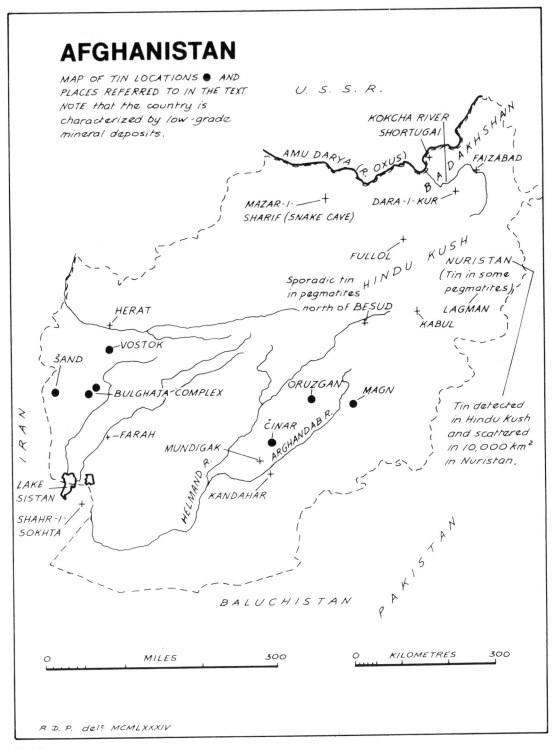

AFGHANISTAN

MAP OF TIN LOCATIONS ● AND
PLACES REFERRED TO IN THE TEXT
NOTE that the country is
characterized by low-grade
mineral deposits.

U. S. S. R.

KOKCHA RIVER
SHORTUGAI

AMU DARYA (R. OXUS)

BADAKHSHAN

FAIZABAD

MAZAR-I-
SHARIF (SNAKE CAVE)

DARA-I-KUR

FULLOL

HINDU KUSH

NURISTAN
(Tin in some
pegmatites)

Sporadic tin
in pegmatites
north of BESUD

LAGMAN

KABUL

HERAT

VOSTOK

ŠAND

BULGHAJA-COMPLEX

ORUZGAN

MAGN

Tin detected
in Hindu Kush
and scattered
in 10,000 km²
in Nuristan.

ČINAR

ARGHANDAB R.

+-FARAH

MUNDIGAK

HELMAND R.

KANDAHAR

IRAN

LAKE
SISTAN

SHAHR-I-
SOKHTA

BALUCHISTAN

PAKISTAN

| 0 | MILES | 300 |

| 0 | KILOMETRES | 300 |

R.D.P. del⁲ MCMLXXXIV

Map 5

Other bronzes from the Snake Cave contained around 19–20% tin, but these were dated to around AD 300. They differed from the contemporary high-tin bronze of China in being low in lead or devoid of it, suggesting that they were made from locally derived ores. So far, 'optimum' tin bronze with about 10% tin has not been found in Afghanistan.

The early Snake Cave tin bronze is in marked contrast to the finds made at Mundigak, a series of mounds some 55 km west-north-west of Kandahar and not far from the Argandah River. Excavations since the 1950s have produced one single artifact from period I$_5$ (*c.* 3500 BC) with about 1% tin – hardly a significant amount but not surprising considering the Soviet discoveries quoted above. In period II (*c.* 3500–2800 BC), tin is still scarce though differing alloys indicate attempts to smelt metals suitable for different purposes. Pins, 'points', and knives still had about 1% tin, but axes and adzes had more, though barely 5%. Wherever the tin came from it was used sparingly, a fact which could well mean that it was being transported a very long distance – perhaps from the north of Afghanistan or from across the border in the Soviet Union. Other objects of long-distance trade found at Mundigak are beads of (?) jade, and lapis lazuli, especially, but not exclusively, in period IV (roughly equivalent to the 3rd millennium).

Excavations in Soviet Central Asia have shown that this area maintained links with the better-known centres of the civilized world. At Altyn-depe, for example, in the period *c.* 2100–1750 BC there was contact with Mesopotamia. Architecture was influenced by the use of a multi-stepped tower, and a golden-headed bull from the priest's tomb is Mesopotamian in style. In the other direction there were links with the Indus civilization, in the shape of ivory sticks and gaming pieces. Metalwork at Altyn-depe is arsenical copper (1·5–8% arsenic), and occasionally up to 12% additional lead. Tin bronze has yet to be found;[38] perhaps it will not be. The important fact is the extent of the trade; it would be surprising if every site in the Near East yielded finds of nephrite, lapis lazuli, or tin. Few archaeological sites have been explored in Iran away from the Iraq border, in Afghanistan, or in Soviet Central Asia. Excavations at such sites as Mundigak, Namazga-depe, and Tepe-Yahya point to even more interesting discoveries in the future. Lamberg-Karlovsky, whose name is welded to Tepe Yahya as firmly as Schliemann's is to Troy, has summed up the emerging picture thus:[39]

> It is perfectly apparent ... that a very large geographical expanse must today be considered in order to understand developments within parts of it ... The interactional patterns which integrate Sumer, Elam, the Persian Gulf, the 'Helmand Civilization' of Seistan, Turkmenistan, Baluchistan, and the Indus Valley remain poorly understood. It becomes increasingly obvious, however, that the urban process within one area cannot be understood in the absence of an understanding of contiguous areas. ...

Evidence from a few sites can be used to illustrate this. Impressive evidence for trade was unearthed in 1966 when peasants chanced upon well preserved gold and silver vessels and other fragments in a narrow gorge 2 km south of Fullol, a small village on the Sohi Azora tributary of the Amur Darya.[40] Unhappily, the treasured objects were broken up when the peasants divided the spoils, but they are sufficiently well preserved to show that their decoration and skill of execution invite direct comparison with the products of Mesopotamia and Elam. Nor were all the items imported along southern routes: two of the silver bowls are characteristic of Turkmenia in Soviet Central Asia. The hoard has been dated to the middle of the 3rd millennium, although Lamberg-Karlovsky[39] favours a date nearer 3000 BC. Whatever the date, the hoard fits well into the pattern of the lapis lazuli trade which has proved to be so important to the town of Shahr-i-Sokhta.

Even more exciting is the evidence from Shortugai. Since the discovery of the first Indus finds at Harappa in 1921, the sphere of influence of this civilization has been greatly extended, first southwards to Gujarat and the Makran coast of Baluchistan, and now into northern Afghanistan. In 1975, French archaeologists discovered on the surface at Shortugai, sherds of Indus pottery extending over more than a millennium – the whole span of the Indus civilization.[41] The sites are clustered above the confluence of the Amur Darya and the Kokcha. Finds also included gold and, not unexpectedly, much lapis lazuli. Particularly important is a Harappan seal bearing an engraved rhinoceros and an inscription which reinforces the belief that the site was a trading post. Shortugai is only 800 km from Harappa, as the crow flies, though the journey involves hundreds of kilometres of mountainous terrain through the Hindu Kush. Other sherds compared well with Turkmenian pottery of the early 2nd millennium (Namazga levels V and VI).

Lyonnet's conclusion was that the most likely explanation for their existence was an interest in 'the mineral resources of the Iranian Plateau and of Central Asia', to which can now be added those of Afghanistan itself. Indus contacts extended well into Turkmenia where the principal bronze age settlements, such as Altin-depe and Namazga-depe, lie close to the Iranian border. Imports here include square and oval gaming-counters of Indian ivory, and decorated sticks, numerous at Mohenjo-daro, related to types described in Sanskrit texts as being used in fortune-telling. The flat daggers of southern Turkmenia also closely resemble Harappan types.

As metal users the inhabitants of the Indus cities were very conservative. As already noted, tin bronze was rare. Their conservatism is also shown by their adherence to the flat axe, often of copper, rather than the adoption of superior types which their technology

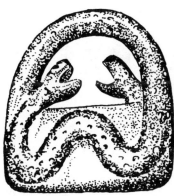

10 **Amulet of black stone decorated with two snakes found at Soch, Fergana (re-drawn from a photograph in *Iran*, 1971 (scale not given), of the object in Tashkent Historical Museum)**

would have allowed. A fine copper axe-adze from Harappa, and similar bronze examples from Chandhu-dara and, in Baluchistan, at Shari-tump, are rare imports of the superior shaft-hole implements developed initially in Mesopotamia before 3000 BC. In northern Iran examples have been found at Shah Tepe, Tureng Tepe, and Tepe Hissar in level IIIc (2000–1500 BC). Just over the Soviet border there is a fine example from Daima near Ashkhabad. This corner of Iran and adjoining Turkmenia show other close links, especially in pottery types going backward in its adoption of good-quality bronze. Tin bronze first appeared at Tepe Hissar in level IIIc and remained at least as rare as in the Indus cities.

Tin was more commonly used in eastern Iran, an area only now emerging from obscurity through the excavation of key sites such as Tepe Yahya and Shadad. In level IV*b* (*c.* 3000 BC) at Tepe Yahya was found a dagger of 3% tin bronze;[42] perhaps the result of using a tin-rich copper ore. However, in later levels tin bronze became a 'significant element in its material culture' comparable with other evidence from southeast Iran where at Shadad bronze shaft-hole axes and bronze vessels were found in graves dated to *c.* 2500 BC.[43] The richness of Tepe Yahya, Shahr-i-Sokhta, and Shadad, are all indicative of trade and 'an accumulation of wealth unsuspected from the area'.[44] Shadad has also yielded lapis lazuli, carnelian, turquoise (perhaps from Nishapur), shell necklaces, silver and bronze cylinder-seals, bronze standards, and bowls carved from steatite and alabaster. Such trade extended into Turkmenia and can be seen at Namazga-depe by 3000 BC (period III). Mason and Sarianidi consider the possibility that 'the most enterprising traders of Sumer and Babylon actually came to these remote outposts of the world of commerce of the day'.[45]

Namazga-depe and neighbouring sites are a long way from the important tin reserves of Fergana, but if it is assumed that this is the region from which most tin must have come – simply because on present evidence tin was far more common here than anywhere in Afghanistan – a thin trail of finds can be followed deep into this metalliferous region. Nearly 480 km northeast of Namazga-depe at Zaman-baba is pottery identical with that from level IV (*c.* 2500 BC) at Namazga-depe. More importantly, the metal types are of southern inspiration, though 'smelted from the local ores, judging from the rare impurities in their composition'.[46] Mason and Sarianidi trace imports further towards Fergana: 'The next such point after Zaman-baba was Pendzhikent, where an adze of the type found at Daina was discovered. At Fergana itself was found a hoard which included some particularly interesting pins with elaborate heads, including a double-spiral and a representation of a cow being milked. The southern influence, or perhaps even the southern origin of these articles seems certain.' Brentjes has also described a black steatite object, now in the Museum at Tashkent, found at Soch in Fergana (Fig. 10). It is regarded as an Elamite import of the mid-3rd millennium BC, a thousand years older than formerly believed; the style is comparable with steatite dishes found in Mesopotamia and India.[47]

Southern influence in Turkmenia extends to a primitive form of writing, though whether the need for it arose as a result of trade is not known since the script has not been deciphered. A number of repeatedly used symbols on Turkmenian figurines are, in some cases, identical with those found on clay tablets at Tepe Godin, Tepe Yahya, Shahr-i-Sokhta, and elsewhere in Iran in the early script known as Proto-Elamite. Some Turkmenian symbols also resemble early Sumerian types, while others show similarities with the Harappan script, which is also undeciphered. Such wide-ranging links will be strengthened as excavations proceed in previously unexplored areas of eastern Iran, Afghanistan, and Soviet Central Asia. The origin of Near Eastern tin remains unproven; the geological evidence would favour the deposits of Fergana and the Tien Shan range. Until more recoverable amounts, more promising than 'indications of tin', are found in western Afghanistan where Strabo believed it was found, it must be assumed that this valuable metal was transported along trade routes as hazardous and long as those that carried lapis lazuli and the rarer nephrite.

A NOTE ON RUSSIAN TIN LOCATIONS

Several writers on tin deposits have admitted that some locations within the USSR are difficult to locate accurately and should be treated with caution. Map references, some question marked, appear in the English edition of V. I. Smirnov's *Ore Deposits of the USSR* (Pitman, 1977), and these may differ from places with the same name in *The Times Atlas of the World.* Thus, there are several places in Soviet Central Asia called Sarybulak. One in Kirgiziya (42°47′ N 78°17′ E), marked on the map of Soviet Central Asia, is the location given in Smirnov's book and is not the town in *The Times Atlas*, also in Kirgiziya, at 40°52′ N 73°54′ E. The more important locations of Karnab and Changalli in Uzbekistan, with their evidence of early mining, are correct.

NOTES AND REFERENCES

1   BURNEY, 1977, 86; MUHLY, 1973, 306–7, 449 note 542
2   CHAKRABARTI, D., in Sherratt (ed.) 1980, 165–6, notes the long trade routes of the late 3rd and 2nd millennia BC with turquoise, lapis lazuli, jade, and tin all reaching the Indus Valley which he includes within 'Meluhha'
3   DAVIES, 1919
4   HOSKING, 1974, map fig. 2; HESS & HESS, 1912
5   WHEELER, 1953, 58
6   ITC, 1964; MALLET, F. R., 1874, *Rec. Geol. Surv. India*, quoted in Hess & Hess, 1912
7   ITC, 1964; OATES, R., 1895, *Trans. Fed. Inst. Mining Eng.*; quoted in Hess & Hess, 1912
8   DAVIES, 1919
9   LAL, 1972
10 RUDRA, S. C., 1904, *Trans Am. Inst. Mining Eng.*, quoted in Hess & Hess, 1912
11 HOLLAND, T. H., 1904, *Rec. Geol. Surv. India*; quoted in Hess & Hess, 1912
12 LAMBERG-KARLOVSKY, 1967, 149
13 TYLECOTE, 1976, 11
14 MOOREY, P. R. S., in Coghlan, 1975, 44
15 WHEELER, 1953, 59
16 SCHLIEMANN, 1882, 171–2, illustrates an axe 50 mm long by 14 mm wide
17 BUSHNELL, 1921, vol. 1, 120–4
18 PIGGOTT, 1950, 175
19 PAYNE, 1968
20 MUHLY, 1973, 314
21 WHEELER, 1953, 59, 88
22 ADAMS, 1954, 25
23 FEARNSIDE & BULMAN, 1961, 207
24 HERMANN, 1968; PAYNE, 1968; LATHAM, 1958, 46
25 BURY & MEIGGS, 1978, 74–5
26 CRAWFORD, 1978
27 RUDENKO, 1970, 207
28 MONGAIT, 1961, 148
29 MATERIKOV, 1977; *Tin International*, January 1970, 20, reported that prospecting in the Tien Shan mountains in 'areas stretching for hundreds of kilometres' had located big tin deposits, while the June issue, 165, noted a considerable tin–wolfram body at Karagayla, Kazakhstan, 300 km north of Lake Balkhash
30 MELLOR, 1966, 249–50
31 GREGORY, 1968, 554, 837
32 LITVINSKII, 1950
33 WOLFART & WITTERKINDT, 1980, 410–11. The ores are generally linked with Eocene and Oligocene intrusives of the Alpine orogeny
34 WERTIME, 1978, 3
35 KOHL in KOHL (ed.), 1981, 110 footnote
36 CALEY, 1972
37 SHAFFER, J. G., in Allchin & Hammond (eds.), 1979, 91, 141–4
38 MASSON, V. M. & KIIATKINE, T. P., 'Man at the Dawn of Civilization', Ch. 5 in Kohl (ed.), 1981
39 LAMBERG-KARLOVSKY, 1978
40 TOSI & WARDAK, 1972
41 LYONNET, 1977
42 LAMBERG-KARLOVSKY & LAMBERG-KARLOVSKY, 1971; MUHLY, 1973, Appendix 11, 347
43 BURNEY, 1975
44 LAMBERG-KARLOVSKY, 1973, reviewing Masson and Sarianidi (1972) in *Antiquity*, 43–6
45 MASSON & SARIANIDI, 1972, 120–9
46 MASSON & SARIANIDI, 1972, 128
47 BRENTJES, 1971

4   Tin in Northern Russia and Siberia

For the archaeologist interested in tin used in the Near Eastern civilizations, the most important Soviet deposits are those in Central Asia already discussed. These are not the only deposits: the USSR is rich in tin. Its undisclosed output is estimated to be in the order of 30 000 t a year, as its imports from Malaysia and elsewhere averaged only 5 600 t a year from 1964 to 1974. Compare the virtually tinless USA, which imported an average of 47 000 tonnes of metal a year during the same period.[1] The USSR has not always made the most of its assets. Tin mining was minimal before the revolution of 1917. Between 1881 and 1896 production reached a peak of 19 t in 1888.[2] Throughout much of its history, European Russia imported its tin from the west, including Cornwall. One of the few records of the Cornish tin trade in the early mediaeval period refers to the metal reaching Russia via Germany and Austria in the 9th century.[3]

The bulk of present-day output is in the extreme east, from Primorskiy Kray to Yakutskaya and Agadanskaya, especially in the Anadyr Khrebet and Chukotskiy Khrebet where there are veins of tin-wolfram and sulphide ores as well as alluvial tin deposits.[4] This remote region has only been opened up in the last half century following successful exploration and development of known reserves in the Nerchinsk region. A mine was opened at Olovayannya on the Onon River in 1933, and others soon followed in the same area.[5] The Onon deposits may be of little importance now, but they were among the few worked in Tzarist Russia. Veins and placers were exploited along more than 110 km of the Onon and Ingoda rivers, and the Malaya Koulinda tributary of the Onon. Tin was found here in 1811 and worked at Olovayannya to about 1835 or 1852, and again by Germans before World War I.[6] Natives are said to have worked the tin before the 19th century, though how far back into antiquity is not known.[7] The area is so remote that it is difficult to imagine that it could have been involved in any prehistoric trade, either with the Near East or with China, but it is unwise to be dogmatic about that. Mongait wrote that bronze-using people in the 2nd millennium BC extended deep within Yakutskaya 'along the Lena and its tributaries up to the Arctic Circle',[8] so it is not impossible that the tin deposits of Nerchinsk were discovered in antiquity.

The only other area of recorded tin production in the 19th century is at Pitkyaranta on the north-east shore of Lake Ladoga near the Finnish frontier. The principal ore is magnetite, but 497 t of tin were recovered between 1818 and the closure of the mines in 1904.[9] This difficult deposit cannot have been exploited before that.

The Ural Mountains are justly famous for their copper ores. No mineralogical collection of distinction is without its polished slabs of malachite from Sverdlovsk, formerly Ekaterinburg. There are also reports of tin which have attracted the attention of archaeologists. Childe, for example, mentions tin from Chelyabinsk,[10] south of Sverdlovsk in Miass Province. What is not usually appreciated, however, is that the authenticated occurrences of tin do not consist of cassiterite, nor is there anything in the archaeological record to indicate that recoverable tin was obtained here in antiquity. The analysis of bronze age artifacts of the Turbino culture of the later 2nd millennium BC shows them to consist of 98 or 99% copper with mere traces of a wide range of impurities – tin, lead, iron, nickel, arsenic, and antimony as well as non-metallic elements – all characteristic of the local copper sulphide ores.[11] Bronze age moulds, crucibles, and hammers for crushing the ore have all been found in the Sverdlovsk region.[12]

In the mining literature, reports of tin concern its occurrence in gold placers in Miass (or Miask) and the adjoining province of South Ural, as well as in Perm and Orenburg.[13] Old reports concentrate on the claim that tin is 'not uncommon' as *native tin* in South Ural, Miass and elsewhere. Native tin has always been regarded with suspicion by mineralogists. It has almost always been attributed to some secondary, usually human, agency as in Nigeria and Cornwall. In the Urals, with no tradition of gold working before 1829,[14] natural explanations have been offered.

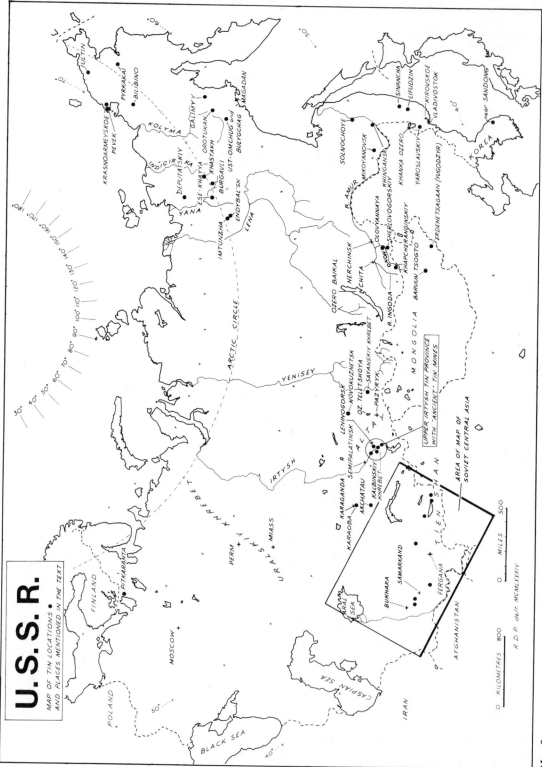

U.S.S.R.

MAP OF TIN LOCATIONS ●
AND PLACES MENTIONED IN THE TEXT

Map 6

Hautpick suggested that cassiterite was smelted naturally during the combustion of vegetation,[15] producing native tin 'in the form of little metallic grains ... combined with a little lead', recorded from the gold washings (*les lavages d'or*) of Siberia as long ago as 1844. Here it was assumed the miners caused their own native tin; when the wet ground was frozen to a considerable depth, shafts were sunk in the gold-bearing alluvials with the aid of fire-setting, the walls of ice serving to support the excavations.[16] As far as the Urals are concerned, the lack of cassiterite makes these attempts at explanation implausible. A far better alternative is offered by Hosking, who believes that the supposed native tin is really associated with stannides of the platinum group of elements, several of which are known from Noril'sk and Talnakh.[17] Hosking accepts only one occurrence of native tin as certain: that reported in 1954 from three veins in the Nesbitt La Bine Uranium Mine, Beaver Lodge, Saskatchewan, Canada, where in the sequence of paragenesis it followed the formation of copper sulphides, zinc blende, and galena. Native tin and native copper are said to have been found in Cornwall 'in a cavity of a lode in Crickbraws [Creegbrawse, Kenwyn] but rare, only a few specimens found'.[18]

Returning to the bronze age, the only objects of the Turbino culture containing a significant amount of tin – up to 6 and 8% – are bronze ornaments from the cemeteries at Turbino itself, a site near Perm, and at Ust'-Gajva. It has not been seriously doubted that the tin was imported from Siberia; evidence for this long-distance trade is quite convincing. A characteristic series of socketed axes in European Russia and Siberia, called Sejma axes after the type-site near Gor'kiy, have been found at places scattered in a broad belt as far east as Irkutsk with a marked concentration in the Semipalatinsk region. It is clear from available analyses that not all the Siberian socketed axes had a high tin content. One from near Minusinsk contained no more than 67% and 'probably considerably less'.[19] Sejma axes correspond in date roughly to the middle bronze age of western Europe, *c.* 1000 BC. Their origin is assumed to have been in eastern Siberia.

Tin deposits occur in the upper Irtysh basin above Semipalatinsk in the Kalbinskiy Khrebet, and it is accepted by Soviet archaeologists that both tin and copper were worked here in the 1st millennium BC. Their discovery by prospectors can be postulated, perhaps initially for gold, in the placer deposits which are widespread in the Altai. It is from here in the Semipalatinsk region that the Scythians obtained much of their gold. The centre of Scythian influence was far to the west in the hinterland of the Black Sea. They had access to seemingly unlimited supplies of gold, which came from the east. Herodotus reported that it was from the Issedones, a tribe living east of the Irtysh, that originated 'tales of the one-eyed men and

the griffins which guard the gold'.[20] When a man died his body was served up with lamb and eaten by his kinsmen.

> The dead man's head, however, they gild, after stripping off the hair and cleaning out the inside, and then preserve it as a sort of image, to which they offer sacrifices.

There was closer contact between the western Scythians and the tribes living west of the Issedones – the Argippaei, whose territory included the upper Irtysh basin. Much was known about them, according to Herodotus:

> from the reports, which are easy to come by, of Scythians who visit them, and of Greeks who frequent the port on the Dnieper and other ports along the Black Sea coast. The Scythians who penetrate as far as this do their business through interpreters in seven languages.

Herodotus does not describe the nature of the business, but it can safely be assumed that it concerned metals since the region contained little else of value and was frozen solid for eight months of the year. The Argippaei were of Mongolian type, shaved their heads (Herodotus quaintly believed them to be bald from birth) and were:

> supposed to be protected by a mysterious sort of sanctity; they carry no arms and nobody offers them violence; they settle disputes amongst their neighbours, and anyone who seeks asylum amongst them is left in peace.

Their 'sanctity' has been compared with that enjoyed by certain tribes of African blacksmiths.[21] It also bears a similarity to Pliny's description of the Cornish tinners as 'especially hospitable to strangers'. The accepted view is that the Argippaei were skilled miners and foundrymen, perhaps also goldsmiths, who worked for all the neighbouring tribes. Supporting archaeological evidence is given by Sulimirski, but he does not go into details. Copper mining as early as the mid-2nd millennium in the mountains south of Karkaralinsk (370 km west of Semipalatinsk) continued into the period of Scythian influence *c.* 600–200 BC, while both tin and copper workings of the period have been found in the upper Irtysh, as well as traces of gold workings at Zmeynogorsk (160 km north-east of Semipalatinsk).

In chapter 3 mention is made of the Andronovo tribes digging subterranean workings for copper, tin, and gold. Rudenko[22] states that ancient mining in the High Altai has been little studied but that in many districts local outcrops of all these metals were worked. He says nothing about alluvial deposits of tin, and it may be that tin was obtained as an accessory in the prime search for gold. Gold was certainly plentiful whereas tin was not. Rudenko can cite only two bronze objects containing 'significant quantities of tin; a mirror and a torc from barrow number 2 at Pazyryk'. Copper varied considerably in its impurities suggest-

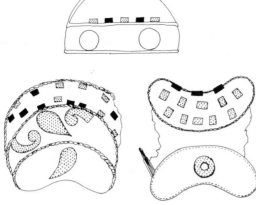

11   Iron age saddle arches from Pazyryk, Siberia, decorated with sheet gold (stippled) and small oblongs of tin foil (solid black); all from barrow 3 (re-drawn from colour plate 163 in S. I. Rudenko, 1970)

ing a variety of sources, but the gold was uniform in composition, an electrum with 20% silver, which Rudenko believed came from a single source. It is possible that the tin was an accessory accompanying this gold, a proposition that conveniently explains the scarcity of tin and its use as a decorative feature in conjunction with sheet gold. Tin was clearly very highly prized. Both metals were used to decorate the arches set at the front and back of saddles (Fig. 11), though sheet gold had wider applications for covering various articles of copper, bronze, wood, and textiles. On the saddle arches from Pazyryk illustrated by Rudenko, the tin foil appears as simple squares contrasting with the more lavish gold to which copper had apparently been added to enhance its colour.

North-east of Pazyryk in the Minusinsk region of the upper Yenisey River are more tin deposits. In the past it has been assumed that the Karasuk culture of this region (*c.* 1200–700 BC) was responsible for introducing a knowledge of tin bronze into China. It is now realized, however, that tin bronze was already established in China,[23] and that the population of the Minusinsk region increased partly 'through an incursion of some tribes from northern China', as proved by anthropological evidence.[24] The bronze knives with characteristically curved blades found here and in China are not, therefore, a Siberian development introduced into China, but the reverse, for the curved knife has a long and respectable Chinese ancestry traceable to stone age prototypes. As will be seen in the chapter on China, the picture that emerges is one of an independent development of Chinese metallurgy.

The bronze age implements of the High Altai are of very high quality, a feature used by Chard[25] to explain the fact that iron did not replace bronze here until the 4th to 2nd centuries BC. The region was doubtless far too remote to have supplied the Near East with tin, but the possibility that it was a source of tin for parts of the Far East remains to be explored, a point touched on in Chapter 7.

NOTES AND REFERENCES

1 ITC, 1964
2 ANON., 1956
3 GAHLNBÄCK, 1928
4 MATERIKOV, 1977
5 ITC, 1964
6 JONES, 1925; DAVIES, 1919. Working at Olovyannaya ceased in 1852 according to Jones, and in 1835 according to Davies
7 FAWNS, 1905
8 MONGAIT, 1961, 150
9 JONES, 1925
10 CHILDE, 1954
11 SULIMIRSKI, 1975, 151
12 GIMBUTAS, 1965, 616
13 ANON., 1958
14 DAVIES, 1901. Gold was first found in quartz veins cutting decomposed granite at Berezovskiy near Ekaterinberg (Sverdlovsk) in 1723. This was the only important underground mine. Sixty-six other sites were worked without success up to 1823, all of them alluvial. Government workings were established on the western side in 1829 and on the eastern side in 1838
15 HAUTPICK, 1912; HERMANN, R., 'Sur l'étain natif', *J. Prakt. Chem.* (1844) and *Ann. Mines* (1845), quoted in Hess & Hess, 1912
16 BROUGH, B. H., in discussion in COLLINS, A. L., 1893
17 HOSKING, 1974*a*
18 MS Catalogue of minerals in the collection of Phillip Rashleigh (1729–1811) at Truro Museum. The specimen (copper ores No. 100), recently analysed at Camborne School of Mines, certainly contains metallic tin, but it is not considered to be native. A possible explanation is that it was caused by an underground fire, intended or accidental
19 COGHLAN, 1975, 125
20 SÉLINCOURT, 1972, 278–9
21 SULIMIRSKI, 1970, 23, 27
22 RUDENKO, 1970, 138, 207–8. For a reconstruction showing the horse harness with saddle arches in place, see Clark, 1977, 196, Fig. 108
23 WILLETTS, 1965, 74
24 MONGAIT, 1961, 149
25 CHARD, 1974, 153

5 Tin in China

For an archaeologist the greatest problem in the Shang period is the explanation of the apparently sudden rise of highly skilled bronze metallurgy.

[William Watson, *The Genius of China*, 1973]

Modern methods of archaeological research established under the direction of the late Li Chi in the 1930s have been pursued with vigour in the People's Republic of China. Results published by the Academica Sinica (Nan-ching) covering the first seasons of excavations at Hsiao-t'un, Anyang, and by the Academy of Sciences, Beijing, embracing a very exhaustive series of excavations, have been impressive, even startling. They have overturned long-cherished beliefs, not the least of which concerns the origin of the Chinese bronze age.

The dominating factor held to have shaped the prehistory of Europe was, until recently, that of 'diffusion': the spread of knowledge and the mass movements of people outwards from the Near East like the ripples on a pond. This view is no longer tenable. Several authorities working on the history of Chinese metallurgy also claim now that Chinese founders invented tin bronze independently of the civilizations of the Near East. The appearance of this useful alloy in western Europe and China in the centuries before 2000 BC does not necessarily point to a common origin at some place midway between the two, attractive as the idea is, for the simple reason that tin-bronze metallurgy developed in China in a fundamentally different way from that in the West. Nevertheless, many contacts between the Near East and the Far East have been cited in the search for a western route for the introduction of tin bronze to China. The similarity between chariot types in the Near East and China has long been appreciated. In particular a 28-spoked chariot wheel from Lahashan in Soviet Armenia (*c.* 1400–1100 BC) resembles wheels from late Shang and Chou China (*c.* 1200–400 BC).[1] But any common influence here, if it exists at all, is too late, for tin bronze was being used in China before 2000 BC.

The commencement of metallurgy in China is now placed somewhere between 2500 and 2000 BC.[2] A link with China through the tin-rich Fergana basin – often called the gateway to China – appears very attractive.

Similarly, it has been suggested that a knowledge of tin reached China through the Minusinsk region of Siberia, but it now appears that the animal-art style of the Siberian Karasuk culture (1290–700 BC) is derived from the art of Shang China, not vice versa, in a one-way traffic flow. There is little evidence of anything reaching China until the late 1st millennium BC, when Minusinsk art returned the compliment by influencing Chinese art of the Warring States period.[3] As Chard put it, 'The Great Wall [of China] seems to have been a distinct cultural frontier and barrier long before it came into physical existence.'[4]

The possibility remains that Siberian tin reached China in late Chou times. The widespread occurrence of Siberian tin has already been discussed, occurring, for example, in the upper Irtish basin. Among the finds from Pazyryk is a fragment of a black-coated mirror dated to the 2nd century BC, and comparable to mirrors made over 2 600 km from Pazyryk at Lo-yang in Honan. Such mirrors are rich in tin (about 25%),[5] so that a trade in Siberian tin is a distinct possibility.

In the Near East, following the use of unalloyed copper, the use of tin bronze was preceded by, and co-existed with, the use of arsenical copper. By contrast in China, a precocious tin-bronze metallurgy apparently sprang, as Watson puts it:[6]

> from the ground fully-fledged technologically. Its earliest bronze-casting demonstrates a skill and sophistication which in Europe equates with the culture of La Tène 1st millennium BC ... and in the Near East would be matched in some extent, but not surpassed by work of the mid-second millennium BC.

The apparent lack of a more elementary technology in China has been a problem stretching beyond credulity the notion that even those proverbially clever Chinese could invent, unaided, such an advanced technology with such speed. But this picture is the result of historical accident, for by chance, the first bronze age site excavated under controlled conditions was at Anyang in the late 1920s – a Late Shang capital founded on this new site around 1400 BC. The annals of Yin record eight such moves of the Shang capital down to the time of T'ang, and five from then until P'an-keng.[7] Two earlier capitals situated near present-

CHINA

NOTE : *Incomplete data available on Chinese tin deposits*

DISTRIBUTION OF TIN LOCATIONS IN SOUTHERN CHINA AND WITHIN 400 KMS. OF ANYANG ●

EARLY COPPER MINES, MAINLY HAN PERIOD ☉

IMPORTANT SHANG PERIOD SITES ○

NATIONAL FRONTIERS – – –   PROVINCIAL BOUNDARIES

U. S. S. R.

U. S. S. R.

MONGOLIA

HWANG HO

TA-CHING

CHENG-TE-CHUAN-CHÜ

KOREA

BEIJING (PEKING)

400 kms

SUIDE

SHILOU
YÜN-CH'ENG

ANYANG

YIDU

ERLITOU

ZHENGZHOU

KYUSHU

TIBET

PANLONGCHANG

YANGTSE-KIANG

SHANGHAI

TUNG-LÜ-SHAN

NAN LING TIN BELT

SHI-CHAI-SHAN

TAIWAN

KO-CHIU

CHINA'S
PRINCIPAL
TIN DEPOSITS

VIETNAM

SOUTHERN TIN BELT

HONG KONG

LAOS

THAILAND

DON S'ON

O   KILOMETRES   600   O   MILES   400

R.D.P. delt MCMLXXXIV

Map 7

day Cheng-chou and Hui-hsien have been found since then to contain marginally more primitive, though still technically advanced bronzes. These discoveries surely anticipate the eventual discovery of an embryonic bronze age somewhere on the plain of Honan between 2500 and 2000 BC.

A number of essentially stone-age sites have yielded a little metalwork among the larger quantities of pottery, bone, stone, and shell in graves. At Chang-chia-tsui a large lump of 'copper' was found on an earthen tile in the neolithic level, while at a nearby site were several awls, ring-ornaments, and small unidentified artifacts of 'copper'. Because of the lack of analyses it is not known whether the Chinese word *t'ung* here means copper or bronze, though some are of copper – an important fact in establishing a copper period before the development of tin bronze. One of the earliest and most important sites is Yen-shih (*c.* 1650 BC) in an area where expectations are high for the eventual discovery of the birthplace of metallurgy. Especially significant here in the earlier phases are *chia* and *chüeh* wine cups of pottery which incorporate such elements as thin supporting rods and knob-stem representations, which are explicable only as copies of metal vessels that have not survived. Barnard and Satō reasonably explain this as an indication of the rarity of metal at this time, so that pottery copies would be adequate in burials.

As a result of more recent excavations several bronze *chüeh* have been found, which exhibit a progression in piece-moulding, ranging from Erh-li-t'ou (Early Shang) through Middle to Late Shang. These were studied by Barnard during a recent visit to mainland China. He noted a simple two-piece mould assembly with its simple, single-unit core from the Erh-li-t'ou period, to complex, later, single-unit cores plus three leg-face units and three intervening units of Late Shang date: that is a technical progression from three to seven pieces for the casting of legs and vessel-bottom.[8] Such a development of mould preparation has an important bearing on the hypothesis of the indigenous discovery of metallurgy in this cultural area.

All the ingredients for the independent discovery of metallurgy existed in late neolithic and Early Shang China. Not least were the advanced experiments by potters, craftsmen held in very high regard. At Cheng-chou their excavated living quarters indicated a 'scale and elaboration not inferior to some of the residences of the Shang nobility'.[9] Among their developments on pre-metal sites are advanced kilns of near-vertical construction, vitrified on the inside by temperatures of *c.* 1400°C. There are also wasters of vitrified pottery which collapsed on itself because of too great a heat. Technically there was nothing to prevent the discovery of copper and tin bronze. On Shang ceramics, pure kaolin (china-clay) was occasionally used, carved with

ornament related to that used on bronzes. Of particular interest is the invention of feldspathic glaze requiring a firing temperature of nearly 1200°C.[10]

Experiments in the neolithic period to find suitable clays and pigments led to the use of materials that turned out to be metallic ores. The use of kaolin in Shang times is of interest, for it is also known that potters in the neolithic were experimenting with kaolin. It should be remembered that kaolin is formed from the decomposition of certain feldspars in granite which, as in Cornwall, may be strongly mineralized and rich in tin. Thus, the discovery of cassiterite and the realization that it melts at only 600°C appears very likely at an early date. Tin bronze was then but a logical step.

According to Cheng Te-k'un there can be no doubt that 'the foundation of the Chinese metal industry was that of existing ceramic practice'.[11] It readily explains the fact that the Chinese technical terms of bronze-casting, still in use today, derive from those of pottery. More importantly, the technique of bronze working was casting, a method with a long ceramic tradition. At Cheng-chou, in levels dating to the early Shang, have been found over a thousand piece-mould fragments for bronze casting, as well as crucibles, copper ore, slag, and charcoal. Metal working techniques used in the West, such as hammering (or 'smithing'), were unknown in China until probably the Chan-kuo period (481–221 BC). This is proved by hardness tests: a tin bronze with 10% tin can have a Brinell hardness of over 220 when hammered, but cast it is only around 88 HB. Chinese cast bronze rarely exceeds 90 HB, and is much less when lead has been added to bronzes of Late Shang and later periods.

The lack of understanding of the western art of smithing is a strong argument against the theory that metallurgical technology, let alone knowledge of tin bronze, was introduced to China through Asia. Further argument, if any be needed, is that the Chinese knew nothing of silver until knowledge of it was introduced in the 6th century BC. Until then it is detectable only as a rare impurity and has led to the suggestion that the lead-carbonate, cerussite, was the lead ore used, and not the sulphide galena, which is often rich in silver. Similarly, gold appeared relatively late after the development of tin bronze, in contrast to its early discovery in the Near East. By the time tin bronze was appreciated in the Near East and Europe, the exploitation of sulphide ores was commonplace and ancient. In contrast, analyses of Chinese bronzes show that copper sulphides were not used and accords with the discovery of copper carbonate, malachite, at sites such as Cheng-chou.

Tin ingots have been found at Anyang (Fig. 12), proving that cassiterite was smelted separately before being alloyed to the copper. Pure copper is difficult to cast (as discussed in the Introduction), in contrast to

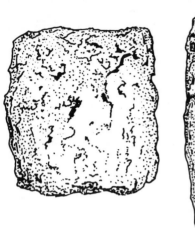

12   **Tin ingots from Hsiao-t'un, Anyang (no scale)**
      (after Li Chi, *The Beginnings of Chinese*
      Civilization, **reproduced in Barnard and Satō,**
      1975, p. 41)

the ease with which it can be hammered into shape when annealed. All Chinese bivalve moulds so far found for casting spearheads or other weapons from the earliest times, contain only one pouring hole (sprue), proving that the Chinese quickly discovered how to obtain an easy-flowing alloy of tin bronze. Different amounts of tin were used for different purposes, so that as much as 16% occurs in ornaments where hardness was subservient to the faithful reproduction of the detail in the mould.

LOCATION OF TIN DEPOSITS IN CHINA

China is rich in cassiterite. In 1946, Chinese output was placed fifth in line behind the Federated Malay States, Dutch East Indies, Bolivia, and the Belgian Congo. Over 14 000 t were produced in 1936.[12] Recent production figures are impossible to obtain, but annual output probably exceeds 20 000 t. The main tin belt is in the south of the country flanking the Nanling Range, and is richest in the western part. Yunnan province contains most with a particularly productive belt at least 50 km long by 20 km wide in the Ko-chiu district.[13] Ko-chiu, together with Fuhochung in Kwangsi province, produced an estimated 20 000 t in 1970, while new placers were found recently at Chiumou in Kwangsi.[14] About 80% of reserves are believed to lie in Yunnan. The most important lodes are here, in limestones associated with granites, occurring as sausage-shaped ore bodies up to 10 m thick following the bedding planes. Tin placers are more widespread; the best are in Kiangsi, typically 2 or 3 m thick with some as much as 30 m, though lode tin is

assumed to account for the bulk of modern output.[15] The placers at Fu-ho-chung, on the border with Kweichow, are particularly important because of the association with wolfram in the alluvials. Tin also occurs at Ta-yu and Nan-k'ang in Kiangsi province.

The tin belt continues through Hunan province where particularly pure cassiterite needing very little refining is mined at Ch'ang-ning and Lin-wu. A minor belt of mineralization hugs the coast from the Fukien-Kwangtung frontier south-west to the island of Hainan, and includes mainland Hong Kong.

All these deposits are a long way from the hub of Shang civilization in the great bend of the Hwang Ho or Yellow River. At present, the use of tin bronze in Yunnan is not known before about 700 BC, as discussed in Chapter 6. Collins[16] stated that mining began in Yunnan in the 16th century when galena lodes were exploited for their silver content, but this is as valueless a statement as the erroneous assertion (discussed in Chapter 12) that tin was not obtained from the Erzgebirge in central Europe before the 12th century AD. The Ko-chiu deposits are not easy to work because of the lack of water for dressing the ores, but even if tin was worked here by 700 BC the source of earlier tin remains a problem. Metalwork is not known from Hunan before the Ch'un-ch'iu period (770–481 BC), or from Kwangtung and southern Kiangsi until the Chan-kuo period (481–221 BC).

An abundant source of tin from the southern tin belt in China is perhaps indicated by finds from tombs of the Warring States Period of the 5th century BC. A good deal of ware at this time imitated the shape and decoration of bronze vessels. Among the variety of techniques used to simulate metal, Thorp has noted the use of tin foil for decoration,[17] though at present it has not been possible to obtain further information about this.

The distribution of bronze artifacts of different periods from Early Shang (1850–1650 BC) to the early 1st millennium AD shows a gradual spread from the cradle of metallurgy on the borders of Honan and Shansi. Where, then, was tin obtained if Siberian and western sources are dismissed? Nowadays the Anyang copper industry uses supplies of copper from around T'ang-yin. Whether cassiterite accompanies the copper is not recorded, but tin sources in Chinese local gazetteers are said to be plentiful in the region of Anyang, although no indication is given of the nature of the ore bodies. Barnard and Satō claim 23 copper and 11 tin deposits within a radius of 300 km of at least one of the three Shang metallurgical centres. Tin at Ch'eng-an is only 50 km from Anyang, and at Ch'-hsien only 55 km away. The nearest deposits to Yen-shih and Cheng-chou are at Lin-jui, a distance of 60 and 102 km respectively.

Many of the major tin-fields of the Soviet Far East consist of sulphide ores, though some contain cas-

a wall plates *b* dividers *c* long end posts *d* end plates
13   Timberwork in a shaft and a windlass (from Agricola's *De Re Metallica* Book V p. 123)

ANCIENT CHINESE COPPER MINES

Early tin mines have yet to be identified in China, but copper mines of the Late Shang and Han periods are known and are mentioned here to illustrate their high degree of technological competence. The earliest dates come from Ta-ching, Lin-hsi-hsien in Liao-ning:[19] 1147 ± 119 BC; 931 ± 99 BC; 911 ± 112 BC; 840 ± 103 BC. Details of this mine are not available at the time of writing. Han period mines are known at Yün-ch'eng in Shansi (*c.* AD 179–259), Ch'eng-te-chuan-chü in Hopei, and T'ung-lü-shan near Ta-yeh-hsien in Hupei (557 ± 129 BC). Of these, T'ung-lü-shan is at present the most important with good accounts in Chinese,[20] which have been discussed in English by Barnard.[21] The ores are malachite, azurite, cuprite, native copper, and among the iron ores magnetite and haematite. Open-cast and underground workings were on a large scale over an area of 2 × 1 km. Over 1 000 pit props were recovered from the open-cast workings alone.

Two areas of underground mining were investigated. In the northern part of the open-cast workings eight vertical shafts and one sloping shaft, averaging 80 cm^2 extended to a depth of about 40 m. In the southern part of the open-cast workings there were five vertical shafts, one sloping shaft, and ten horizontal galleries. Galleries branched off the sloping shaft at three different levels. Details of the construction need not be given here, suffice it to say that the timber work was of a very high order and compares very favourably with the 16th century work illustrated in Agricola's *De Re Metallica* (Fig. 13). The tools found in the workings merit similar comparison and include a variety of picks, axes, wedges, hammers, mallets, spades, scoops, ladles, buckets, hooks, rope, and baskets, some of which were filled with ore. Wooden troughs up to 260 cm long were used for drainage.

Barnard re-assesses, in particular, the remains of a windlass (Fig. 14) and believes it employed a length of rope wound around the windlass with a hook at each end allowing a full bucket to be raised while at the same time lowering an empty one, a device also illustrated by Agricola. Barnard interprets the 'rundles' at each end of the windlass shaft as evidence for a form of gearing. Hsia Nai, however, has since then proposed a different function for the 'rundles', following constructions given in the *T'ien-kung k'ai-wu*, a Ming period (late mediaeval) account of various agricultural and industrial technologies (Barnard). Whichever interpretation is correct, the device is a sophisticated one. Also of interest is a wooden *batea* (Spanish for a wooden tray), precisely like one illustrated by Agricola and described by him as 'generally two feet long and one foot wide', used for carrying ore on the shoulders or hung about the neck with a rope (Fig. 15). Pliny in his *Natural History*

siterite. The deposits of the Anyang region may be similar, and sulphide ores are known in the southern Chinese tin belt. Slessor[18] described the ores at the Da Soh Fu mine near Pai-sha in Hunan as:

> a mixture of sulphides of tin, iron, copper, and arsenic, with a quartzite and limestone gangue. The surface ores of the past were decomposed and rich, and filled large caves in the limestone. ... Old workings show that large quantities of ore had been taken out in the past.

Similar ores were found about 30 km to the north at the Kyang Yu tin mine, which were also worked 'in the early days'. The Da Soh Fu ores (which included stannite) yielded 1·7% tin, 3·8% copper and 3·5% arsenic.

If this typifies the nature of the tin in northern China it explains why Shang China had a readily available supply of tin for bronze from the birth of its metallurgy, while in the Near East trade routes of enormous length resulted in a fitful and unreliable supply of tin.

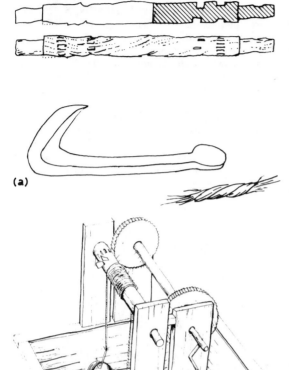

(a)

(b)

a small batea *b* rope *c* large batea

15   Illustration from Agricola's *De Re Metallica* (Book VI p. 157)

14a The remains of a wooden windlass shaft with grooves or 'rundles' at each end suggesting some form of gearing; from the mine opening at T'ung-lü-shan (length 250 cm, barrel dia. 26 cm) (*Wen Wu*, 1975)
  ***b* Hook and rope from the mine with Noel Barnard's suggested reconstruction of the T'ung-lü-shan windlass (Noel Barnard, 1981)**

(XXXIII, 21) noted how ore was carried on work-men's shoulders: 'night and day each passes the material to his neighbour, only the last of them seeing the daylight'. Tools at T'ung-lü-shan include an iron-tipped wooden shovel with a central square hole for inserting a handle, similar to many recovered from Cornish tin streams and described in 1602 by Carew as being used in Elizabethan days.

The advanced nature of the T'ung-lü-shan workings in the 6th century BC point to an earlier long de-velopment, and it can only be hoped that the rapid advances of Chinese archaeology will soon reveal much older mines, for tin as well as copper.

NOTES AND REFERENCES

1   PIGGOTT, 1974
2   BARNARD, 1983, 238 note 2
3   GROUSSET, R., 1959, *Chinese Art and Culture*, 47; quoted in Barnard & Satō 1975
4   CHARD, 1974, 208
5   CHASE, 1979
6   WATSON, 1974, 35
7   BARNARD & SATŌ, 1975. Unless stated to the contrary, the greater part of the information in this chapter is taken from this sumptuously produced book
8   BARNARD & CHEUNG, 1983
9   CHÊNG, 1975
10 WATSON, 1973. Fig. 113, p. 88 is the glazed pottery vase excavated at Ming-kung-lu, Chen-chou, in 1965 and dated to *c*. 1500 BC. The temperature required for firing was nearly 1 300°C
11 CHÊNG, 1975
12 VEI, 1946; ITC, 1964; 'Mineral Industries of China', *Mining Annual Review*, USA Bureau of Mines, 1979, 14
13 JONES, 1925
14 WANG, 1971
15 VEI, 1946
16 COLLINS, 1909–10
17 THORP, 1980, 59
18 SLESSOR, 1927
19 BARNARD, 1981
20 ANON., 1975
21 BARNARD, 1981

6 Tin in South-West China and Vietnam

The great bronze age of northern China had no links with the south of the country until the Shang period and later. Here they advanced in their own way and developed a regional style of art in Shang times along the River Yangsi from Hunan in the west to Jiangsu province on the seaboard of the East China Sea.[1] However, this leaves out the greater part of south-west China which includes the tin-rich province of Yunnan. The first contacts between Yunnan and Ch'u China were not until the 4th century BC when the General Chuang Ch'iao considered Yunnan culture on a par with Han civilization.[2]

The culture in Yunnan has been named Tien after the lake of that name. On its eastern shore lies the important late bronze age site of Shi-chai-shan on the top of Shi-chai (Stone Fort) Hill. Excavations began here in 1955 on tombs identified as belonging to the royal household of the Kingdom of Tien.[3] Copper and tin are both plentiful in Yunnan, and since Shi-chai-shan is only 160 km north of China's most important tin-field of Ko-chiu, it is not surprising that one of the outstanding features of the site's tombs is the abundance and variety of bronze weapons and tools, including socketed implements decorated with characteristic geometric patterns. Various items of Western Han origin, a lack of Eastern Han objects, and a gold 'Seal of the Kingdom of Tien' date the tombs to the approximate duration of the Western Han period (206 BC–AD 24). The relationship of Shi-chai-shan to earlier neolithic and chalcolithic sites found elsewhere in Yunnan is not yet fully understood, but the contemporary art style and metallurgical techniques 'tell of exchanges with the realm of the nomadic art style in the north-west, and with the Dong-son civilization to the south-east'.[4]

Dong-son near Hanoi in north Vietnam, famous in the west for its remarkable bronze drums, is about 480 km south-east of the Ko-chiu mining area. There is also tin in Vietnam near the Chinese border west of Cao Bang. It is sometimes found with wolfram and there have been recent operations to expand the Tinh Tuc mine.[5] The mineralized region is in the Pia Ouac mountains and includes rich tin alluvials in the Tinh Tuc river.[6] A bronze-using civilization in north Vietnam is, therefore, not unexpected, and it is possible that the Ko-chiu deposits in Yunnan were located by prospectors from Vietnam about the 7th century BC. Shi-chai-shan has already been described as 'an important Dong-son site'.[7]

The subject remains full of questions. As demonstrated by the richly furnished tombs in such Yunnan sites as Shih-chai-shan, Ta-p'o-na, Li-chia-shan, and Ch'u-hsiung, of which several have been radiocarbon dated, the use of tin bronze goes back to at least 700 BC,[8] though as Barnard says, 'It is a little premature yet to present a studied impression as to the range of time involved as suggested by the presently available ^{14}C dates.' The Heger type 1 drums, used as containers of cowrie shells and often surmounted by lively scenes of humans and animals, are almost the only ones excavated in Yunnan, and it may be that the type developed here and was diffused to Vietnam and elsewhere in south-east Asia.[9] Barnard suggests that the metallurgical development that led to the Dong-son and Tien drums began as early as 1000 BC somewhere in the frontier regions of Yunnan, Thailand, and Malaysia.

The earliest appearance of tin bronze in Vietnam and northern Thailand was probably around 2000 to 1800 BC, as discussed in Chapter 9. This is roughly the same period as it developed elsewhere, but with the scanty knowledge of the archaeology of south-east Asia, it is too early to say whether this dating is correct, and pointless to speculate on whether the advance to a bronze-using society was aided from outside the region or not. Thailand is relatively well excavated, while work started by the French in Vietnam continues under local patronage, but very little is known about Cambodia, Burma, or Laos.[10] Bronze development in south-east Asia may have owed little or nothing to outside influence, but further research is needed in this part of the world where archaeological research has only just awoken to the late Chairman Mao's famous dictum: Let the past serve the present.

NOTES AND REFERENCES

1 BAGLEY, 1980
2 DEWALL, 1967
3 WANG, 1960
4 DEWALL, 1967
5 FISHER, C. W., 1983, 'Viet Nam' in *The Far East and Australasia 1982–1983*, Europa Publications
6 ITC, 1964
7 RUDOLPH, 1961
8 BARNARD, 1980
9 BARNARD, 1980
10 BAYARD, DONN, 'East Asia in the Bronze Age' in Sherratt (ed.) 1980, 168–73

7 Tin in Korea and the Soviet Maritime Territory

Tin was discovered in South Korea in 1959 6 km from the famous Sandong tungsten mine, the third largest in the world. Resources have not been thoroughly prospected and production is tiny, averaging only 76 t of tin in concentrates annually from 1964 to 1972 and 24 t in 1974.[1] Two mines operated 8 km apart in Kangwon-Do (Do = province) and Kyongsang Pukto respectively, some 48 km east of Yongwol, in the Taeback Mountains. The tin occurs as cassiterite erratically distributed in pegmatites cutting the Precambrian Taeback metamorphosed shales.[2] How much tin, if any, is produced in North Korea has been impossible to find out. Kangwon-Do is a rugged mountainous area 'traditionally regarded as poverty-stricken and remote' and only recently opened up for the mining of, principally, tungsten, iron ore and anthracite.[3] On present evidence it is unlikely that South Korean tin was known in antiquity.

However, these are early days to be dogmatic about the Korean bronze age. Advances in archaeology have been so rapid since 1970 that Korean prehistory has to be rewritten, not only regarding its internal elements but also its relations with neighbouring countries.[4] The very existence of a Korean bronze age has only recently been established, and just when scholars were beginning to accept its birth at around 1000 BC, new data have pushed back the date to c. 2000 BC. Information is difficult to obtain from North Korea, but a recent government publication quoted a date of 2000 BC for the start of the bronze age.[5] In the South there are ^{14}C dates that early in the south-east and c. 1300 BC in the central area. The distribution of bronze age sites as yet probably reflects the sphere of influence of archaeologists based in Seoul, though the greatest concentration of sites would be expected in the lusher areas of west-central Korea. Kyonggi-Do has more lowland below 300 m than any other province except Chungchong Namdo, which borders it to the south.

Contrary to what Western scholars might expect, there was little Chinese influence in Korea until the Han period (from c. 200 BC). For the bronze age it is important to realize that Korea was not limited by present political boundaries but had strong links with the Liaotung peninsula, Liaoning, and the Maritime Provinces of Siberia. For example, the typical Korean bronze swords originated in Liaoning.[6] There were close ties with Mongolia (which also contains tin deposits) and Siberia, notably with the area of the Karasuk culture, which developed in about the 13th century BC out of the indigenous Andronovo culture.[7] This region embraces the tin-bearing regions of the upper Yenisei and Onon rivers.

A good deal more archaeological research is needed before the full importance of the Siberian tin deposits can be evaluated. This applies equally to the extreme Soviet Far East. Charcoal from a burned wooden house at Kirovskoe, near Vladivostok, produced a ^{14}C date of 4150 ± 60 BP (c. 2200 BC), a particularly early date for the establishment of the bronze age in the Primorskiy Kray or Maritime Region.[8] As is well known, the Soviet Far East is rich in tin, providing most of its modern output.[9] For example, deposits discovered in the 1930s are worked 80 km inland from the port of Tetyukhe Pristan, and a large open-cast mine has operated at Yaroslavskiy, 100 km north of Vladivostok, since 1957.[10] Many of these Soviet deposits are rich in sulphides such as stannite, and similar minerals predominate. It is not certain if these were exploitable in antiquity, but cassiterite is widespread in alluvial deposits as well as in lodes. Placer tin is found at Pevek, Deputatskiy, and Ege Khaya, all well north of the Arctic Circle, probably much too far north to have been available in the bronze age. Yet tin was being obtained from somewhere.

The idea that the Soviet Far East remained in a neolithic culture until the arrival of the Russians in recent times is quite untrue. As Mongait pointed out,[11] 'Bronze using cultures extended northwards along the Lena and its tributaries up to the Arctic Circle'. There is no suggestion here that their northward trek was in search of metal ores; all the tin used in the bronze age could have been obtained from a number of deposits

45

scattered all along Siberia's southern border from Fergana in the west to Vladivostok in the east.

Tin also occurs in Mongolia, though it is not known to be mined. There is little information on the deposits, but Taylor[12] has published cross-sections of greisen–quartz–cassiterite associations in the east of the country at Barun Tsogto, Tumen Tsogto, and Yugodzyr. It remains to be seen whether these deposits could have been of any importance in antiquity.

NOTES AND REFERENCES

1   ITC, 1968; ITC *Tin Statistics 1964–1974*
2   UNITED NATIONS *Mineral Resources Development Series*, No. 23, 1964
3   BARTZ, 1972
4   NELSON, 1982
5   SIHANOUK, 1980
6   KIM, 1975
7   CHARD, 1974
8   CHARD, 1962, item RUL-177; Sulimirski 1967 with map of sites p. 81. As well as Kirovskoe, Sulimirski draws attention to Krasino, just south-west of Vladivostok, where bronze objects were being cast in the 2nd millennium BC
9   MATERIKOV, 1977, esp. 290f for placer deposits
10 ITC, 1968
11 MONGAIT, 1961, 150
12 TAYLOR, R., 1979, 60

8   Tin in Japan

There are widespread minor deposits of tin in both Kyushu and Honshu. Exploitation reached a peak, as it so often does, during wartime with nearly 2 200 t produced in 1941. Current production is around 800 t.[1] Many of the lodes contain copper as well as other sulphides and wolfram. Bronze and iron were introduced to Japan from Han China at the same time in the last few centuries BC. Early bronze weapons for ceremonial and religious use fall into two types: firstly those with a good tin content and superior workmanship imported from China, and secondly those with a lower tin content copied by local craftsmen.[2] Chinese bronze mirrors have been found at several sites in Kyushu and are thought to date between about 100 BC and AD 100. Some occurred with local pottery vessels of the middle Yayoi period. A recent study of mirrors from the Yayoi and Tumulus period tombs showed the two distinct types of metal.[3] The Chinese mirrors (Hakusai-kyō) are generally rich in tin: 60–75%Cu; 17–30%Sn; 3–7%Pb. The Japanese copies (Bōsei-kyō) are twice as numerous with generally less tin, though the amounts can still be considerable: 70–90%Cu; 2–20%Sn; 3–8%Pb. The locally made mirrors only occur in the later Kofun period c. AD 300–700.[4] The question is whether the local mirrors, and locally produced bronzes, used Japanese tin or metal imported from China. With increasing Japanese involvement with Korea and a weakening of ties with imperial China from the 5th century, there is also the possibility of tin reaching Japan from other far eastern sources, perhaps through Korea from sites now in the Soviet Far East. As yet the question is unanswerable.

According to Jones,[5] the important Taniyama tin deposit was discovered in 1655, and it is to the 16th and 17th centuries that all the earliest discoveries are assigned: the Toroku and Mitate mines in Miyazaki, Kiura in Oita, Suzuyama in Kagoshima,[6] while an area of about 30 km² of alluvial tin on the banks of the Kiso-Garva at Gifu was worked in the 18th century or earlier.[7] The largest tin deposit is in the Ikuno-Akenobe district of Hyogo. Copper and silver were mined here in the 16th century, though tin was not discovered at Akenobe until 1908.[8] The lack of early evidence does not mean that tin could not have been recovered in the Kofun period, for as will be seen in Chapter 12 the lack of archaeological finds in the tin deposits of the Erzgebirge does not mean that cassiterite was not mined there in the 2nd millennium BC.

There is evidence for 'ancient' working at Taniyama, and we can do no better than quote Gowland's eye-witness account of his visit to the mines.[9]

> The excavations of the old miners here are of a most extensive character, the hill sides in places being literally honeycombed with their burrows, indicating the production in past times of large quantities of the metal. No remains, however, have been found to give any clue to the date of the earliest workings. But, whatever may have been their date, the processes and appliances of the early smelters could not have been more primitive than those I found in use when I visited the mine in 1883.
>
> The ore was roughly broken up by hammers on stone anvils, then reduced to a coarse powder with the pounders used for decorticating rice; the mortars being large blocks of stone with roughly hollowed cavities.
>
> It was finally ground in stone querns (Fig. 16), and washed by women in a stream to remove the earthy matter and foreign minerals with which it was contaminated.
>
> The furnace in which the ore was smelted is exactly the same as that used for copper ores, excepting that it is somewhat less in diameter. The ore was charged into it wet, in alternate layers with charcoal, and the process was conducted in precisely the same way as in smelting oxidised copper ores ... The tin obtained was laded out of the furnace into moulds of clay (Fig. 17) ...
>
> The entire yield of each smelting charge was only about thirty pounds of metal, much being lost by volatilisation and in the slag.
>
> Yet, with this primitive process and its rude appliances, the Satsuma smelters were producing tin at a cost not greater than that of the imported metal.

Almost all Japan's tin came from this mine in the 19th century. It was a sizeable concern comprising 21 veins varying from a few cm to 1·20 m wide with an average of 0·45 m.[10] This accounts for Gowland's warren of burrows. The lodes trend NW–SE cutting slates, sandstones, soft tufas, and occasional hard, dark blue quartzite. Most of the cassiterite is almost microscopic in the quartz gangue, though Monroe found one specimen to contain over 56% metallic tin compared with the usual 12 or 13%. Monroe also noted how the ore was crushed between the millstones: the first grinding yielded 15% of concentrated ore, the

47

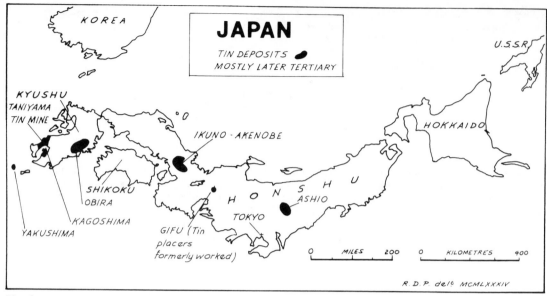

Map 8

second 5%, the third 2%, and the fourth 'after two years exposure to the weather' a mere 0·5%. The *ita* boards used for washing the ore were circular, a little over 0·6 m in diameter and less than 25 mm deep. Methods may have improved by Gowland's day, for Monroe stressed the laboriousness and expense of the method, necessitating more than 80 days' work to treat a single ton! Small wonder that in 1870 the mine, which employed about 120 men and boys, produced only 8 tons of tin, an amount Monroe considered to be the annual average yield.

16   Grinding ore in a quern of the type used at the Taniyama tin mine when William Gowland visited the site in 1883 (from *Archaeologia*, 1899)

17   Tin smelting and casting the metal into bars (*Archaeologia*, 1899)

NOTES AND REFERENCES

1   ITC, 1964
2   KIDDER, 1959, 110
3   SAWADA, 1979
4   CHARD, 1974
5   JONES, 1925; ROLKER, 1894, states production of stream tin at Nakatsugawa, Gifu prefect, in 1890 amounted to 175 Singapore piculs (10·4 tonnes). Total Japanese production in that year was 791 piculs (47 tonnes)
6   KANEKO, 1960
7   JONES, 1925
8   KANEKO, 1960
9   GOWLAND, 1899, 294–5
10 MUNROE, 1876, 298–9

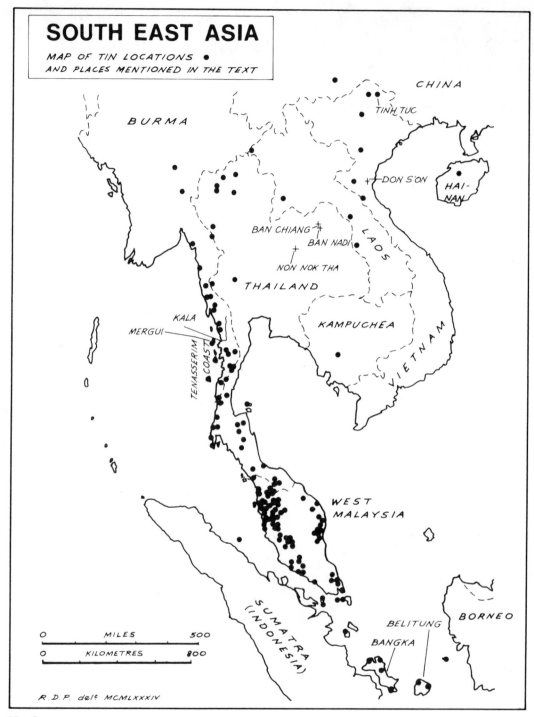

SOUTH EAST ASIA

MAP OF TIN LOCATIONS ●
AND PLACES MENTIONED IN THE TEXT

Map 9

9 Tin in South-East Asia

Malaysia is the world's largest producer of tin averaging over 80 000 t annually (*see* Fig. 18). The possibility that here was a source of tin for the West in prehistory is thus an attractive one. There is evidence from Thailand – itself no mean producer with around 20 000 t annually – for the early production of tin bronze, though whether this was an independent discovery is impossible to say on present evidence. There are numerous tin sources in the Annamite Mountains bordering north-east Thailand and Vietnam, and it was at Ban Chiang in that part of Thailand that bronze artifacts associated with pottery were found, including socketed axes. Early thermoluminescent dates for the pottery caused a stir by placing some of it as far back as 4600 BC, an unprecedented and apparently nonsensical date,[1] especially worrying as ^{14}C dates of skeletal remains buried with the bronzes varied between 3600 and 2000 BC. Technically the bronzes are of high quality without 'primitive' forms. This fact, plus the occurrence of similar tin bronzes found with iron artifacts at Don Son in Vietnam contemporary with the development of the iron age in China around 600 BC,[2] demanded further enquiry. The problem with the ^{14}C chronology at Ban Chiang is that the charcoal was not always of certain provenance within the site.[3]

To resolve the difficulties, Higham and Kijngam excavated an undisturbed site at Ban Nadi (pile dwellings) close to Ban Chiang. Bronze was cast at Ban Nadi throughout its period of occupation. Finds included pieces of partially smelted ore, bronze casting furnaces, crucibles, and bronze arrowheads, bangles, and chisels. All the furnaces yielded charcoal samples, which have now been dated by [14]C. Ten dates vary between 3440 ± 200 BP and 1575 ± 85 BP (*c.* 1500 and 400 BC). The finds are comparable to those from Ban Chiang, and Higham concluded that 'There is no evidence to sustain the exceptionally early chronology which has been advanced for Ban Chiang, but rather the data parallel recent chronologies obtained for the Vietnamese sites where bronze was first found in the Phung Nguyen culture during the early 2nd millennium BC.'[4] Layers at Ban Nadi dated between *c.* 1500 and *c.* 600 BC contained abundant evidence for local bronze smelting and casting. It is probable that

the evidence from here and Vietnam points to the establishment of a bronze industry by about 2000 to 1800 BC.[5]

In spite of the early use of tin bronze in south-east Asia, there is no evidence that tin formed part of any international trade. Voyages across the Indian Ocean from the Near East to India and China were a development of the 1st millennium AD.[6] Indian trading interests intensified in south-east Asia during the same period for a variety of reasons. It is probable that Vespasian (AD 69–79), in prohibiting the export of precious metals from the Roman Empire, intensified a scarcity of gold in India. This had been initiated in the previous century by nomadic disturbance in Central Asia closing the Bactrian trade routes over which gold had reached southern Asia from Siberia. Other reasons were better ship design and the expansion of Buddhism. After Islam, Arab interests also spread further than hitherto. The China run became very lucrative, especially in the 9th century until 878 when the rebel Huang Ch'ao massacred thousands of foreign merchants at Canton – Muslims, Christians, Jews, and natives alike.[7]

It is during this period that the Malaysian tin deposits were opened up to trade. The centre of commerce was the island of Kilāh, described by Ibn Khurdādhbin in his 'Books of the Roads and Kingdoms' (AD 844–8) as the usual meeting place of Muslim ships from Siraf and Oman with ships from China. The island itself contained 'famous mines of tin and plantations of bamboo'. Abū Zaid, who lived at Siraf on the Persian Gulf, gleaned information from sailors and traders, and listed Kalāh as the centre of trade for 'aloeswood, camphor, sandalwood, ivory, tin, ebony, baquamwood [brazil wood], spices of all kinds, and a host of objects too numerous to count'. Abū Zaid did not visit Kalāh, but Abū Dulaf did; he lived at Bukhara at the court of the Sāmānids. In about AD 940, he journeyed to China and wrote an account that is preserved in later Arab texts. After leaving China he wanted to visit Kalāh which he found to be 'very great, with great walls, numerous gardens and abundant springs. I found there a tin mine such as does not exist in any other part of the world except in its *Qal'ah* [fortress] ... In the entire world there does not exist

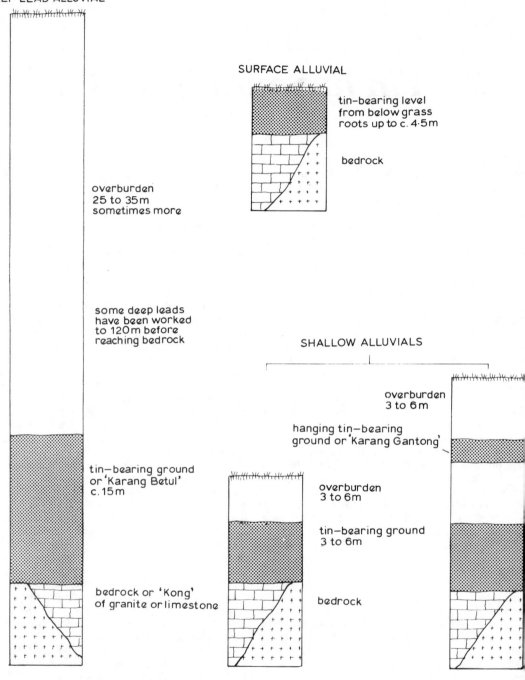

DEEP LEAD ALLUVIAL

SURFACE ALLUVIAL

tin-bearing level
from below grass
roots up to c. 4·5m

bedrock

overburden
25 to 35m
sometimes more

some deep leads
have been worked
to 120m before
reaching bedrock

SHALLOW ALLUVIALS

overburden
3 to 6m

hanging tin-bearing
ground or 'Karang Gantong'

tin-bearing ground
or 'Karang Betul'
c.15m

overburden
3 to 6m

tin-bearing ground
3 to 6m

bedrock or 'Kong'
of granite or limestone

bedrock

18   Simplified sections illustrating the variety of alluvial tin deposits in Malaysia (based on Fawns, 1905)

19   The laborious method of hand excavation used by the Chinese in Malaysia to reach the often deep-seated 'karang' or tin ground. The sloping face is the edge of the valley workings. The overburden was carried out in small baskets, a pair hanging from a pole slung across the shoulders (photograph from Fawns, 1905)

such a tin mine as the one in this *Qal'ah.*'

The location of Kalāh is disputed, but Wheatley, from whose book all the above quotations are taken,[8] demonstrates convincingly that it was on the Tenasserim coast of Burma, an area still rich in tin. The Mergui district fits the description well because of the bamboo plantations. Furthermore, the largest island off the mouth of the Tenasserim River is today called Kala.

Before the rise of Arab seafaring, literary evidence points to trade over the Indian Ocean from west to east, at least for India in Roman times. Pliny refers to the trade in his *Natural History* (book VI, 26), as does the anonymous Greek author of the *Periplus of the Erythraean Sea* (i.e. the circumnavigation of the Indian Ocean) written in Egypt, probably in the early 2nd century.[9] Wheatley suggested that it was a compilation of material gleaned mainly in the second half of the 1st century with a few sections on India referring to AD 110–15.[10] Roman trade with India is well known. Periods of intensity and decline are reflected in the age of coin hoards. More impressive evidence includes the temple of Augustus at Maziris on

the Malabar coast, a major port. Pliny gave the sailing time from Ocelis near the mouth of the Red Sea to Maziris as 40 days. As well as tin, gold, bronze, wine, and other manufactured goods were taken from the Mediterranean port of Alexandria up the Nile to Coptos, thence overland to the Red Sea ports of Myos, Hormos and Berenice for shipment to southern Arabia, Somaliland, and India.

Spain may be a source for this tin, though a Cornish origin cannot be ruled out in later centuries. The famous story in the life of St John the Almsgiver, patriarch of Alexandria (AD 611–19), quoted at the end of this book, indicates that the idea of Cornish tin reaching Egypt in the early 7th century was not exceptional.

West European countries did not import Malaysian or Indonesian tin until the 16th century. The Portuguese imported small amounts 'from the Indies' from 1513, the Dutch following a century later.[11] In the 18th century the Dutch regularly imported tin from Bangka, though only as ballast rather than as commercial cargoes, because the European market was

20   **Malaysia: winning shallow alluvial tin ore from a dried-out stream bed (photo *Tin*, June 1957, p. 131)**

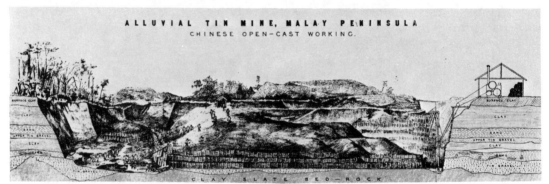

21   **Chinese hand-working two tin-bearing strata in Malaysia. In the centre the waste is being dumped behind retaining fences of 'bakau' or 'kayu kachu' (poles resistant to water when kept wet). Chinese women interlaced 'fallang' grass between the uprights. The ore is being carried in baskets to the dressing floors (from the lithograph in Fawns, 1905)**

still dominated by Cornish production. The Bangka trade was mainly confined to the East. Even Cornish tin was shipped to China, though Cornish trade with the East ceased in 1817.[12]

The Portuguese on their first expeditions to Bangka found local ships laden with tin, and it is said that Bangka tin reached China in the 9th century.[13] In 1817 Raffles gave an excellent account of the primitive manner of mining the alluvial ores on Bangka.[14] As in Cornwall, methods of streaming can have changed very little over a considerable period. Raffles' account may be the closest appreciation possible of the way tin was won in South East Asia to supply the metalworkers of Ban Chiang and Ban Nadi in the 2nd millennium BC (*see* Figs. 19–21).

The process of mining in Banca [Bangka] is remarkable for its simplicity. It consists in an excavation made by digging perpendicularly to the beds of the ore, and a proper application of the water to facilitate the labours of the miners, and the washing of the ore. A favourable spot being selected, the pit is sketched out, a canal conducted from the nearest rivulet, and then the miners excavate the soil until they arrive at the stratum containing the ore, which is next deposited in heaps near the water, so as to be placed conveniently for washing; the aqueduct is lined with the bark of large trees, and, a stronger current being produced by the admission of more water, the heaps are thrown in, and agitated by the workmen; the particles of the ore subsiding through their gravity, and those of common earth being carried away by the current.

This description might have been culled from the pages of Agricola's description of winning tin in the Erzgebirge, or from Pryce's account of Cornish tin streaming. It admirably demonstrates that wherever and whenever alluvial tin was mined, the same methods were universally employed.

NOTES AND REFERENCES

1 CARRIVEAN & HARBOTTLE, 1983
2 WILSON, 1978, 306–7; CLARK, 1977, 346–7
3 HIGHAM, C., personal communication, 1982
4 HIGHAM, 1981; HIGHAM, 1982
5 HIGHAM, C., personal communication, 1982
6 WHEATLEY, 1975, 233
7 HOURANI, 1951
8 WHEATLEY, 1961
9 SCHOFF, 1912, esp. 77–9
10 WHEATLEY, 1961, 130 footnote
11 HEDGES, 1964, 16
12 WONG, 1965, 3–4
13 BAGNOLD, 1820
14 RAFFLES, 1817, esp. 192. A simple method of working still employed by the Chinese, frequently women, is the washing of tailings using wooden bowls or 'dulongs'; HAMPTON, 1886, notes how the Malays, who were 'rather averse to hard work', mined tin from 'lampan' workings, excavations in the sides of hill which were self-draining and needed no machinery.

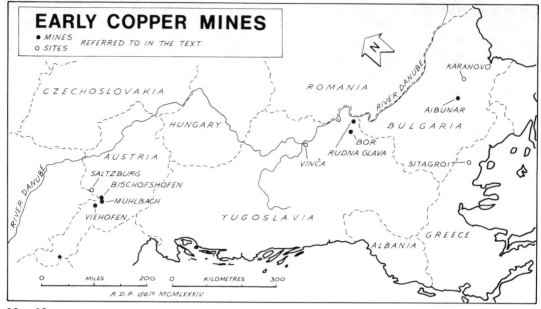

Map 10

PART 2

EUROPE
(excluding south-west England)

꧁꧂꧁꧂꧁꧂꧁꧂꧁꧂꧁꧂꧁꧂꧁꧂꧁꧂꧁꧂꧁꧂꧁꧂꧁꧂

10 *Early copper mining in Europe*

The realization that tin bronze was discovered in China independently of influences from the Near East may have come as a surprise, but it has not produced the same degree of shock as the proposition that European metallurgy may also have developed largely on its own. This notion is not accepted unreservedly; even its strongest proponents admit that 'no conceivable find can ever show that contact did *not* take place'.[1] The precise relations between the Near East, the Aegean, and Europe are imperfectly understood. What is not in dispute is that the once invincible theory of the gradual diffusion of knowledge from the civilizations of the Near East to Europe via Greece and the Aegean is no longer tenable. No one denies that copper was first used in the Near East, certainly by 6000 BC in Turkey, and possibly as early as 7000 BC. Powdered malachite was used at this time for personal adornment by the inhabitants of Aby Hureyra, a pre-pottery settlement on the Euphrates where traces of malachite were found adhered to the inside of a cockle shell.[2] This early beginning was not followed by any rapid development; there was no breakthrough into a new metallurgical technology. For 3 000 years copper was no more useful than a host of other pretty stones – a long time for such dallying. During this period much

of Europe was inhabited by people competent at exploiting stones of one sort or another and consequently very much aware of the geological diversity of Nature. It would be odd in the extreme if during those three millennia various groups of people did not investigate one or two of Europe's rich copper deposits without any outside stimulus.

Copper metallurgy had begun by 3000 BC in south-east Spain. The traditional view is that it was introduced by colonists from the Aegean or Anatolia, a view based on the superficial resemblance of architectural features, notably the bastioned walls at Los Millares in Almería, Spain, and Chalandriani on the Aegean island of Syros in the Cyclades. But metal objects or, indeed, pottery from Iberia are not identifiable as imports from the Aegean or from anywhere else. On present evidence it is safer to assume an independent discovery of copper in the El Argar culture of Spain. The origin of tin bronze in Iberia is another matter, as will be discussed in Chapter 15.

The evidence for copper mining in the Balkans is even earlier, probably not long after 5000 BC. This is still later than the earliest mining at Ergani Maden in eastern Turkey, and as the Balkans are virtually next

AEGEAN	BALKANS	Date BC
middle bronze age		
		2100
early bronze age (and Troy)	early bronze age	
		3000
late neolithic	late Vinča and Gumelnitsa	
		4500
middle neolithic	early Vinča	
		5200
early neolithic II	Starčevo (early neolithic)	
		6000
early neolithic I		
		6500

door to Asia Minor, a diffusion of knowledge from Asia Minor to the Balkans via the Aegean and Greece seems an eminently sensible theory. Unfortunately, this is not supported by ^{14}C which dates metalworking in the Aegean and Greece a millennium after its widespread use in the Balkans. The key site is Sitagroi close to the Aegean sea near Drama in northern Greece. Here in 1968 and 1969, Gimbutas and Renfrew excavated levels containing articles of the Balkan Vinča culture and the Aegean bronze age. As the excavators hoped to find, they were not contemporary. The Aegean bronze age levels were on top and younger than the Vinča levels. The result of this and other evidence is to have produced the equivalent of a geological fault in which the strata of the Balkans have been thrust downwards and shown to belong to earlier horizons. The effect is best seen in the diagram above in which the divisions with similar shading were once believed to be contemporary.[3]

In moving away from the old diffusionist theories, archaeologists are reverting to some extent to the 19th century idea of history as an evolutionary process in which stages of technological development are achieved through a natural process in different regions at varying times, or even at the same time in unconnected localities. An example is Gabriel de Mortillet's universal Laws on the antiquity and progress of humanity drawn up after the Paris exposition of 1867:[4]

Loi du progrès de l'humanité
Loi du développement similaire

Whatever the mechanism by which metallurgy was born in Europe, the importance of the new chronology is to demonstrate the great antiquity of mining, so that by the time tin bronze came on the scene, whether through diffusion of ideas or independent evolution, there existed in Europe a tradition of hard-rock mining extending over several millennia. Although the main

subject is tin, it is not irrelevant to look at the techniques of mining as revealed by excavations of early copper mines, because there are no known prehistoric tin mines, as distinct from open-cast or streaming operations.

The Vinča culture of the Balkans, named after a settlement mound or 'tell' near Belgrade, is essentially neolithic. Copper ores are extensive in the southern part of Vinča territory, especially in the Bor region of eastern Serbia. Here at Rudna Glava in Yugoslavia between 1968 and 1974 early copper mines were revealed during open-cast mining for iron. The prehistoric mine shafts followed narrow veins of cuprite and chalcopyrite to depths of 20–25 m. In the shafts were vessels, some complete, of Vinča pottery. The ore was obtained by fire-setting: heating the rock with fire then cooling it quickly with water, a highly efficient method in crystalline rocks which survived long after the introduction of gunpowder. At the famous Norwegian silver mines at Kongsberg the method was used until the 1880s and abandoned only because modern steel drills, and dynamite invented by Nobel in 1867 made the traditional method too expensive of labour.[5] In the prehistoric Balkan mines the ore was broken up with wedges of bone and wood. Lumps were then smashed with heavy hammers of gabbro and similar rocks. The hammers, like similar Cornish finds of uncertain date, had a shallow groove around the middle facilitating the attachment of a rope. Conventional wooden handles could not be used through lack of space in the shafts. Copper finds from several localities, such as a copper awl from Bolimar in Romania, show that local ores were being smelted in the later Vinča phases around 4900 to 4500 BC.[6]

More mines have been excavated in Bulgaria, the most famous being at Aibunar near Stara-Zagora.[7] Here veins of malachite and the rarer azurite, 0·5–5 m wide, cut the limestone and dolomite hills. Eleven mostly wedge-shaped, open-cast workings were identified. Associated coarse pottery was typical of the Gumelnitsa culture found throughout Bulgaria and belonging to level VI of the great tell at Karanovo, dating it to around 4000 BC. Tools included over 20 antlers used for prising apart the limestone blocks surrounding the copper veins. Stone hammers were not used for mining, only for crushing the extracted ore. Particularly interesting are a number of copper axe-adzes, typical of those found widely in the Balkans and Carpathians. An axe-adze and an axe-hammer from site 1 show strong traces of wear. These and other tools are of very pure copper with only minute traces of a wide range of metallic impurities: the adze contained 0·25% arsenic, while tin was a negligible 0·000 7% in the hammer. Impurities in the Aibunar copper ores are about 1% zinc, 0·1% lead, and lesser but 'significant amounts' of arsenic, silver, bismuth, and nickel. Ores varied sufficiently to show that copper found at various neighbouring settlements had come from no. 2 working, and at other settlements had come from no. 3, all of which helped to confirm the earliest working between 4000 and 3700 BC.

Mining at Aibunar was more active than at any time until Classical Antiquity and the Middle Ages. The total length of open-cast shafts was nearly 0·5 km. Although the miners lived on the site, no smelting was done there. Where it was done is uncertain, for although ore was found at neighbouring prehistoric settlements, no metal tools appeared to be made of Aibunar ore. The conclusion is that there must have been a considerable trade in copper ores, not only from Aibunar, but from other mines elsewhere in Bulgaria. Objects of Aibunar copper have been found in other parts of Bulgaria, with some turning up in the USSR at Tripole sites on the Dnepr below Kiev.

The Bulgarian excavations throw light on a huge mining and metalworking industry, the earliest yet found in Europe, in the south-east Balkans and throughout the Carpathian province. Technically the mining process is simple, either open-cast pits or single uncomplicated shafts. By the late bronze age, central European miners were sufficiently skilled to work vast underground deposits involving massive timbering and penetrating hillsides for up to 100 m. Simple operations are seen at Viehofen near Salzburg, but particularly extensive are those working chalcopyrite in the Mühlbach-Bischofshofen region of the Mitterberg,[8] where a bed of copper ore 2 m thick dipped into the hillside at about 25°. It was worked along a front 1 600 m long. The miners used some bronze picks with blades about 0·3 m long, but for the most part the ore was extracted using the by now ancient art of fire-setting. Thick soot still hangs on some of the gallery walls. Foul atmosphere was one of the drawbacks of this method, and the Mitterberg miners doubtless allowed the smoke from the extinguished fires to clear overnight before entering to extract the loosened ore. Not only was the copper followed down its sloping course, but the roof was also 'stoped' or dug away by building an elaborate wooden staging on which waste rock was back-filled. This then served as a firm platform for attacking the roof, as well as providing upper and lower galleries in which the air could circulate. Passages in the waste itself were constructed to prevent too much weight resting on the wooden staging. By these means passages sloped down for up to 160 m while stopes reached a height of 30 m.

It is assumed that each of the 32 Mitterberg mines were worked for about seven years, allowing a working life for the whole group of over two centuries, assuming continuous consecutive work. When each working was at its peak development, Zschocke and Preuschen estimated that 180 men were employed, most of them used in timbering. As in a modern mine few men worked at the face.

Miners	40
Timbermen	60
Ore workers	20
Smelters	30
Porters	10
Cowmen	10
Supervisors and guards	10

The total quantity of copper produced in the Mühlbach-Bischofshofen region is estimated to have been 20 000 t, two-thirds of it from the Mitterberg ore body. Production of this kind must have sustained bronze smiths over a wide area of Europe. The sophisticated mining methods are those described from the Rammelsberg mines by Cancrinus in 1767. Most of the ore was stoped by fire-setting and Cancrinus gives an illustration showing how the waste or 'deads' was piled up and covered with wood so that the fire could attack the roof to best effect.[9]

There are no comparable workings for tin in prehistoric Europe. Hard-rock mining in bronze age Cornwall is highly likely, but there are no identifiable lode workings. Copper was mined open-cast at Alderly Edge, Cheshire, where malachite impregnates the soft Keuper sandstone.[10] Until recently it was confidently asserted that early bronze age copper workings existed at Mount Gabriel, Co. Cork, where 'old men's workings' or 'Danes' workings' were uncovered by the Geological Survey in the 1920s.[11] These have attracted much attention in recent years and given rise to the assertion that Ireland was an important exporter of copper in the early bronze age. However, Briggs[12] has highlighted the dangers of trying to date a hole in the ground when the complete mining history of a region has not been adequately researched and concludes that the workings at Mount Gabriel are more likely to be the otherwise apparently lost 'superficial workings' of 1862. Jackson[13] has reasserted his claim that the workings date within the period 1710–1275 BC. The debate will certainly continue.

NOTES AND REFERENCES

1   RENFREW, 1969
2   MOORE, 1979
3   RENFREW, 1971
4   DANIEL, 1950, 120
5   COLLINS, 1893
6   JOVANOVIĆ & OTTAWAY, 1976; JOVANOVIĆ, B., 'Primary Copper Mining and the Production of Copper' in Craddock (ed.), 1980
7   CHERNYCH, 1978; *see also* the same author's *Mining and Metallurgy in Ancient Bulgaria*, Sofia, 1978 (in Russian). A lengthy review and summary is given by T.A.P. Greeves in *J. Historical Metallurgy Soc.*, 1981, 45–9, and is reprinted in *P.P.S.*, 1982, 538–42

8   ZSCHOCKE & PREUSCHEN, 1932; PITTIONI, 1951, esp. 22–6 for mining details and reconstruction of the operations. Plates I–VI include photographs of the *Pingen* (ancient workings) and of the artifacts found

9   CANCRINUS, F.L., 1767, *Beschneibung der Bergwerke,*

quoted in Collins, 1893

10   ROEDER, 1901

11   JACKSON, 1968

12   BRIGGS, 1983

13   JACKSON, 1984

11 The advent of bronze in Europe

The precocious Balkan copper age of the 4th millennium BC is regarded as a local development free of outside influence. This independence parallels the rise of Chinese metallurgy in one respect. In both regions the technological breakthrough was due to the expertise of potters. At Karanovo in Bulgaria ovens in the earliest levels were capable of firing pots, and by about 4000 BC (level VI) black graphite-decorated ware was fired under reducing conditions at 1 050°C (in an oxidizing atmosphere the pots would be red). Copper melts at 1 083°C, and although malachite can be smelted at only 700°C, it also needs a reducing atmosphere.[1] The technology for successful copper working was thus well established in the Balkans, but here the similarity with China ends. The speedy appearance of tin bronze in China is to be contrasted with the protracted use in Europe of copper and alloys not containing tin. Simple copper bracelets from the earliest Vinča culture and contemporary Tisza culture of Hungary are dated to well before 4000 BC, whereas tin bronze first appeared not in the Balkans but in the central European Únetice culture around 2400 BC. The reason for the difference between China and the Balkans is simple. In China the metalliferous regions within easy reach of Anyang and related settlements contain both tin and copper, but in the Balkans tin appears to have been unknown in antiquity.

Knowledge of the Yugoslavian bronze age is inadequate, but ^{14}C dates place the earliest knowledge of tin bronze at about 2100 BC. Dates from the settlement at Crnobuki near the borders with Albania and Greece are 1650 ± 175 bc and 1710 ± 150 bc, several centuries later than the early Únetice dates, e.g. 1895 ± 80 bc (c. 2300 BC) for Kroměříz south of Olomouc in Czechoslovakia.[2] The inference is that the knowledge of tin bronze spread to the Balkans from the Únetice culture centred on the tin-bearing Erzgebirge.

However, tin does exist in Yugoslavia. Since 1970 occurrences have been discovered in Tertiary granites of the Serbo-Macedonian metallogenic province at Cer, about 60 km west-south-west of Beograd, and at Bukulja, about 50 km south-west of Beograd. In the hill at Cer cassiterite is associated with a greisen zone with more widespread disseminations of fine-grained cassiterite. Alluvial deposits have also been located here with cassiterite accompanied by some niobium–tantalum minerals. In the Bukulja granite hill, cassiterite is mainly in quartz veins and smaller amounts in pegmatite dykes. Alluvial tin also occurs here and is of potential economic importance.[3] In 1972 it was optimistically reported that enough tin existed to ensure Yugoslavia's self-sufficiency sometime in the future.[4] It remains to be seen whether Europe's newest tin-field escaped the searching eye of prehistoric prospectors.

A small amount of cassiterite is said to have been found in Romania between Reşiţa and Caransebeş[5], but no tin has ever been mined in the famous Banat copper region; the most that could ever be hoped for is a rare trace of tin in coppers mined here. Rare tin-bearing minerals of purely academic interest have also been found in southern Bulgaria; hemusite (Cu_6SnMoS_8) is reported from the Chelopech deposit associated with enargite and a host of other copper sulphides.[6]

Where tin was lacking other alloys were used, the most important being arsenic–copper, well named 'natural bronze' since it would have been difficult not to produce this alloy in many areas. In Hungary, antimony-bearing coppers, the so-called 'grey coppers' such as fahlerz, and the usual suite of generally scarce antimony minerals such as stibnite are common. At the bronze age foundry of Velem St Vid near Köszeg on the western frontier, ingots of copper contain up to c. 18% antimony. Other ingots contain some tin which must have been imported, though the two alloys are exceptional in the same piece. Clearly the local antimony was exploited to supplement scarce imported tin. Ore from the mines at Velem assayed at 16·6% antimony and 17·4% copper. Working here dates back to around 1800 BC if not earlier.[7]

The flowering of the European bronze age depended on the discovery of the tin deposits of the Erzgebirge,

61

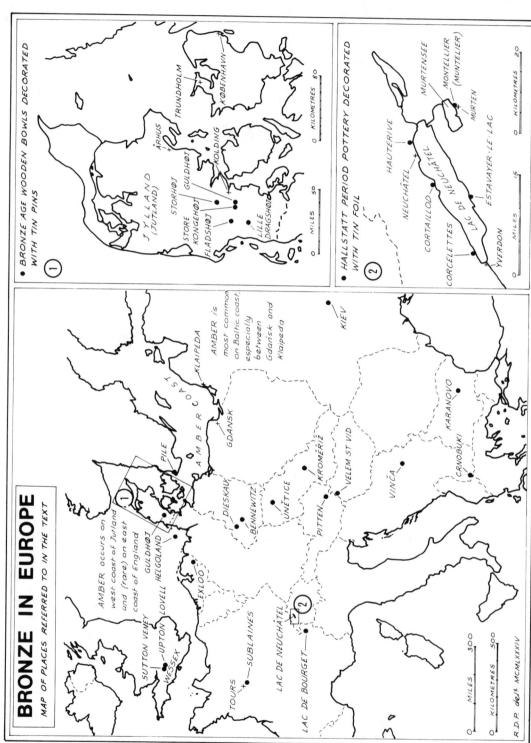

BRONZE IN EUROPE

MAP OF PLACES REFERRED TO IN THE TEXT

AMBER occurs on
west coast of Jutland
and (rare) on east
coast of England

AMBER is
most common
on Baltic coast,
especially
between
Gdansk and
Klaipeda

SUTTON VENEY
UPTON LOVELL
WESSEX
GULDHØJ
HELGOLAND
EXLOO
SUBLAINES
TOURS
LAC DE NEUCHÂTEL
LAC DE BOURGET
PILE
KLAIPEDA
GDANSK
AMBER COAST
DIESKAU
BENNEWITZ
ÚNĚTICE
KROMĚŘÍŽ
PITTEN
VELEM ST VID
VINČA
KARANOVO
CRNOBUKI
KIEV

MILES 300
KILOMÈTRES 500

R.D.P. del. MCMLXXIV

① • BRONZE AGE WOODEN BOWLS DECORATED
WITH TIN PINS

JYLLAND
(JUTLAND)
ÅRHUS
TRUNDHOLM
KØBENHAVN
KOLDING
STORHØJ
GULDHØJ
STORE
KONGEHØJ
FLADSHØJ
LILLE
DRAGSHØJ

MILES 0 50
KILOMÈTRES 0 80

② • HALLSTATT PERIOD POTTERY DECORATED
WITH TIN FOIL

HAUTERIVE
NEUCHÂTEL
CORTAILLOD
CORCELETTES
MURTENSEE
MONTELLIER
(MUNTELIER)
MURTEN
LAC DE NEUCHÂTEL
ESTAVAYER-LE-LAC
YVERDON

MILES 0 15
KILOMÈTRES 0 20

Map 11

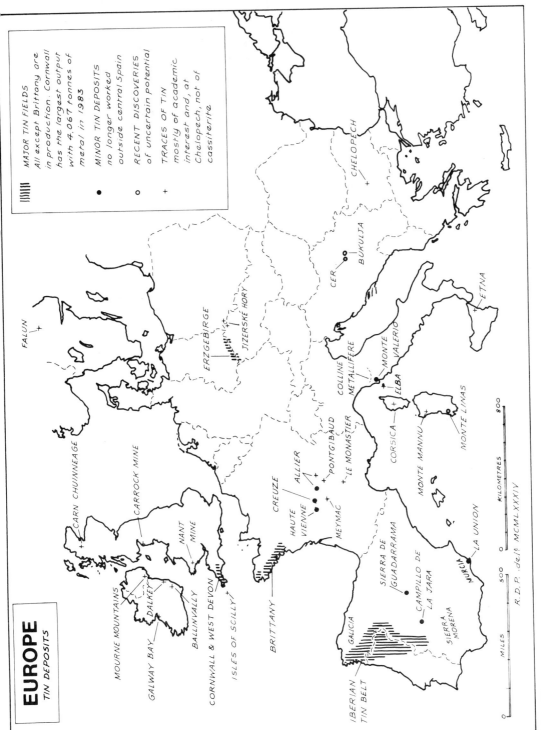

EUROPE
TIN DEPOSITS

MAJOR TIN FIELDS
All except Brittany are in production. Cornwall has the largest output with 4,067 tonnes of metal in 1983

MINOR TIN DEPOSITS no longer worked outside central Spain

RECENT DISCOVERIES of uncertain potential

TRACES OF TIN mostly of academic interest and, at Chelopech, not of cassiterite

● MAJOR TIN FIELDS
○ RECENT DISCOVERIES
+ TRACES OF TIN

FALUN

CARN CHUINNEAGE

MOURNE MOUNTAINS

CARROCK MINE

NANT MINE

GALWAY BAY

DALKEY

BALLINVALLY

CORNWALL & WEST DEVON

ISLES OF SCILLY

BRITTANY

CREUZE

HAUTE VIENNE

ALLIER

MEYMAC

PONTGIBAUD

LE MONASTIER

ERZGEBIRGE

JIZERSKÉ HORY

CER

BUKULJA

CHELOPECH

COLLINE METALLIFERE

MONTE VALERIO

ELBA

CORSICA

MONTE MANNU

MONTE LINAS

ETNA

SIERRA DE GUADARRAMA

CAMPILLO DE LA JARA

SIERRA MORENA

MURCIA

LA UNION

GALICIA

IBERIAN TIN BELT

MILES 0 500

KILOMETRES 0 800

R.D.P. delt MCMLXXXIV

Map 12

the mountain chain now forming the border between East Germany and Czechoslovakia. When and under what circumstances tin was discovered here can only be guessed at for there is no direct evidence, in the form of old mine workings or artifacts in tin streams, to prove the deposits were known before mediaeval historical records. At Graupen tin was supposedly first exploited in 1146, while at Schönfeld and Schlackenwald the 13th century writer Matthew Paris claimed that a Cornishman discovered the deposits in 1241.[8] Probably working had fallen off or completely stopped before our era, for the tin does not seem to have been known to classical writers. There is a growing belief, too, from the analyses of bronze age metalwork, that late bronzes of the Erzgebirge region contained less tin than at an earlier period, while attempts were being made to use the alloys of copper and antimony, suggesting that tin production was greatly diminished (P. J. Northover, pers. comm., 1982). Trade with southern Scandinavia was affected in the mid-1st millennium and the northward flow of metal objects from central Europe 'seems to have ceased entirely'.[9] In the mediaeval period when German production was at a high peak, the superior quality of Cornish tin was still acknowledged,[10] and the increasing tin output from Cornwall as the bronze age progressed may have adversely affected the Erzgebirge tinners.

But tin mining there must have been in the Erzgebirge during the bronze age, for without it the achievements of European metallurgists before the discovery of the Cornish ores cannot be explained. The great influence was the Únetice culture which grew up in the hinterland of the Erzgebirge and eventually penetrated Poland, a country largely devoid of metalliferous minerals.[11] Dating the earliest bronze age is uncertain, depending upon a handful of ^{14}C dates. When calibrated they suggest a transition from the copper age around 2500 BC, with tin bronze appearing by 2300 BC, passing into the middle bronze age Tumulus cultures *c.* 1700 BC.[12] The early bronze age of central Europe is dominated by the name of Únetice (Aunjetitz in German orthography), named after the type-site of 60 graves a few kilometres north-west of Prague. Among the characteristic bronze types are torcs, often called ingot torcs on the assumption that they were a convenient way of transporting metal as well as serving as a sort of currency. Similar torcs are found in the Near East, as at Ugarit (Ras Shamra) and Byblos,[13] leading to the now discredited theory that they were introduced to Europe by a migration of 'torc bearers'. If they were, the other metal types were left in the Levant. There is a variety of pins, rings, and other ornaments, flanged axes, daggers, and in some areas 'halberds', of which more below. From the earliest date Úneticïan hoards contain some arsenical bronze, 'but we begin to see the beginning of tin bronzes during this period', with arm rings and ingot torcs often

containing 8 or 10% tin.[14] By 2000 BC tin bronze was widespread in southern Germany and Bohemia.

The famous hoard from Dieskau, 60 km north-west of Leipzig, contains 293 cast flanged axes, as well as halberds, arm rings and other ornaments; 40% are tin bronze, mostly the heavy rings. The hoard dates to around 2000 BC and contains objects found over a wide area. The bronze-hilted daggers with the hafts almost, or entirely, encased in bronze are abundant in Saxo-Thuringia, East Germany, the Polish border, and were traded as far as Lithuania, the Baltic shore, and south along the Danube.[15] The hoard also contained 120 amber beads, which have a special significance.

Amber is a fossil resin exuded around 30 million years ago in the Oligocene forests of the extinct pine *Pinus succifer*. It is washed up on the west coast of Jutland and on the Baltic shore especially east of Gdansk; some is dug up inland. Not only is its golden colour attractive, but when rubbed it creates static electricity so that it was sometimes classified as a magnet. Of particular interest in modern times are the remarkably preserved Oligocene insects it contains. Whatever its special attraction to bronze age man – to the extent that it was traded to the Mediterranean – it was highly valued, though maybe its theft did not incur the ultimate penalty meted out in the 16th century AD.[16]

> The Amber that is brought from these parts lies in great quantity scattered on the sand of the sea, yet it is as safe as if it were in warehouses, since it is death to take away the least piece thereof. At Dantzic I did see two polished pieces thereof, which were esteemed at a great price, one including a frog with each part clearly to be seen, (for which the King of Poland then being there offered five hundred dollars) the other including a newt, but not so transparent as the former.
>
> Fynes Moryson (1566–1630) *Itinerary*

The frog and newt were undoubted forgeries; even fish were introduced to the amber which is easily softened. Such jiggery-pokery need not be levelled against Únetice traders and their intermediaries, who must be credited with introducing the Mediterranean civilizations to northern amber and their own tin. The distribution of bronze age metal types is patchy throughout Europe. The pattern is clearly the result of trade developed on a vast scale, though unsurprising compared with the far longer routes crossing the Near and not so Near East.

> It was amber, lovely, magical, prophylactic, which ensured that Jutland would be the starting-point of Bronze Age culture in the northern lands.[17]

Slaves, furs, hides, fish, cereals, and doubtless other goods besides, from southern Scandinavia must have been as attractive to southern markets in the bronze age as they were to the far-flung markets of the

22 One of two birchwood bowls decorated with pins of tin from Guldhøj; dia. 17 cm, height 11·5 cm (photo National Museum of Denmark, Copenhagen)

Christian and Moslem worlds in the heyday of Viking trade. It is highly significant that the transition from the bronze age to the iron age in southern Scandinavia was marked by a change in burial practice with the introduction of boat-shaped graves outlined with stones. The keel of Viking trade was well and truly laid in the bronze age.

Trade is the hallmark of the period. The high standard of bronze age metalwork in Scandinavia is remarkable because of the region's distance from metal sources. Cassiterite of purely mineralogical interest was found in Sweden in 1806 less than 0·5 km east of Falun. It occurred as small black grains and rarer octahedral crystals in a granite vein in Finbo quarry.[18]

23 Underside of wooden bowl from Guldhøj, showing the star pattern formed by pins of tin (after Vilh. Boye, 1896)

Yet tin was obtained in the early bronze age in plenty, initially probably from the Erzgebirge since much Danish pottery shows close ties with central Europe. Nor was there a shortage of contacts with the British Isles, so the opening up of Cornwall's tin reserves may have been of great importance in stimulating bronze production over a wide area of northern Europe. Sufficient tin reached Denmark to be used on its own for ornamental purposes, though surviving examples are few and confined to rows of tiny tin pins decorating wooden bowls. Examples are known from Store Höi, Flödhöi, Store Konhehöi, Lille Dragshöi, and – best known of all – two handled bowls of birch, each with pins forming star patterns on the undersides, from Guldhøj west of Kolding in Jutland (Figs. 22, 23). The primary burial, excavated in 1891, contained the skeleton of a man dressed in the now-familiar bronze age cloak, cap and leather shoes, accompanied by a bronze dagger in a wooden sheath, a bronze palstave with a wooden handle, a rare horn spoon, a round, lidded box of bark, a remarkable folding stool formerly covered in otter skin, as well as the wooden cups.[19]

The bronze work of Denmark rivals – occasionally surpasses – the work of southern artisans. Who cannot be impressed by the famous chariot ploughed up in 1902 in a peat bog at Trundholm in north-west Zealand? This remarkable contraption represents a horse pulling a bronze disc gilded on one side to represent the sun, and is good evidence for the widespread belief in sun worship in Europe in the middle of the 2nd millennium. Although a northern product, it is accepted that the techniques of manufacture, such as the hollow-cast bronze horse, are southern, perhaps Greek in inspiration. Indeed, the four-spoked wheels are usually compared with varieties current in Mycenaean Greece and Egypt.[20]

The influence of both central Europe and Britain can be seen in Scandinavian metalwork. The hoard from Pile in southern Sweden contains some flanged axes of tin bronze with decorated bands that follow the curve of the cutting edge. This is a British feature, but the hoard also contains ingot-rings of Únetiče type. An idea of the dominating influence of Únetiče culture can be gained from the enormous hoard, perhaps of items intended for export, found at Bennewitz near Halle. It contained 297 flanged axes, more than is known from the whole of Sweden.[21] Many metal objects can be used to trace Únetičian influence well outside the homeland. The ceremonial halberd, a dagger-shaped blade set at right angles in a long handle, is widespread in central Europe, though the distribution pattern is complicated by the presence, notably in southern Scandinavia, of halberds of Irish type. Únetičian halberds are of tin bronze, but Irish ones are of copper and continued to be made of copper even after the introduction of tin bronze to Ireland. Formerly this was one of the reasons taken to indicate an

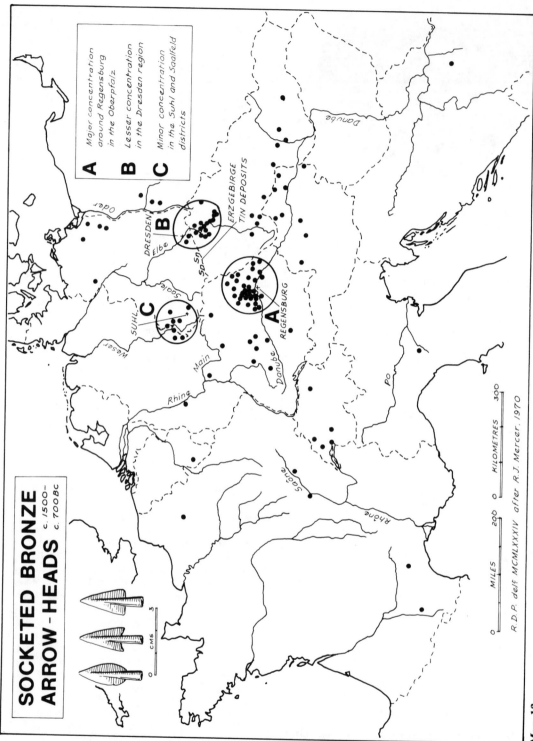

SOCKETED BRONZE
ARROW-HEADS c. 1500–
c. 700 BC

0   CMS   3

A Major concentration
   around Regensburg
   in the Oberpfalz

B Lesser concentration
   in the Dresden region

C Minor concentration
   in the Suhl and Saalfield
   districts

Oder

DRESDEN

Elbe

ERZGEBIRGE
TIN DEPOSITS

SAALE

SUHL

Saale

Weser

Main

REGENSBURG

Danube

Rhine

Po

Soône

Rhône

Danube

MILES 200 0

KILOMETRES 0 300

R.D.P. delt. MCMLXXXIV after R.J. Mercer. 1970

Map 13

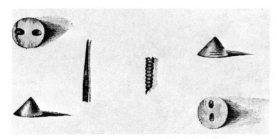

24 Perforated conical bone buttons, bronze awl, and segmented bead of metallic tin from Sutton Veny (from Plate XII of Colt Hoare's *Ancient Wiltshire*, 1, 1812)

independent invention of copper halberds in Ireland, just as Chinese halberds are assumed to be unconnected with European types.[22]

A characteristic item of Únetičian metalwork is the bronze socketed arrowhead. The socketed spearhead is a common European artifact well known in Britain, but its smaller relative is unknown there, and on the Continent examples become progressively more common towards their centre of origin. The major concentrations are at the foot of the Erzgebirge and at the heads of the adjoining river systems. The distribution pattern clearly illustrates how the arrowheads were traded along river valleys. All are made of tin bronze, and Mercer concluded that their distribution reflected the 'superfluity of tin' from the Erzgebirge.[23] Elsewhere tin was too precious to be squandered on weapons that might easily be lost after one shot. This is not a reflection on the skill of bronze age archers but on the merits of the bronze arrowhead. Practical experience within historic times has shown that socketed arrowheads of bronze tend to break on impact with any resistant target. This alone was sufficient to deter their use elsewhere in Europe, even in Britain where there was no shortage of tin. In the Únetičian homeland these arrowheads must have been for prestige only, perhaps used by the rising chieftain class for hunting game. They are a feature of princely burials accompanied by another typical Únetičian piece, the bronze halberd with elaborate bronze handle.

If a single object were chosen to epitomize trade in the European bronze age nothing would be more appropriate than the beautiful necklace dug up in 1881 by a peat-cutter at Exloo in Holland (*see* Frontispiece). It is composed of beads made of faience, amber, and tin. The 25 tin beads are almost unique. A lone tin bead, now sadly lost, was found with typical early bronze age 'Wessex' bone buttons (Fig. 24) in a barrow on Sutton Veny Downs 'not far from Pertwood', although the exact location of the barrow is not now known.[24] A reconstructed necklace from a burial at

Upton Lovel in Wiltshire consists of amber, shale, and blue faience beads which, among other British examples, closely resemble the Exloo beads. The British amber might have been collected on the Norfolk coast, but it is safer to see it as a European export, perhaps from the island of Basileia (Helgoland) where Diodorus Siculus (Bk V, 23) said 'the waves of the sea cast up great quantities'.

The Exloo beads of tin and faience are likely to be of British origin. The tin beads so closely resemble the British segmented faience beads in shape that they may be products of the same atelier. British faience is distinguished by its high tin content, so eventual analysis of the Exloo faience would help prove, or disprove, this theory. The peculiarity of the composition of British faience can be explained by assuming a deliberate addition to the faience melt of tin metal otherwise used in the manufacture of tin beads. Tin beads were a rarity, perhaps a speciality of one manufacturer, with a good supply of Cornish tin, but some could have disappeared because of the tendency of tin metal to reconvert to the oxide and crumble away.

Elsewhere in this book there are references to the rare use of tin foil for decorative purposes. The finest expression of this art form is to be found on Hallstatt pottery, transitional between the bronze and iron ages, in Switzerland and adjoining areas. Its simple beginnings may, perhaps, be seen at Pitten in Austria, where curious sherds of pottery in graves 10, 11, and 151 had folded over them thin, narrow strips of metallic tin, though for what reason is far from clear. Grave 151 was ^{14}C dated to 1085 ± 212 bc.[25] The tin strips resemble those used to decorate pottery found in the lakeside dwellings of Lac de Neuchâtel and elsewhere, which were occupied up to the 8th century BC when climatic change forced their abandonment as lake levels rose to 9 m in Lac Léman.[26]

Geometric designs were incised on vessels made without the use of the potter's wheel. After firing the designs were enhanced by the application of thin strips of tin (occasionally alloyed with silver and lead, but not in Switzerland) stuck in place with resin or pitch.[27] Examples are rare and occur mainly on small vessels up to 8·5 cm high. When newly made, the shining tin formed a striking contrast to the black graphite-coated pottery. By far the most impressive surviving examples are the fragments of a dish 35 cm in diameter from Cortaillod on the northern shore of Lac de Neuchâtel (*see* Figs. 25–27). As on other vessels, the tin foil has a geometric decoration derived from lines incised in the pots before firing. Obviously firing took place at too high a temperature for the tin to have been applied to the wet clay, as Keller assumed.[28] A sherd from Montellier was technically impressive since the foil had been stuck to raised bands of clay.

Tin-decorated pottery occurs in France as a western

25   Plate XCI from Keller, 1878; a reconstruction from the fragment in Fig. 26

extension of the 'lake-side civilization' around the Lac de Bourget between Geneva and Grenoble. There are fine examples from Châtillon with very narrow strips of tin and matchstick-type figures, and other examples from Grésine and Conjux.[29] Sanders also lists examples from Chilly-sur-Salins, Jura; Scey-en-Varais, Doubs; Pommard, Côte d'Or; Grotte Deroc near Vallon, Ardèche; Malgoirès, Gard, all sites within the Rhône–Saône basin.

If, as mentioned earlier, tin production in the Erzgebirge was greatly diminished by the end of the bronze age, a western source for the metal is probable. Recently, Piggott[30] has drawn attention to a handled burial urn, *c.* 800–700 BC, from a barrow at Sublaines near Tours (Fig. 28). It is clearly of Alpine type with a band of geometric designs and a schematic four-wheeled wagon with polychrome and tin-foil inlay. Sublaines lies conveniently on a line between Switzerland and Brittany, and there may well have been a reciprocal trade in Alpine copper and Breton tin. Brittany is poor in copper and some late bronze age ingots from there contain traces of nickel, suggesting to Briard[31] an Alpine origin.

A Breton link suggests the possibility of Cornish tin reaching lakeside dwellings. Finds from Estaveyer include two tin ingots 16 and 15 cm long (Fig. 29), a little gold (from a tin stream?), and some jet rings. Nowhere has jet been more commonly exploited than at Whitby in Yorkshire. Jet occurs in several localities in France, Spain, Switzerland, Germany, and Austria. Little

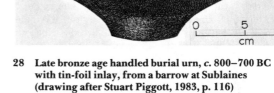

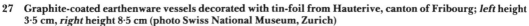

28 Late bronze age handled burial urn, *c.* 800–700 BC with tin-foil inlay, from a barrow at Sublaines (drawing after Stuart Piggott, 1983, p. 116)

29 Ingot of tin, *c.* 15 cm long, from Estavayer, Lac du Neuchâtel (after Keller, 1878, Plate XCVII Fig. 1)

26 Fragment of earthenware plate with graphite coating and tin-foil decoration from Cortaillod; dia. of plate 35 cm (photo Swiss National Museum, Zurich)

27 Graphite-coated earthenware vessels decorated with tin-foil from Hauterive, canton of Fribourg; *left* height 3·5 cm, *right* height 8·5 cm (photo Swiss National Museum, Zurich)

work has been done on distinguishing jet from archaeological sites, even though it was realized in 1875 that Spanish jet was inferior to that from Whitby and could be distinguished by scratching it with a knife, 'the Spanish giving a very irregular scratch, whilst that from Whitby gives a fine groove'.[32] Childe believed Whitby jet was imported into Germany,[33] but this is considered unlikely in Hallstatt Switzerland according to Dr V. Rychner (per G. D. Engel, pers. comm., 1983), and trade in jet is probably unconnected with trade in cassiterite.

A good specimen of Cornish 'sparable' tin is readily distinguishable from a Bohemian cassiterite in a zindwaldite matrix, but once smelted they are at present indistinguishable and we must await the results of Professor George Rapp's 'finger-printing' experiments to see if the position can be changed. The extent to which tin from different ore-bodies was traded remains a matter of guesswork. Gimbutas has described the rich metallurgy of the Chernoles complex south of Kiev *c.* 750–500 BC.[34] It is easy to envisage tin reaching this part of the southern USSR after being shipped from the Erzgebirge down the Danube, across the Black Sea and up the River Dnepr. From what has been said above, this now seems unlikely. Should a Cornish origin be ventured? But to suggest that Cornish tin reached this part of Russia in the early 1st millennium BC just as it did in the 9th century AD, is carrying speculation too far. On present evidence it is safer to assume an eastern origin for Chernoles tin. Some of their bronze objects were influenced by Scythian types, and it has already been noted that the Scythians obtained their gold from Siberia, so their tin may have come from the same remote location.

Whatever the extent of trade in tin from central Europe, Brittany, or Cornwall, the wealth and splendour of European bronze age metalwork is adequate testimony of the division of labour promoted by the advent of metal. Mining, smelting, the manufacture of multifarious ornaments and artifacts, and a widespread, efficient trade in manufactured goods and raw materials, all depended on specialist workers. It has been suggested that tin mining in the Erzgebirge was a part-time occupation,[35] and it may have proved impossible to stream tin, as distinct from mining it, in the winter. Something of the difficulties imposed on streaming in summer and winter in climatically equable Cornwall will be discussed in Chapter 24, but it must be assumed that tin mining in the Erzgebirge was

as efficient as copper mining at Bischofshofen and elsewhere in the bronze age. It cannot be doubted that for at least a millennium in the Erzgebirge, from well before 2000 BC, there was an unknown but substantial number of people whose principal reason for being there was the winning of tin ore.

NOTES AND REFERENCES

1   RENFREW, 1968
2   COLES & HARDING, 1979, 201
3   DUNNING, MYKURA & SLATER (eds.), 1982, 186
4   ANON., 1972, *Tin International*
5   DAVIES, 1935, 206, noting that 'ancient workings are unknown'
6   TAYLOR, 1979, 373–4
7   CLARK, 1965, 195–6; TYLECOTE, 1976, 8
8   PARIS, (died 1259), *Historia Major Angliae* (London, 1571); quoted in Hoover & Hoover, 1950, 283
9   JENSEN, 1982, 155–67
10 HEDGES, 1964, 16
11 JANECKA & STEMPROK, 1967, who note tin in skarns (altered limestones) and schists on the Polish side of the border, as at Kopaliny. This is a modern discovery of no economic importance
12 COLES & HARDING, 1979, 67–9
13 BURNEY, 1977, 106
14 TYLECOTE, 1976, 23–4
15 HAWKES, 1940, 330
16 SEAGER, 1896, 7
17 JONES, 1968, 157
18 WÖHLER, 1826, 45–8
19 GLOB, 1974; BOYE, 1896
20 POWELL, 1966, 160
21 CLARK, 1965, 257
22 O'RIORDAIN, 1937
23 MERCER, 1970
24 HOARE, 1812
25 HAMPL & KERCHLER, 1981
26 SAUTER, 1976, 107–8
27 TRACHSLER, 1966, 140–51, 152–60
28 KELLER, 1878
29 SANDARS, 1957, esp. 369 and map XIII
30 PIGGOTT, 1983, 116
31 BRIARD, J., in Giot, Briard & Page, 1979, 31
32 HUNT, 1875, **3**, 10
33 CHILDE, 1939, 165, 255, 260, 291
34 GIMBUTAS, 1971
35 NEUSTUPNY & NEUSTUPNY, 1961

12 Tin in the Erzgebirge

The great Bohemian massif is one of the classic mining regions of the world. It is especially famous for its tin, which occurs from the Vogtland and Slavkovsky Les region in the west, through the 'Ore Mountains', the Erzgebirge of Germany and the Krusne Hory of Czechoslovakia, on both sides of the East German–Czechoslovakian border, to the Altenberg region (see Fig. 30). A little tin of no consequence also occurs in the Jizerské Hory on both sides of the Czech–Polish border.[1]

A tradition that tin was discovered in the Erzgebirge by a Cornishman can be traced back to Matthew Paris, a Benedictine monk of English birth who died in 1259. In his *Historia Major Angliae* (not published until 1571) he relates how a Cornishman forced to flee his native country discovered tin in Germany in 1241 with the result that the price of tin fell markedly. The story is retold by Camden, Borlase, and Pryce among others, and may well be believed by some modern archaeologists who, in spite of the evidence presented in Chapter 11, have denied that tin mining could have taken place in the Erzgebirge in antiquity. Because the cassiterite occurs in hard granitic rocks it is claimed the tin would 'have been completely inaccessible to the metal workers of the Bronze Age', and that the tin used must have been imported from Cornwall.[2] This claim not only overlooks the fact that tin bronze was known in central Europe before the discovery of the Cornish deposits, but denies to the bronze age miners of the Continent those skills which would have allowed them to work the Erzgebirge ores with the same success which attended their operations in contemporary and earlier copper mines. Moreover, greisen-bordered tin veins are often found in soft kaolinized granite.

In hard-rock mining the value of 'fire-setting' should not be underestimated; this method was already ancient by the time tin bronze came on the European scene in the middle of the 3rd millennium BC. Collins[3] studied the use of alternately heating and quenching stones; the effects of this method must have become apparent through cooking or by accident when attempting to put out fires. The controlled use of fire for splitting obsidian, flints and other siliceous material in axe manufacture extends back to the palaeolithic age, and has also been recorded as recently as 1875 amongst the Shoshoni of Colorado. Collins' own experiments with cherts demonstrated that 'percussion flaking is greatly facilitated by heating' without any increase in the brittleness of the resulting artifact. To use fire would have been an obvious ploy by the first hard-rock miners of copper and, later, of tin.

There are few accounts of the use of fire-setting in Cornwall, but Tonkin,[4] writing in 1733, asserted that the Cornish miners:

> When they meet with rocks and very hard ground, as sometimes they do with such as require not only three weeks, but three months, to hew so many feet through the same, they formerly burnt furze, faggots, &c. to break the rocks; but that proving insufficient, and very often fatal to the workmen, by the sudden changing of the wind, which drove down the smoke upon them, and suffocated them, they have of late had recourse to gunpowder...

a method which Tonkin admitted is 'likewise attended with many sad accidents'. Hardness of rock was no deterrent to fire-setting – in fact quite the opposite, for according to Agricola:[5]

> As for the hardest kind of metal-bearing vein, which in a measure resists iron tools, if the owners of the neighbouring mines give them permission, they break it with fires... As I have just said, fire shatters the hardest rocks, but the method of its application is not simple.

Agricola may well have been speaking of the mesh of small veinlets (a 'stockwork') of tin at Altenberg, or the copper and lead mines in the Rammelsberg where fire-setting was used until the 1860s (Fig. 32). Simonin gives an excellent concise account of the method employed in the Rammelsberg Mine.[6]

> Horizontal layers of billets of firewood disposed crosswise above one another, are piled up in a nearly vertical position so as to present four free vertical faces (whence the pile has been called a *chest* by the miners), and set fire to. The flame plays on the face of the ore, which becomes shattered and traversed by cracks, and when cooled is very easily detached with the pick, or long iron forks...
>
> At the Rammelsberg Mine, in the Harz, to the south of Goslar, this old method is still practised, and the fires are lighted on Saturday night, and kept burning until Monday morning when the miners return to the mine to resume their usual labours, after the fireman and his

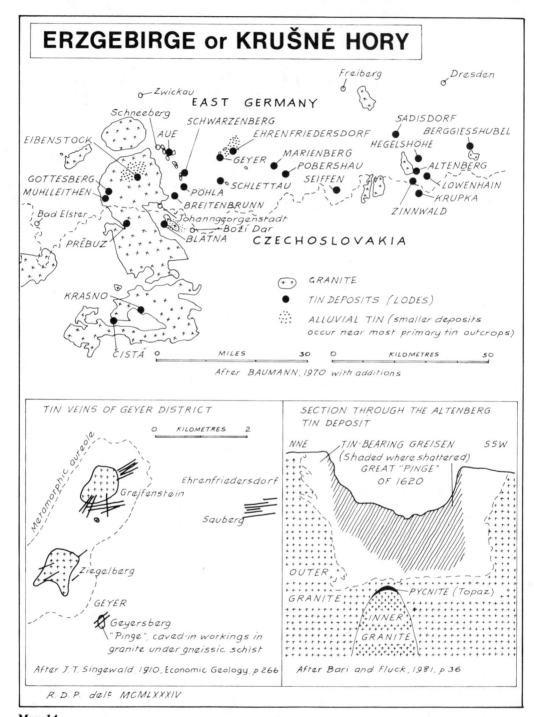

ERZGEBIRGE or KRUŠNÉ HORY

Freiberg

Dresden

Zwickau

EAST GERMANY

Schneeberg   SCHWARZENBERG   SADISDORF

EIBENSTOCK   AUE   EHRENFRIEDERSDORF   BERGGIESSHUBEL

HEGELSHÖHE

GEYER   MARIENBERG   ALTENBERG

POBERSHAU

GOTTESBERG   SCHLETTAU   SEIFFEN   LÖWENHAIN

MUHLLEITHEN   PÖHLA   KRUPKA

Bad Elster   BREITENBRUNN   ZINNWALD

Johanngeorgenstadt

Boži Dar

PŘEBUZ   BLÁTNA   CZECHOSLOVAKIA

GRANITE

KRASNO   TIN DEPOSITS (LODES)

ALLUVIAL TIN (smaller deposits
occur near most primary tin outcrops)

ČISTÁ   0   MILES   30   0   KILOMETRES   50

After BAUMANN, 1970 with additions

TIN VEINS OF GEYER DISTRICT

0   KILOMETRES   2

Metamorphic aureola

Greifenstein   Ehrenfriedersdorf

Sauberg

Ziegelberg

GEYER

Geyersberg

"Pinge", caved-in workings in
granite under gneissic schist

After J. T. Singewald, 1910, Economic Geology, p 266

SECTION THROUGH THE ALTENBERG
TIN DEPOSIT

NNE   TIN-BEARING GREISEN   SSW
(Shaded where shattered)
GREAT "PINGE"
OF 1620

OUTER   PYCNITE (Topaz)

GRANITE   INNER
GRANITE

After Bari and Fluck, 1981, p 36

R. D. P. delt MCMLXXXIV

Map 14

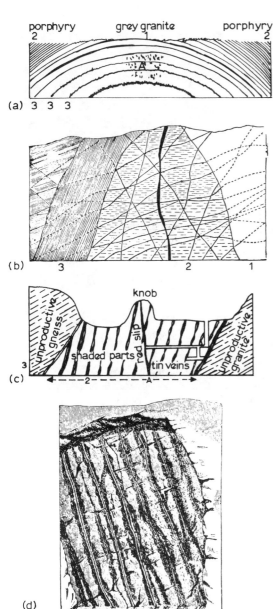

porphyry 2   grey granite 1   porphyry 2

(a) 3 3 3

(b)   3   2   1

knob

unproductive gneiss

shaded parts

red slip

tin veins

unproductive granite

3

(c) ←— — 2 — —A— — — →

(d)

30 **'Stockwerks' and tin-bearing strata at various locations in the Erzgebirge (from E. H. Davies, 1901)**

31 **Fire-setting in progress (Agricola, *De Re Metallica*, bk V, 1556)**

work, mining was done partly by the use of gunpowder followed by underhand and overhand stoping, and partly by the use of fire followed by overhand stoping (*förstenbau*), i.e. mining the rock above the miners' heads. Gunpowder was used on the softest rocks, 'generally on the veins themselves', while fire-setting was employed principally where hard stanniferous granite separated the small veins of tin, a commonplace in a stockwork. The piles of wood (*bûchers*) were prepared for fire-setting once or twice a week. Two days after firing, the fractured portions could be worked loose with pick-axes or with gunpowder. By these means a cubic fathom yielding 700 to 800 cwt of ore required twenty-four days work, two or three *cordes* of wood and 15 lbs of powder.

Fire-setting was extremely skilful. Good ventilation was of the utmost importance. A set of lodes would be attacked at several points from working platforms 11–18 m above each other and linked by shafts. At Gayer work proceeded upwards by overhand stoping from each level on the working face. Manès comments that for a long time it had been proposed to try working by descending stopes, that is by underhand stoping, but gives no indication that it had ever been tried. The way in which the bundles of wood, or 'pyres' as Manès called them, were placed was all important. Upon them depended the action of the fire as it could be arranged to attack the rock face horizontally or vertically. In driving levels, either along the lode or crosswise, the force of the fire had to be directed horizontally. To achieve this the 'pyres' were stacked against the working face at an angle of less than 45° and were placed on top of bundles of wood arranged in a cross-shape to allow the passage of air currents. The

assistants have extinguished the remains of the bonfires. While the fires are burning it is impossible to remain in the galleries, for a thick smoke fills all the working places of the mine.

It is fortunate that an excellent account of the use of fire-setting in the tin mines was published by Manès.[7] At Langenzeche and elsewhere in the Geyer stock-

32   **'Attack of rocks by fire' in the Rammelsberg copper mine (Fig. 122 from Simonin, 1868)**

bundles of wood were also surrounded by a thick high wall (*muraille*) of large pieces of ore already extracted 'so that the flame can have no action in the rear and only a little against the roof'. Not only did this method attack the base of the working face, but also shattered the huge pieces of ore and subjected 'them to a roasting'. This use of a *muraille* could be extremely ancient, and the illustration by Simonin of an 'Attack of rocks by fire' (Fig. 32) probably shows a more recent innovation in which a shield covers the sides and top of the fire in order to direct the flames to the base of the working face; the miners are clearly driving a level and not overhand stoping.

In overhand stoping the need is simply to attack the roof. At Geyer this was done by stacking all the wood cross-wise, much as Simonin described at the Rammelsberg Mine, forming a lattice with channels through which the air currents direct the flames upwards. The way the wood is stacked in the illustration in Agricola's *De Re Metallica* (Fig. 31) suggests overhand stoping, the proximity of the bundles to the surface of the earth being merely an artistic device. Note the refinement of shaving the wood, performed by the man at the surface, so that the curled-back shavings 'easily take light, and when once they have taken fire communicate it to the other bundles of wood'. For the best results at Geyer, the roof

might need prior preparation. If as a result of the previous fire the surface had become too smooth it was made irregular again, in Manès' day by using gunpowder, 'so that the flame can touch it at many points, and penetrate the interior by many crevices'.

Fire-setting may have been slow but it would certainly have been the best method available in antiquity. Its efficiency could be improved by quenching the fire with cold water, a practice alluded to by classical writers, though the supposed use of vinegar, referred to by Pliny in his *Natural History*, is probably the result of a wrongly transcribed text in Pliny's own time. Experiments have shown that quartz rocks heated to *c.* 600°C and subsequently quenched with water are rendered so friable as to turn to powder when rubbed with the fingers.[8] When Collins visited the Norwegian silver mines at Kongsberg about 1893 to see the last use of fire-setting, he found the method used only for the occasional driving of levels 'in hard siliceous rocks'.[9] Until the introduction of steel borers, dynamite and other high explosives, the cost and efficiency of fire-setting equalled the use of double-handed iron borers and gunpowder. If 19th century figures are any guide to bronze age capabilities, fire-setting could drive 1·5–6 m a month. The methods of mining in the Mühlbach-Bischofshofen copper mines, referred to earlier, are strikingly similar to descriptions of modern methods of fire-setting and should allay fears that it is irresponsible to compare recent practice with mining 4 000 years ago. Cancrinus could well have been describing Mühlbach bronze age miners when, in 1767, he described how at the Rammelsberg mine, when the ore was stoped by fire, the 'deads' or waste materials were built up and wood placed on top under the roof of the stope so that the fire would act to best effect. Collins rightly saw here the origin of 'overhand stoping'.[10]

In the Erzgebirge, tin is not everywhere disseminated through solid granite. Some veins are in mica-schists, as in the Ehrenfriedersdorf region. Even in the granite the country rock varies considerably in hardness, so that 'near the veins it is tender, and offers but slight resistance to the miner'.[11] Kaolin (china clay) is not an important economic mineral in the Erzgebirge, but decomposed feldspars are widespread, so that soft white layers of kaolin and bands of soft grey-green chlorite often accompany the stanniferous veins.

As in Cornwall, there is a wide range of minerals in the Erzgebirge. At the most important tin deposit centred on the Altenberg, the country rock is a dark 'greisen' composed of lithium-rich biotite–mica, quartz, and topaz. The ore minerals include cassiterite, wolframite, molybdenite, bismuth, chalcopyrite, and sphalerite or zinc blende. Chalcopyrite is not important here, but it also occurs at Ehrenfriedersdorf and in the west of the range at Mühlleithen. On the Czech side of the border there is a good deal of chalcopyrite, sphalerite, galena, and silver at Hora Blatna (Platten), Přebuz (Fruhbuss), and Hora Slavkov (Schlaggenwald). Manès also referred to copper and silver veins at Schneeberg. It is not known whether tin was found when prospecting for copper in these regions. It may not be too fanciful to suggest that fire-setting could have produced smelted tin by accident.

Collins records an interesting tale by Löhneyss,[12] that at the St Georg mines at Schneeberg the heat from the fires was intense enough to melt silver which is said to have flowed from the working-faces. Argentite, or silver glance, occurred in abundance here with native silver, and argentite melts readily in a candle flame. Cassiterite is not immediately recognizable as a metal-bearing mineral, so fire had to be employed by chance or design to discover its true nature.

Stannite, a tin–copper–iron sulphide rare in workable amounts, was formerly recovered in commercial quantities in the Erzgebirge, notably south-east of Freiberg. Exploitation of these lodes in the bronze age might have resulted in the production of a natural bronze without the separation of the tin from the other elements. Such separation would, in any case, have been impossible in the bronze age according to some metallurgists,[13] and the importance, if any, of stannite in prehistory remains open to debate.

Having established that hard-rock mining would have been possible in antiquity, it is deflating to have to restate that there is no proof of it in the Erzgebirge before the historical working at Altenberg in the 12th century AD. Altenberg is Europe's largest tin working, an open-cast pit some 300 m in diameter and 100 m deep. Ancient galleries are visible in the upper part of the quarry face,[14] but it can only be surmised that this ancient pit began as a small openwork in the bronze age. The importance of the Únetiče culture that straddled the Erzgebirge in the early bronze age has been mentioned in Chapter 11. Forman and Poulik[15] refer to Etruscan goods which reached Czechoslovakia, and since it is known that the Etruscans had to import tin, cassiterite from the Erzgebirge may have been one good reason for the trade.

The tin extracted in antiquity could all have come from alluvial sources, though here again proof is lacking. Prehistoric finds from tin streams are virtually confined to Cornwall and Brittany, but at least the erroneous belief that there is no stream tin in the Erzgebirge can be quashed (Fig. 33). Indeed, with the rising value of this scarce metal nowadays it may well be of economic importance in the future. The nature of the alluvials has been summarized by Dr Ludwig Baumann of the Mining Academy at Freiberg:[16]

> For centuries cassiterite placers in the Erzgebirge have been among the most important tin producers. They occur in the vicinity of almost all primary tin ore deposits of the Erzgebirge, both as eluvial [i.e. resting where eroded from the parent lodes] and fluviatile concentrations. With regard to geological age, fossil placers [Permian to Tertiary] and Quaternary placers can be distinguished. It is mainly the Quaternary placers that have become commercially important. Because they are easily mined, placer deposits should be given special attention in any future geological exploration for tin deposits. Above all, the great placer districts of Steinbach-Eibenstock and Ehrenfriedersdorf-Thum deserve investigation in view of their probable reserves and their favourable situation as regards transport facilities. The favourable conditions of extraction and treatment of tin placers contribute to promising future prospects.

The placer deposits were well known to Georg Bauer (1494–1555), better known by his latinized name Georgius Agricola, whose treatise on mining *De Re Metallica*, published in the year after his death, was not superceded as a textbook until the appearance in 1738 of Brunswick's *Grundlicher Unterricht von Hutte-Werken*. Agricola's oft-quoted classic has been made especially famous by the meticulously annotated translation, embellished with all 289 woodcuts, by President C. Hoover and his wife in 1912 and still in print with more recent editions (Fig. 34). The end of book VIII deals with alluvial tin, the metalliferous material 'usually found torn away from veins and stringers and scattered

33   Late mediaeval tin streaming area west of Boží Dar near Rozhrani (Halbmail) (photo Jiří Majer 1980, Plate 2 Fig. 2)

far and wide by the impetus of water'. Eight methods of washing the tin-bearing sands are described, six of them of great antiquity in Agricola's day.

A valuable and little known account of the mediaeval and later tin industry is given by Arqué[17] who specifically mentions stream tin near villages 'half-an-hour from Weissenstadt'. This town is in the Fichtelgebirge, an area that Schmidt[18] asserted 'played a considerable rôle in the earlier historic and prehistoric times as a source of tin ... most of the former mining operations were confined to the working of placers'. By far the best account of the method of working the placer deposits is given by Manès, whose description of the environment is a reminder that however 'easy' the exploitation of such deposits may be today, they were not so in the past.[19]

> The climate is rigorous. There are frequent violent winds and long, heavy rain. In winter, snow covers the ground for four or five consecutive months. The higher parts are covered with wretched pine trees, and those parts given to cultivation only offer a meagre soil carrying grain hardly able to push through. Only in the more productive valleys does the traveller see a few pleasant meadows, a few beautiful crops.

Such a picture may have induced Neustupny to claim that tin working in the Erzgebirge would have been seasonal work undertaken by the agricultural population in the neighbourhood of the deposits.[20] The great importance of the Cornish deposits is that they lie in an area of equable climate allowing mining more or less

continuously, though even here, as will be discussed later, there were difficulties with too much or too little water. In the Erzgebirge, as in Cornwall, the 'tin ground' (to use a Cornish expression) was buried beneath an overburden of sediments 'three, five fathoms and more deep' and several thousand fathoms in length. These occurred in the low-lying parts of the valleys, what the Cornish tinner called 'bottoms'. Pebbles of tin also occurred on hill slopes, notably covering the lower half of the eastern side of the Auersberg. This is the equivalent of the Cornish 'shode tin', but since the tin veins in the Erzgebirge usually outcropped in a mass of fine veins or stockworks, the resulting shode (or eluvial tin, to use a modern term) would have been of greater value than in Cornwall where it was of little commercial value in itself, but merely an indicator of where the parent lode lay. In the Erzgebirge, therefore, shode tin may have been of the utmost importance in prehistory, and its exploitation may in part account for the lack of prehistoric finds in the streamworks. The pebbles of tin-stone are described by Manès as:[21]

> formed of granitic and schistose rocks ... more or less rounded. They range in size from a 'ligne' [*c.* 1·5 mm] to a foot, and sometimes even three feet in diameter. They usually rest in a sand of fine quartz grains and are particularly rich in tin. The granite pebbles most often contain the tin in veins; those of tourmaline-schist, which are very abundant, are thoroughly impregnated [with tin]...
> One finds, besides, pebbles of massive tin-stone, and fragmented crystals of 2 to 4 'lignes' diameter disseminated in

a stream *b* ditch *c* mattock *d* pieces of turf *e* seven-pronged fork *f* iron shovel *g* trough *h* another trough below it *i* small wooden trowel.

34 Tin streamers at work in the Erzgebirge; the nearest man in the stream, shod in high leather boots, is using a seven-pronged fork, another of which is shown on the bank (Agricola, *De Re Metallica*, Bk VIII)

the sand. Some pebbles contain iron oxide as well, some tourmaline, and Charpentier said that one can find there opal, and a few little flakes of gold.

Manès then gives an account of the method of working the streamworks or *seyffenwerk* which it is worthwhile quoting in full:

> The greater part of these *seyffenwerk* have been exploited with vigour for more than a century, always advantageously until recently, for the low price of tin has forced them all to be abandoned.
>
> The *seyffenwerk* were usually situated in the bottoms (*bas-fonds*). On the slopes of escarpments which rose up here and there were constructed channels to conduct the water, and these channels led it, at the correct gentle

slope, whereby the water separated with its strength of flow the rocks from the soil, and carried them to the bottom in a channel which followed the valley. Men standing in this last channel, shod with large wooden boots and with water up to their calves (*jarrets*) separated the large pebbles from the small by means of an iron rake (*râteau*): the smallest [pebbles] passed through the teeth and were deposited further on according to their weight: the largest [pebbles] rested on the rake, were thrown over the sides and collected by other workers. The pebbles, large and small, judged rich enough and recovered, were then carried to the stamps (*bocards*, for crushing) and dressing-floors (*laveries*), then smelted. By this means of exploitation, a hundredweight (*quintel*) of smelted tin always fetched 32 to 36 *écus*. [An *écu* equalled three francs].

In the year 1817, one *seyffenwerk* near Eybenstock produced 38 hundredweight of tin which occasioned the following expenses –

		écus	gros
1	For overseers and master streamers	162	0
2	For sorting pebbles and transport to the stamps	75	20
3	For pay of the streamers and upkeep of the channels	463	4
4	For expense of the smelter, purchase of wood and other materials	109	4
5	For expense of mechanical preparation	82	20
6	For expense of smelting crushed ore	105	0
7	For general expenses	342	0
	Total	1 340	0

From which it follows that a hundredweight of tin realised that year about 35 *écus*.

It is hardly surprising that the wealth of the Erzgebirge is summed up in a proverb 'preserved in every village': 'The pebble with which you hit your cow is more precious than the cow herself.'[22]

Manès might have been describing Cornwall, so similar were the methods of working. Maybe the Cornish climate was kinder to streamers who frequently worked barefoot, but the use of the rake is recorded once in Cornwall in working the unusual stream wolfram deposit at Buttern Hill on the north side of Bodmin Moor as recently as the early 1900s.[23] More striking in the present context is the similarity to the streaming methods described by Agricola, even to the rake and leather boots in favour of clumsy wooden ones, as depicted in one of his woodcuts (Fig. 34).

The discovery of alluvial tin is credited traditionally to gold seekers. But gold is scarce in the Erzgebirge, and it is not known how long prospecting went unrewarded before a streamer, who perhaps learnt the art of gold panning in Transylvania, realized that the heavy but undistinguished brown and black pebbles of the Erzgebirge would prove to be more valuable to the advance of civilization than the gold he had hoped to find.

NOTES AND REFERENCES

1   BAUMANN, 1970; JANECKA & STEMPROK (1967) assume bronze age working in their introductory remarks, as does MAJER 1965
2   MUHLY, 1973, 256
3   COLLINS, 1973
4   TONKIN, T. (early 18th century), footnotes to the 1811 ed. of Richard Carew's *Survey of Cornwall*, London, 37
5   HOOVER & HOOVER, 1950, 118
6   SIMONIN, 1868
7   MANÈS, 1824, 292–4
8   HOLMAN, 1927
9   COLLINS, 1893
10 CANCRINUS, 1767, *Beschreibung der Bergwerke*, 95ff, quoted in Collins 1893
11 DAVIES, 1901, 97
12 LÖHNEYSS, J. E., 1617, *Béricht vom Bergwerke*, Clausthal, quoted in Collins, 1893
13 COGHLAN, 1975, 24
14 BARI & FLUCK, 1981
15 FORMAN, FORMAN & POULÍK, no date, *c*. 1950. Note that plate 107 illustrates a Hallstatt handled cup from the Náklo hoard found in 1884. It is said to rest in an openwork container of pure tin. This is incorrect; the Czech edition describes the whole vessel as normal tin bronze (Dr Vít Dohnal, Olomouc Museum, pers. comm.)
16 BAUMANN, 1970
17 ARQUÉ, 1906
18 SCHMIDT, ALBERT, 1906, 'Das Vorkommen von Zinnstein im Fichtelgebirge dessen Gewinning in Mittelalter', *Zeitschr. Berg. Hütt. Sal. preuss. Sta.*, **54**, 377–82: quoted in Hess & Hess, 1912
19 MANÈS, 1824, 653–6
20 NEUSTUPNY & NEUSTUPNY, 1961, 101–2
21 MANÈS, 1824, 653–6
22 ARQUÉ, 1906
23 BARROW, 1908

13 Tin in the Mediterranean Islands and Italy

Tin was by far the scarcest base metal in constant demand by the Mediterranean civilizations. Homer, an Ionian poet whose disputed date may have been *c.* 850 BC, presents us with a picture of Greece 400 years or so before. He furnishes proof in the *Iliad* of the high regard for tin, whose source Herodotus still puzzled over in the 5th century BC. Homer describes tin not only as a constituent of bronze but as an adornment in its own right. The shield of Atreides was 'made of parallel strips, ten of dark blue enamel, twelve of gold, and twenty of tin', while Hephaestos cast for Achilles a shield five layers thick, two of them of tin, with such lavish decoration as only a god could produce, using tin to depict a fence and, with gold, a herd of long-horned cattle. His greaves, too, were 'pliant tin', and the chariot of Diomedes was overlaid with gold and tin.[1]

Classical writers regarded tin as an exotic product, for there is none in Greece. The alleged mine near Kirra, the port of Delphi,[2] is a figment of the imagination; this limestone country is in the wrong geological setting for cassiterite, a point blissfully ignored by many archaeologists until spelled out by Benton.[3]

Several Mediterranean islands contain traces of tin mineralization; it has been detected in Corsica and Elba. Elba is a famous mineral locality best known for its iron, especially the lustrous crystals of haematite from Rio Marina in the east of the island. Diodorus Siculus (Book V, 13) called the island Aethaleia because of the smoke (αἴθαλος) caused by smelting the iron which 'they possess' in 'great abundance'. At the west end of the island in the Monte Capanne granite are gem-bearing pegmatites containing more than 30 mineral species. Cassiterite, in stout, prismatic, twinned crystals, occurs in vein pockets at Grotta d'Oggi, Fonte del Prete, I Canili, and Facciatoia, in the vicinity of San Piero in Campo a few kilometres west of Marina di Campo.[4] However, belief that the Etruscans appropriated the island for its tin is erroneous because the cassiterite does not occur in economic quantities and appears to have been

unknown before the discovery of the pegmatites in 1825. Since 1850, these have been exploited specifically for gem-quality specimens. The presence of Etruscan bronze slags on Elba probably indicates the import of tin, perhaps from Tuscany though proof is lacking.

Sicily has recently added cassiterite to the list of its minerals, for granite blocks 'included in the lavas of Etna contain cassiterite, testifying to the presence of tin-bearing granites in the substrata of Sicily'.[5] Whatever the economic geologist may want to make of this, there is nothing here to attract the attention of the archaeologist.

Sardinia is highly mineralized, and the island may well prove to have been even more important as a metal producer in antiquity than formerly believed. Copper was probably exported in the 2nd millennium BC. Ox-hide ingots occur on the island; two bear Minoan-like signs suggesting that they were imported from an eastern source.[6] Alternatively, it has been argued that these ingots were the product of local mining cast into an ox-hide form for the benefit of east Mediterranean traders accustomed to such a shape.[7]

The difficulty with tin is that insufficient has been found to make it appear likely that the island could have been a producer in antiquity. None of the known deposits are of any economic importance though trial workings continue. Recently tin has been discovered off the south coast,[8] while on the island itself tin is known from two areas.[9] In the north at Monte Mannu (Canali Serci) is an Hercynian intrusion, a 'stockwork' of sub-vertical veins in Silurian schists and quartzites. The mineralization is varied with cassiterite associated with zinc blende, chalcopyrite, iron pyrite, and galena. Even if this were proved to be economically workable today, it would have presented prehistoric mineral dressers with problems their technology might not have coped with. In the south tin occurs in the Iglesia district. Mineralization of a similar age in Silurian strata occurs at Perda Maiori in sub-vertical veins trending NE–SW containing molybdenum, wolfram, and subsidiary tin, lead, and zinc. Prof. R. F. Tylecote

79

has examined cassiterite from Monte Linas (1 236 m) between Gonnosfannadiga and Fluminimaggiore in the northern part of Iglesiente, where the quartz veins were being tested for their economic viability. The cassiterite was very pure containing around 83% tin metal. Tylecote found it to be virtually identical to Cornish tin, and indeed to any other tin (*pers. comm.*), a point worth remembering when attempts are made to distinguish the source of tin from the analyses of bronze age artifacts.

The bronze age Sardinians were skilled metallurgists and smiths well known for their very fine bronze statuettes which give a rare insight into contemporary dress. The Nuraghi people, named from their massive stone towers or *nuraghi*, may have mined local copper sulphide ores in the 2nd millennium BC and certainly did so in the 1st millennium. Such ores would be responsible for the traces of tin found in the very pure cakes of smelted copper (*c.* 98% copper) from Abini near Teti. They contained about 0·15% tin with equally small or lesser amounts of lead, silver, nickel, arsenic, antimony, bismuth, iron, and zinc.[10]

Corroded tin metal, formerly wrongly thought to be cassiterite, was found with ten cakes of copper at the smelting site of Ferroxi Nioi near Nuragus, ancient Valeria, 60 km north of Cagliari.[11] For Cambi[12] the source of tin had to be Cornwall or Spain. Through lack of evidence Cornwall can be discounted, and there is sufficient proof of contact with Iberia. A double-looped palstave was found at Monte Arrubbio. Better known is the hoard from Monte Sa Idda, probably a nuraghic foundry, containing bronzes of about the 6th century BC with 'a remarkable proportion of Iberian types', fragments of carp's tongue swords, double-looped palstaves, trunion adzes, and sickles.[13] The traders may have been the Phoenicians who had, from the 8th century, settlements in southern Sardinia,[14] though before this the Sardinians or the Iberians must have been the traders. Iberian exports of tin continued into the Roman period, for south of Capo Bellavista a wreck was discovered with 32 tin ingots; these are discussed in Chapter 16.

Such trade must also account for the tin used in bronzes on the Balearic Islands. A Beaker period workshop in a rock shelter at Son Matge, Mallorca, was occupied around 2100–1700 BC. Among the finds were an awl of bronze containing 10% tin, part of a bronze bead with 5 or 10% tin, and a local Talyotic bronze pin with 6 or 7% tin. The conclusion of the excavators was that true tin bronze was established in the Balearic Islands *c.* 2000 BC.[15]

Chapter 15 opens with a quotation from Ezekiel that lists tin among the trade goods of Tarshish. Various locations for Tarshish have been proposed: Tarsus in south-east Turkey, and Tartessos in the south of Spain are the most popular, with the latter being generally favoured. Another view is that it refers to Sardinia.[16]

This does not alter the view of the importance of Iberian tin, but if it is true it does emphasize further the pre-eminent status of Sardinia as an island trading in Iberian tin, as the islanders' meagre supplies would have been insufficient even for their own needs were they able to exploit them.

Traces of tin have been found at several localities in Tuscany in peninsular Italy. The claim that it has been found with copper and iron in the Tolfa range[17] needs confirmation, but there is no doubt about traces of tin mineralization in the Valle Pozzatedo and at Monte Spinosa in the Colline Metallifere.[18] The only tin deposit that has proved to be of any economic worth is at Monte Valerio near the small town of Campiglia Marittima (Fig. 35). In 1875 miners examined the extensive ancient iron workings known as Cento Camerelle ('100 chambers').[19] In following up a vein of 'brown haematite', an east-west striking tin lode was revealed, the tin ore being 'very compact, of a yellowish-grey colour, and of granular fracture. Specimens yielded from 58 to 72 per cent.'[20] Tin production was small. Various authors give 20 tons of cassiterite in 1877 and 1911, 350 tons in 1912, 25 tons in 1913, and 4 tons in 1914.[21] Total production to the end of the first world war is estimated to have been 200 kg of 29% tin, 700 kg of 3–5% tin, and 850 t of 2% tin. During the most recent period of working, 1936–47, a little over 1 500 t of tin metal were recovered from 400 000 t of ore.[22] The tin deposit is not rich, only the constraints of war encouraging modern production. The tin ore is sparsely disseminated through a largely iron-bearing deposit, which also contains a little zinc blende, galena, and chalcopyrite. Mineralization is linked to a granite pluton intruded into Liassic limestones as recently as the Pliocene, thus giving it the distinction of being Europe's youngest tin deposit.

Whatever small amounts were dug out in prehistory, the tin at Monte Valerio does not alter the view that tin was permanently deficient for the Mediterranean civilizations. Even supposing that the Minoans were aware of Tuscan tin, there was certainly not enough for there to have been an extensive trade with Crete, as Sir Arthur Evans envisaged. Nor was there any likelihood of Tuscan tin reaching 'across the Alps, to the Rhône, and to southern France'.[23] Such trans-alpine trade as there was in bronze age metals was in the reverse direction.

The metalwork of the Polada culture north of the River Po owes nothing to peninsular Italy; it is, instead, linked to the Únětice and related cultures. The flanged-axes, round-heeled riveted dagger-blades, torcs, and various decorated pins, are all closely paralleled north of the Alps. Metal objects are scarce in much of the Polada territory, and flint retained an importance that indicates that metal objects were not freely traded through the Alpine passes. But contact there was, perhaps even with Hungary, as some metal-

35 'Working by the Excavation of a Mass, at Campiglia, in Tuscany' (Fig. 113 in Simonin, 1868)

(a)

scale = 75%

36a **Etruscan lamp from a mine at Monte Valerio in the Colline Metallifere, Tuscany (Fig. 132 in Simonin, 1868)**

(b)

above Triens (third part of an As) struck at Populonia in copper from the mines of Campiglia; the head is of Vulcan *below* half As found at a mine in Monte Catini, Tuscany; actual size. (Figs. 131 & 133 in Simonin, 1868)

36b **Etruscan coins**

work contains less than 0·5% tin but over 14% antimony in finds associated with ingots of tin bronze in the Val Seriano near Bergamo.[24]

In peninsular Italy throughout most of the bronze age, metalwork was very restricted, limited to "the area of copper and tin ores" in Tuscany and, for copper, in northern Lazio as well. Throughout this area technological development was remarkably unaffected by the metal industry of northern Italy[25] – supposed tin buttons from Monte Bradoni turn out to be made of local antimony.[26] Even so, it is a reasonable assumption that local tin ore was used in the bronze age. The transition from copper to bronze in Tuscany has been dated at La Romita to around 1800 bc (*c.* 2100 BC).[27]

The Etruscans, whose culture becomes recognizable about the mid 8th century BC, exploited copper in Tuscany. Extensive slag and mine tips are attributed to them, while a remarkably preserved, brick-lined shaft-furnace excavated at Fucinaia in 1936 contained chalcopyrite and smelted copper ores.[28] Simonin described the mines before tin was known to occur. The vast copper workings contained chambers large enough to 'hold a six-storied house with ease' linked to each other by narrow galleries scarcely large enough for a person to crawl through. Many workings contained wooden pit props in situ and 'fragments of vases, lamps and amphorae connected with Etruscan art.'[29] Simonin illustrates coins struck from local copper and an Etruscan lamp found at Monte Valerio (Fig. 36). Without going into details, Simonin also refers to the Cento Camerelle iron workings as Etruscan in origin.[30] The Etruscans, according to some authorities immigrants from the Black Sea coast around Trebizond,[31] were known for their energy and drive. As Finley puts it:[32]

> If one had to hazard a guess as to what the immigrants from Asia Minor contributed which made the new amalgam [of populations] so dynamic, mine would be in the first instance their ability to exploit the rich metal deposits of the region ...

Yet whatever the Etruscans may have extracted at Monte Valerio it proved insufficient to quench their thirst for tin, which they found necessary to import.[33] The similarity in composition of bronzes found in Etruria, Sardinia, and in the Huelva district of southwest Spain, all with *c.* 7% tin,[34] suggests a link in metalworking traditions which may token a common use of Iberian tin. After the foundation of Massalia *c.* 600 BC, it is likely that Cornish tin also reached the Etruscans. Even if trade relations were not amicable, tin could have been acquired by illegitimate means for, as Cristofani says, 'The predominant form of trade ... must have been piracy.' The Etruscans patrolled the Tyrrhenian Sea as far west as the Balearic Islands, bringing the French coast well within their territory – a point proved by the discovery of a wrecked Etruscan ship of *c.* 178–138 BC near Cap d'Antibes between Nice and Cannes.[35]

Any mining for tin at Monte Valerio cannot have continued during the Roman Republic later than the time of Sulla (*c.* 80 BC), for according to a senatorial decree all mining was halted on the Italian mainland. Pliny makes frequent references to this in his *Natural History* (*lib.* iii, xxiii, xxvii) and leads to the inference that it was to encourage emigration to provinces such as Spain, Sardinia, Greece, and Asia Minor, and even to encourage Italian agriculture, which might be injured in places by the working of mines at home.[36]

NOTES AND REFERENCES

1 RIEU, 1966. References to tin are on pp. 197–8, 349, 370, 396, 425

2 DAVIES, 1929

3 BENTON, 1964

4 ORLANDI & SCORTECCI, 1985

5 SCHUILING, 1967, 532–3

6 GUIDO, 1963, 158–9; SANDARS, 1978, 100–1 and Fig. 58

7 SANDARS, 1978; HARDING, 1975, stresses the importance of Sardinian trading contacts. The copper trade was at its height in the 12th century BC. Some local finds, such as large picks, are no longer regarded as imports from the Aegean but are seen as Sardinian products; for the local mining of copper ores see Zwicker, Viridis & Ceruti, 'Investigations on copper ore, prehistoric copper slag and copper ingots from Sardinia', in Craddock (ed.) 1980, 135–64

8 BRAMBATI, 5 Sept. 1981, 'Dove sono le miniere sottomarine italiane', *Corriere della Sera*

9 CASTALDO & STAMPANONI, 1975, items 315 (Perda Maiori) and 332 (Monte Manu)

10 CAMBI, 1959; he does not believe that Sardinian tin was worked in antiquity or that any was imported from the Italian mainland

11 TYLECOTE, BALMUTH & MASSOLI-NOVELLI, 1983

12 CAMBI, 1959

13 TRUMP, 1981, 286–7 and Fig. 64

14 HARDEN, 1971, 34–5 and 62, Fig. 14 for map of sites

15 WALDREN, 1979

16 ALBRIGHT, 1941

17 GRANT, 1980

18 A LA SPADA (pers. comm.) quoting from a report of the *Societa Azionaria Minero-Metallurgica*, Rome, Sept. 1981; Bergeat 1901, refers to stanniferous copper ores at Boccheggiano and Massa Marittima

19 FAWNS, 1905

20 HUNT, 1879, Supplement vol. 4, 903

21 FAWNS, 1905; JONES, 1925; DAVIES, 1919

22 CASTALDO & STAMPANONI, 1975, item 159 (Monte Valerio)

23 BARFIELD, 1971, 75–7

24 TYLECOTE, 1976, 22

25 BARKER, 1972; BARKER, 1981, stresses the rarity of bronze in Apennine settlements

26 PIGGOTT, 1977

27 BEER, 1962, Ch. 11 'Who were the Etruscans?'

28 COGHLAN, 1975, 38 and Fig. 6

29 GRANT, 1980, quoting Boni & Ippolito, *Contributi introduttivi allo studio dell monetazione etrusca* (Atti del V Convegno del Centro Internationale di Studi Numismatici, 1975), Naples, 1976, 53

30 SIMONIN, 1868

31 BEER, 1962, Ch. 11 'Who were the Etruscans?'

32 FINLEY, 1977, Ch. 6 'Etruscheria', esp. 103

33 GRANT, 1980; he notes (pp. 12 and 192) that a 'heap of selected tin ore' was found on the Etruscan site of Rusellae, modern Roselle; the implication is that it is Etruscan tin, but it could have been imported

34 CRADDOCK, 'The composition of the non-ferrous metals from Tejada' (Huelva, Spain), in Rothenberg & Blanco-Freijeiro, 1981, 279

35 CRISTOFANI, 1979, 64

36 SIMONIN, 1868

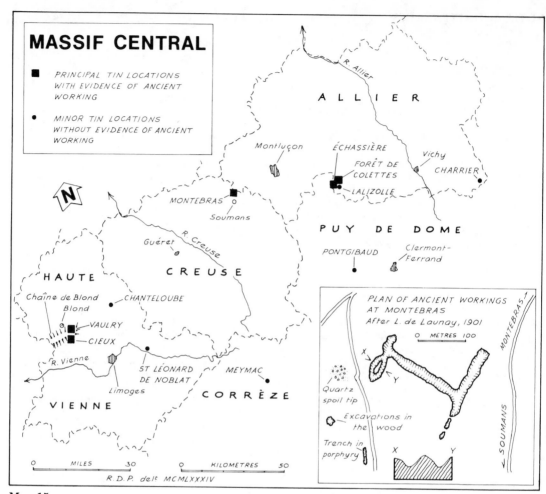

Map 15

14   Tin in France

When the Revd William Borlase published his *Antiquities of Cornwall* (1754) and Sir Christopher Hawkins his *Observations on the Tin Trade of the Ancients in Cornwall* (1811), no mention was made of the possibility of the Phoenicians – then invariably credited with initiating the Cornish trade – also participating in the trade of Breton tin. The reason is that Breton tin was not then known to exist. Mining there certainly was in 18th century Brittany; lead was worked as early as 1729 at Poulawen and Huelgoit (Huel is same as Cornish Huel or Wheal, a mine), and again with the aid of Saxon miners by about 1750. The Saxons also tried to interest the government in the importance of Breton coal,[1] but the tin even eluded these experienced miners. Men of letters were not ignorant of the use of French tin 'by the ancients'. They were also aware of the similarity of the terrain in Brittany, Cornwall, and the Erzgebirge, and entertained the hope that tin would be met with 'in the maritime departments of the north-west' of France.[2] In spite of this there was no prospecting for tin, and its eventual discovery was left to the whim of Fate.

Tin was first discovered in France in 1809 by Cressac as a rare accessory mineral in a wolframite deposit discovered when quarrying for road metal at St Léonard de Noblat, 16 km east of Limoges.[3] Breton tin was found in 1817, but before dealing with these deposits, the other French tin-fields should be looked at. Davies,[4] quoting from Buffon's *Histoire Naturelle des Minéraux* (1749), mentions tin in the Ercé and Uston tributaries of the upper Garonne in Foix, and notes a law of 1483 dealing with the export of tin and gold from Couserans, modern St Girons. But no tin is reported in the modern mineralogical literature in the Pyrenees.

TIN IN THE MASSIF CENTRAL

Tin is widely distributed in the Massif Central, particularly in the departments of Haute-Vienne and Creuse. Not all is of economic importance and some may not have been known in antiquity, such as the tin at Monastier, Lozère, where a stanniferous quartz vein was found cutting a large barytes deposit early in the 20th century.[5]

The first workable deposit was found by Villeluna and Alluaud in 1812 at Vaulry, Haute-Vienne, on the

northern flank of a range of low hills called la chaîne de Blond. It was worked with some success in 1813–26 and 1856–9. In 1856 stanniferous veins were also found on the southern flank of the hills at Cieux, with gold-bearing alluvial tin deposits 'in nearly all the valleys flowing off la chaîne de Blond'. In the centre of the Cieux valley the stanniferous sands were up to 2 m thick. There was further working at Vaulry in World War 1,[6] and in 1925–7 on the La Garde lodes which proved to be too complex and difficult to treat.[7] The veins at Vaulry do not exceed 5 cm wide and form a stockwork of tiny veins in a much kaolinized granite. As well as cassiterite and a little gold they contain wolfram and a host of iron and copper sulphides, native copper, molybdenum, uranium, and non-metalliferous minerals.

In spite of the difficulties encountered in treating these ores, there is plenty of evidence of ancient working in the form of open-cast pits in rows between the hamlets of Garde and Tournerie. One row follows the line of mineralized veins (5° to 10° E of N), the other crosses them at right angles, evidently to establish the full extent of the deposit. The largest pits reached 60–70 m across and 10 m deep. According to Mallard,[8] Cressac concluded 'perhaps not without exaggeration' that 400 000 m^3 of rock had been extracted. Cressac also found in the neighbourhood of the workings, fragments of slag scattered over the fields. On analysis they contained up to 21% tin, proof of the difficulty of treating the ores, and proof also that they were smelted on the spot.

Tin has been found in old feldspar quarries near Limoges at Chanteloube, but is unlikely to have been known in antiquity. Apart from Vaulry, workings assumed to be ancient are known at Beaune with 7 000 m^3 of spoil heaps, and Ribière with 10 000 m^3 of spoil. These do not contain tin, but something was being sought after over a wide area – logically gold and tin. Some mineral was worked, perhaps iron at Beaune. The presence of gold most probably accounts for the names Aurières, Laurière, and Auriéras, hamlets between Millemilange, Creuse, and Couzeix, Haute-Vienne, adjoining a stream called the Aurance. Here the etymology is not in doubt for the stream contains flakes of gold, sufficient to have attracted panners at the end of the 18th century. Mallard and others assumed that gold, as well as tin, was important in the Limousin region in the pre-Roman or Roman periods. Mallard came to this conclusion in 1859 when he turned his attention to the geological mapping of Montebras, Creuse, and found that remains, formerly believed to have been those of a Roman camp, were the weathered remnants of open-cast workings in a tin-bearing granite elvan much decomposed near the surface, hence easily worked.

At Montebras is a series of funnel-shaped depressions each up to 200 m long and 20 m deep. About 400 m to the west a large spoil heap of quartz still contains cassiterite. Similar workings in a wood at Feuillade were in a 'lightly stanniferous' alluvial deposit.[9] The Montebras workings were large enough to cause drainage problems; on the edge of a wood one trench 130 m long by about 60 m wide and 8–10 m deep was linked at its northern end by a smaller trench that followed the slope of the ground and was 'without doubt, dug to allow the water which could fill the workings to flow away'.[10] The alluvial deposits of Montebras extend along the Petite Creuse stream for more than 10 km. A Cornishman reported in 1908[11] that the 'possibilities of Montebras are immense'. Samples from 'the mud of a wood, the surface of a flower bed, and the dirt attached to grass roots, all showed appreciable quantities of tin on the shovel'. More than a shovelful of Cornish optimism here, for after the main period of working between 1860 and 1877, little tin was recovered, though at the end of 1955 a new prospecting campaign was considered.[12] The granite at Montebras is rich in the rare lithium-bearing phosphate mineral montebrasite, isomorphous with amblygonite.[13] Davies[14] believed that beads recognized in a bronze age context must have come from Montebras as a by-product of tin mining. In fact, the beads and pendants from Breton burials which he must have had in mind are of variscite and neolithic in date (P-R. Giot, pers. comm., 1984). It is unfortunate that there is no better evidence of prehistoric working at Montebras than two Gallic coins and an early Augustan one.

At Charrier in the department of Allier, tin is found with slightly argentiferous copper pyrites. These are complex ores and the successful treatment of 135 tons in 1952 was suspended the following year.[15] There is no tradition of early working here, but further west at Échassière and Colettes, where cassiterite and other minerals occur in a kaolinized granite, there is said to have been working 'at an extremely remote period' at Échassière.[16]

Through lack of archaeological evidence it is difficult to evaluate the importance of prehistoric tin working at Montebras and other sites in the Massif Central. Clearly it should be sufficiently remote in time for folk-memory or historical record to have left

no trace beyond, perhaps, the place-name evidence already mentioned. Whatever activity there may have been during the Roman period, it was not enough to have attracted the attention of classical writers. If there was a heyday of Limousin tin mining that accounted for the massive spoil tips, then exploitation in the bronze age and pre-Roman iron age is a better suggestion than most. Unfortunately, archaeological evidence does not support the idea of an important bronze age metalworking centre in the Massif Central where, in the 2nd millennium BC, there is no evidence of long-range contacts with other communities 'beyond the ordinary pursuit of desirable objects through seasonal or random relationships'.[17] Such contact is sufficient to account for such imported items as a Rhône-type dagger from Puy-de-Dôme. More telling is the poverty of the local economy in which grave goods tend to be rather sparse 'and metal rarely wasted'.[18] This is not the hallmark of a rich mining district and is in stark contrast to the archaeological evidence from bronze age Brittany.[19]

TIN IN BRITTANY

The modern history of Breton tin working began with a chance discovery in 1817, a discovery not without a Cornish connection. A French naval officer had been held prisoner in Cornwall during the Napoleonic wars. Such prisoners were allowed a good deal of freedom, and it is likely that he was held at Roscrow or Kergilliack near Penryn, not far from the Carnon valley tin stream which was working at that time. After the war he returned to his native sea port of Piriac, Loire Atlantique, and:[20]

> Going out sea fishing one day, and wanting some weights for his lines, he picked up a pebble on the shore, which appearing to him to be unusually heavy, he took it home to compare with a piece of Cornish stream tin which he had brought back from the place of his captivity, and found it to be the same substance. He gave notice of his discovery in the proper quarter, and M. Dufrenoy, now a distinguished French geologist, then a young aspirant of the school of mines, was sent with another to investigate the matter, and the report they made shows a remarkable uniformity of structure between that part of Brittany and the tin district of Cornwall, on the opposite side of the Channel. The country between the mouth of the Loire and Piriac is composed of granite and slate, and it is at the junction of these rocks that the tin ore is chiefly found. They met with veins traversing the rocks, and a considerable quantity of stream tin, both in the form of pebbles and of sand; and their impression was, that this stream tin was produced by the wearing of the rocks containing the veins, by the action of the waves, the same action going on now, as in Cornwall in remote ages. The continued large importation of tin from England into France shows that this discovery has not yet been attended with any great results.

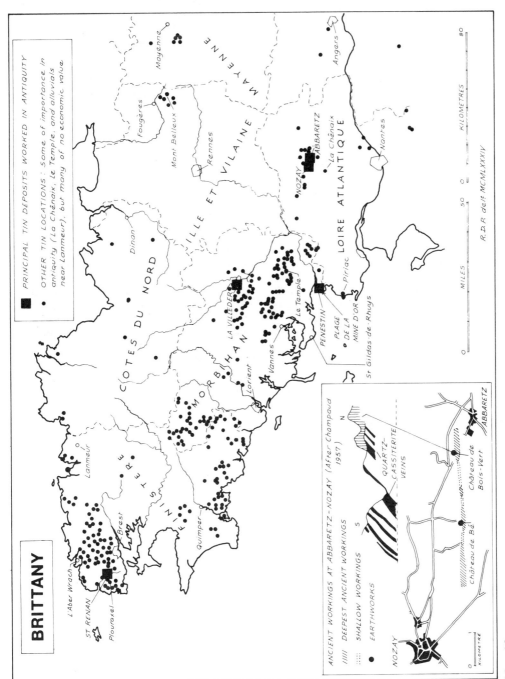

BRITTANY

PRINCIPAL TIN DEPOSITS WORKED IN ANTIQUITY

OTHER TIN LOCATIONS: Some of importance in antiquity (La Chênaix, Le Temple, and alluvials near Lanmeur), but many of no economic value.

L'Aber Wrach

St RENAN

Plouraxel

Brest

Lanmeur

FINISTÈRE

Quimper

Lorient

Vannes

Le Temple

LA VILLEDER

MORBIHAN

Dinan

CÔTES DU NORD

ILLE ET VILAINE

Rennes

Mont Belleux

Fougères

Mayenne

MAYENNE

Angers

Nantes

LOIRE ATLANTIQUE

NOZAY

ABBARETZ

La Chênaix

PENESTIN

Piriac

PLAGE DE LA MINE D'OR

St Gildas-de-Rhuys

MILES   50

0   KILOMETRES   80

R.D.P. del: MCMLXXXIV

ANCIENT WORKINGS AT ABBARETZ-NOZAY (After Champaud 1957)

////// DEEPEST ANCIENT WORKINGS

::::::: SHALLOW WORKINGS

• EARTHWORKS

QUARTZ-CASSITERITE VEINS

NOZAY

Château de Bé

Château de Bois-Vert

ABBARETZ

KILOMETRE

Map 16

A report in the *West Briton* for 26 May 1848 probably refers to Piriac and is hardly encouraging:

> On Monday the 14th instant a French vessel arrived at Penzance with forty or fifty tons of ore that was supposed to contain a large quantity of tin. It was shipped from the coast of Brittany by persons who fancied that the sand on the beach there contained tin, but in a portion of the sample tried there was not in the proportion of half a ton of tin in as much ore as the ship would carry. In a small sample, very little of which was imported, there were better appearances; but it is said that the speculation was so far unsuccessful as to give little encouragement for another importation.

Hope springs eternal and Cornish tinners were encouraged to have a go, as the same newspaper reported on 26 April 1850:

> Monday last, about thirty miners and tin streamers, from the parishes of Madron, Paul, Sancreed, Ludgvan, and Buryan, embarked at Penzance in a French lugger, for Pirac [sic], that having been engaged by a company to prosecute tin-mining in France.

However, in 1878 three workmen struck it rich, collecting between 600 and 700 kg of alluvial tin from the beach at Piriac 'in a matter of a few weeks panning'.[21] The beach has been sampled many times with small amounts recovered in 1903–7, 1930, 1950, 1952, and 1960–3.[22]

The deposits at Pénestin cover about 15 ha and have yielded on occasions for more than a century between 10 and 15 kg tin per m³ sand, as well as 0·005 g gold in the same amount.[23] The presence of gold accounts for the name La Mine d'Or ('the gold mine') being applied to the beach. The name Pénestin is more interesting as it is recorded as early as the end of the 11th century as Pennestin. The neighbouring parish contains a tiny settlement called Lestin, doubtless from *l'estin*, both words coming from the Breton *staen* or *stean* meaning, as in Cornish, tin.[24] Pénestin, ('tin headland') is close to deposits of tin at Péaule on the right bank of the Vilaine, and near the important deposits of Abbaretz-Nozay which will be discussed below. Small amounts of tin were probably obtained in antiquity from the beach at La Mine d'Or as the cliffs backing the beach were being eroded away, as they still are.[25]

The exploitation of tin and gold at Piriac has been limited. 'Limited' is the operative word for the story of Breton tin production as a whole since 1817, for it is not one of unqualified success. The Breton deposits must be seen in perspective. Brittany is no Cornwall, nor indeed a Saxony. Nothing comparable to the 'strong masterly lodes' of Cornwall has ever been found under French soil. Consequently, alluvial deposits, while comprising locally rich pockets, are limited in extent. Notwithstanding the measure of success that attended the working of the St Renan deposits in the 1950s, 'all

attempts at mining tin ore on a commercial scale have proved to be disastrous failures'.[26] This does not undervalue France's contribution to tin production in antiquity. What the geological evidence implies is that once the seemingly limitless stream tin deposits of Cornwall were in full exploitation, the limited extent of French alluvial tin and the difficulties encountered in smelting some of its lode material reduced the rôle of French tin to, for example, making up any shortfall in Cornish supplies owing to natural or other disasters. The very proximity of Cornwall and Brittany makes it sensible to see the latter as an annexe of the former, as much concerned with the shipment of Cornish tin across the Continent as with trying to produce vast amounts of its own.

Breton tin deposits have been surveyed in great detail, and only those localities revealing evidence of mining in antiquity are discussed below. The deposits at La Villeder, Morbihan, were worked with a little success from 1845 until 1886 with stoppages owing to the 1848 Revolution among other causes. About 250 people were employed in 1882 when it was optimistically hoped that there might be 'untapped veins at depth'.[27] The ore body consists of four principal lodes aligned roughly 8° west of north and underlying (dipping) to the west, the biggest being Le Gros Filon, 2·5 m wide and traceable for 6 km. More important, however, are the many small stringers forming a stockwork. Alluvial tin also occurs, being especially concentrated at La Ville Clément and Poudelan where, moving up the valley, 'the mineral grains become bigger and more angular', the cassiterite varying in size between 'a pinhead and that of a nut'. Much of it was the beautifully banded variety 'wood tin', originally known only from Cornwall and called appropriately *étain de bois* by the French.[28] As at Vaulry, these placers contained gold, a feature which probably did not go unnoticed in antiquity. Old workings and spoil-heaps had long been apparent at La Villeder, but they only served as a quarry for road metal until 'an intelligent land owner' took the trouble to have the brown stuff in quartz lumps analysed by mining engineers from Rennes in 1834. As at Vaulry and Montebras there was no memory or historical record of their working, a pointer to great antiquity. In the 1850s and 1860s artifacts were unearthed in the workings (Fig. 37):

> There have been found an axe of polished stone, an axe of bronze, fragments of tiles, of pottery, and the conduits which carried water to the placers for washing the metalliferous sands; not to mention the enormous heaps of spoil, the surface excavations everywhere, nor the mounds, found here and there, of slag speckled with grains of tin.

Baudot mentions several stone axes, one 'strongly blunted and bearing numerous signs of percussion'.[30] This sounds like a miner's axe-hammer of universal

37 Polished stone axe-head and middle bronze age bronze palstave from the tin streamworks at La Villeder (Figs. 138 & 139 in Simonin, 1868)

type and virtually undatable, but the polished axe, illustrated by Simonin, is of fibrolite. It was found only a few paces from the outcrop of Le Gros Filon and strongly suggests an early bronze age interest in the deposit. The metal axe, also illustrated by Simonin, is a middle bronze age palstave with well marked stop-ridge between blade and handle. The other items, tiles and pottery, are more suggestive of Gallo-Roman working, indicating a mining history spread over more than a millennium. Limur described the ancient workings as they were uncovered in the mid-19th century: three large trenches, which he attributed to the Romans, were up to 100 m long, 10 wide and 4 or 5 deep, strewn with debris and blocks rounded with hammer blows. The main lode was picked out and exposed rather like a wall standing up out of the surrounding gravel.[31]

This bears a striking resemblance to the ancient workings between Abbaretz and Nozay, Loire Atlantique. Archaeologically they are arguably the most important of any European tin deposit being the only ones that have been excavated, if not under ideal conditions, at least under close supervision – thanks to the co-operation of the mining company, the *Société Nantaise des Minerais de l'Ouest*, which began work there in 1949 and produced tin from the end of 1951 to the end of 1957. Production averaged *c.*450 t of concentrates a year and ceased through rising costs and the exhaustion of the main deposit.[32] The pre-

historic workings were written up by Champaud in an award-winning essay.[33]

The deposit is a typical stockwork, of minute quartz-cassiterite veins, rarely exceeding 100 m wide and traceable east-west over 8 km. The ancient workings were found along 5 km west of Abbaretz. The country rock is a variety of schists, brown, red, and softer grey. As long ago as 1778 attention was drawn to a line of 'disturbed ground' between Abbaretz and Nozay, assumed to be some kind of frontier work until in June 1882, Davey, a mining engineer, not only proved them to be open-cast tin workings but sent slag to Mallard for analysis. The latter, having already proved the importance of Limousin as a tin bearing region, was delighted to find that the slag contained 'metallic tin in tiny scattered grains'.[34] Further discoveries had to wait until the 1950s. Even so, some questions remain unanswered. In particular, what is the nature of the curious 'castles', the *château de Bois-Vert* and the *château de Bé*? They were thrown up where the stockwork was richest, but excavation revealed little, and it remains speculative whether they originally held miners' quarters or were defensive sites, or had something to do with smelting – a suggestion prompted by the discovery of charcoal. A military presence was suggested by the discovery of many stone 'projectiles' all over the site, many stacked in pyramidal piles. Of varied form, they are mostly 'olive-shaped' and regarded as sling-stones used against the Romans.

The one object which proved beyond all doubt that tin was mined here is the bottom part of an amphora (perhaps broken during the excavation) that contained cassiterite. It cannot be dated closer than the 1st century BC–2nd century AD, but this period is more than covered by coins from the site.

AUGUSTUS	Bronze, as minted at Nimes *c.*14 BC
ANTONINUS PIUS	Bronze sestertius minted in Rome AD 145–61. Another, perhaps a surface find, minted in Rome between 10 December 156 and 9 December 157
LUCILLA	Bronze sestertius minted in Rome *c.* AD 164
MARCUS AURELIUS	Bronze sestertius minted in Rome between December 172 and December 173
POSTUMUS	Silver-washed antoninianus minted at Cologne in AD 263
MAURICE TIBERIUS	A 10 nummi piece struck at Carthage *c.* AD 585

Two gold Merovingian coins may have been surface coins accidentally incorporated into the excavated

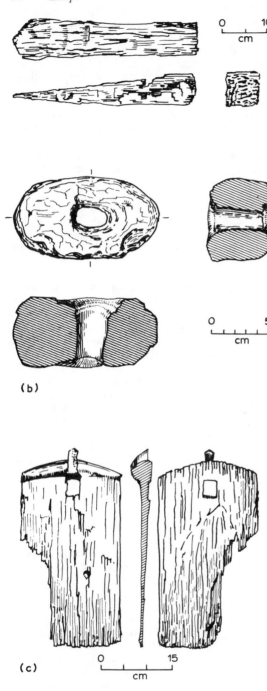

a wooden wedge *b* iron hammer head *c* wooden 'hoe' or shovel (after Champaud, 1955)

38   Tools recovered from ancient workings at Abbaretz-Nozay

material, but it is reasonable to accept them as evidence for contemporary mining. They are gold *triens* minted in Vannes and Nantes *c.* AD 550–60, earlier than the Tiberius coin noted above. The suspected discovery of 30 kg of presumed 'mediaeval' coins in a well on the side of the chateau de Bois-Vert, which found their way into the hands of a dealer, is too vague to have any historical value. Equally tantalizing, at the other end of the time-scale, is the discovery in 1878 of a bronze age torc of twisted gold, of a type known from other Breton sites, but best known as a British and Irish feature of the so-called 'ornament horizon' *c.* 1200 BC. The farmer would not reveal the find-spot, only that it was roughly at the foot of the Château de Bé. Another is said to have been found at the beginning of the century in a well at the foot of the Château de Bois-Vert. Authenticated finds are known from near Quimper, at Cesson near Rennes, and one from Nozay itself. Optimists will certainly interpret this as a bronze age interest in the tin of Abbaretz-Nozay.

Of particular interest at Abbaretz are the remarkably well preserved tools, found where they had been left in the Gallo-Roman period at a depth of 18 m in contact with the old working surface (Fig. 38). The deepest working produced two wooden wedges, and at least a dozen pieces of wood which certainly resemble the wooden mediaeval shovels found in Cornish tin streams. A detailed examination by Champaud proved that the handles would have been attached at an angle of about 75° to the blade, suggesting to him their use as hoes. Another piece, not illustrated by him, may have been 'part of a very large shovel', possibly something comparable to a Cornish vanning-shovel for giving a rule-of-thumb assay of the ore, or it might have been the base of a wooden tub for carrying ore. Great quantities of plank fragments and other pieces of wood with various possible uses were recovered, as well as over half-a-dozen iron hammers. With these finds and the exposed old working surfaces, Champaud was able to reconstruct the probable method of mining.

The tin-bearing veins were attacked in step fashion, being lowered from the edges towards the centre of the quarry. Because the lodes dipped to the south, 30–40°, the northern edge of the quarry was often revealed as a long sloping edge of a vein. The schists, especially the micaceous ones, between the quartz veins are very soft and easily transformed into a whitish clay which presented no difficulties in removal with pick and shovel and carting away in baskets or panniers. Great spoil heaps still line the edge of the excavations. The quartz–cassiterite veins were then left in high relief; these were attacked by fire-setting. Some large blocks of quartz still carried traces of reddening by fire. Into the fissures created by fire-setting, wooden wedges were driven. When soaked with water they swelled, splitting the quartz further. Many apparently barren

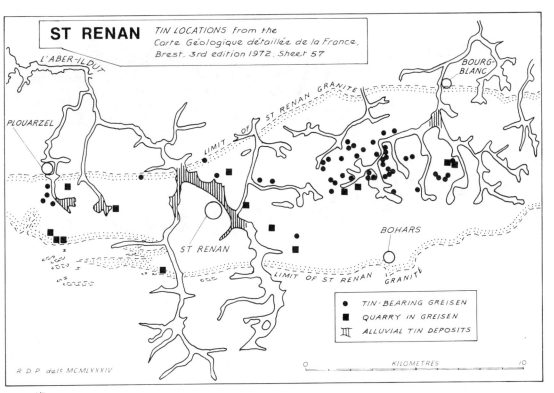

Map 17

quartz blocks were then thrown out and still litter the countryside where more recent inhabitants have used them for building walls and field boundaries. The metalliferous veins were next broken down with iron hammers: microscopic flakes of iron in the residual mud of the dressing floors may have been derived from these tools. Some of this ore dressing could have taken place in the *châteaux de terre* and might account for the layer of stones and charcoal in them.

To separate the cassiterite from the crushed ore, it is assumed that running water must have been employed, a view supported by the presence in the old workings of lenticles of very fine clay interspersed with, but distinct from, coarser beds. This disposition of material could not result from the natural weathering of old spoil heaps but had to be the controlled use of water by man. The many pieces of broken planks and other bits of wood are best interpreted as the remains of sluices constructed to direct the flow of water over the washing floors.

About 3 000 t of tin-in-concentrates were recovered in the 1950s. Champaud estimated that *c.* 1 000 t would have been recovered in antiquity – not much by modern standards but enough to make *c.* 9 000 t of good 10% tin bronze.

After Abbaretz closed at the end of 1957 another previously unsuspected deposit of alluvial tin was found in 1958 at St Renan in the valley of the Aber-Ildut. Small amounts are known from tributaries as far west as Plourazel, and at Bourg-Blanc in the valley behind L'Aber-Wrach. Reserves at St Renan were estimated to contain 8 000 000 m^3 cassiterite-bearing sands in a deposit some 3·5 km long, up to 600 m wide and averaging 4·5 m deep. The St Renan granite contains many small veins of quartz with tin and tungsten, both wolframite and scheelite, but none seem to be economically workable. The placers, however, were exploited from 1960 by the *Compagnie Minière de St Renan*, producing around 300 t of tin-in-concentrates each year; 477 in 1964 and 219 t in 1974 (Fig. 39).[35]

It was while 'outlining the deposit with light drills' in 1963 that the first evidence was found that Gallo-Roman tinners knew all about the potential of St Renan.[36] The finds consist only of 'a few *fibulae* and some slag',[37] but they speak most eloquently of the astonishing expertise of iron age prospectors and, by inference, of the skills of their bronze age forebears. A section through the St Renan deposits shows the similarity with some of the smaller Cornish tin streams.

39 General view of the newly opened alluvial tin workings at St Renan in October 1960 (photo *Tin*, June 1961, 137)

St Renan

Thickness	Layer
0·5–1 m	Humus etc. passing into clay usually devoid of any tin
0·5–1·5 m	Recent alluvials; clay and sand weakly mineralized (50–300 Sn g m^{-3})
0·5–2 m	Grey gravel with granite and quartz pebbles, 500–3 000 Sn g m^{-3}
1–2 m	Ancient alluvials; compact sands and clay more weakly mineralized than the Recent alluvials

The depth at which the fibulae and slag were found is not recorded, but they did not have to be far down to be on the most profitable layer.

In the deposit at Bourg-Blanc, 2 km east of St Renan, the recent exploitation of tin produced a large perforated disc of fibrolite, almost circular, 28 cm max. diameter and 4·5 cm max. thickness.[38] The find circumstances and the relation of the disc to the 'tin ground' are not recorded, so that its value as an indicator of previous tin working is limited. Nor is the age of this curious piece known. Fibrolite, a hydrated form of sillimanite widespread in this part of Brittany, has been used since the neolithic age for the manufacture of polished stone tools, but the disc cannot be assigned at present to any period.

France, like Cornwall, is more fortunate than other tin-bearing regions in having deposits which concealed prehistoric evidence of their exploitation. The French evidence is sparse, especially for the bronze age, but it is backed up by the whole corpus of bronze age discoveries. Metalwork is widespread in France, though not uniformly spread. The paucity of finds in the Massif Central has already been mentioned. There is also a marked concentration of finds in Brittany.

While this could be accounted for by the proximity of Cornish tin and Irish or Cornish copper, it is not likely since there is little metalwork in Normandy where imports could be received just as easily. Breton metalwork has a consistently high tin content; 10–15% in palstaves and over 20% in some socketed axes,[39] a sure sign of a local source. This is not to deny cross-Channel links. The close correspondence between early bronze age Brittany and Wessex is clear enough; the bronze daggers with handles decorated with gold pins from Kernonen, Finistère, would not look out of place in Wiltshire. Even more striking is the evidence from early bronze age gold lunulae. Of the three from Kerivoa, Côte-du-Nord, one is exactly like the less ornamental of the two from Harlyn Bay in Cornwall. So similar are they that the same hand, even the same tools, must have created them. Whether these impressive ornaments were made by an itinerant goldsmith close to sources of native gold, or whether they were exported from a single workshop elsewhere is not important here. What is relevant is that there were close links, whatever those links might imply in human terms, between the metal-producing regions of north-west Europe, between essentially copper rich Cornwall and Ireland and tin-and-gold-producing Cornwall and Brittany.

There were close technological links, too, between the western metalliferous centres and central Europe. The gold pin decoration already mentioned (15 000 such minuscule pins were recovered from the tumulus at Kernonen), the silver quoit-headed pins from the Côte-du-Nord, are all ultimately of Unetičian inspiration. There is no need to elaborate on such links here; the bronze age 'common market' is discussed, for example, by Briard[40] at a quite popular level, as well as in more heavyweight tomes. The conclusion drawn from such studies is that the knowledge of tin bronze, with its associated skills of prospecting and mining for tin, spread to both Brittany and Cornwall from the Erzgebirge mining region around 2000 BC. Once minerals were discovered in Brittany, it cannot have been long before they were found in Cornwall. The similarity between the two peninsulas can hardly have escaped for long the attention of prospectors familiar with granite outcrops, quartz veins, and the range of colourful secondary minerals that characterize most metalliferous regions. The scarcity of copper in Brittany must have been a spur to external contacts. No copper deposits of economic merit are known, though the tin lodes at Lanmeur near Morlaix do contain a little.[41] In contrast to copper, lead is widespread in Brittany with over 70 recorded deposits, some of which must have been exploited in the later bronze age to account for the tremendous quantity of socketed 'Breton' axes, commonly of pure lead and useless as tools. Supporting evidence comes from the Île d'Er near Donges, Loire Atlantique, where 40 such axes were found with the same composition as the ore from the nearby Pont-du-Gué mine.[42] These axes were extensively traded. Their dating is controversial but should fall within a period of about a century either side of 600 BC.

The lack of references by classical authors to the mineral wealth of Brittany is unfortunate and has led to wild speculation about the early abandonment of the Breton tin mines. Cary, for example, thought they fell into disuse soon after 500 BC.[43] On the contrary, links with Massalia (Marseille) were maintained and must account for the discovery, among other numismatic evidence, of a silver drachma in the parish of l'Aber-Wrach in Finistère minted between 250 and 200 BC. The subject of Greek contacts is dealt with in Chapter 23. Far from showing an abandonment of the mines, finds from the mines themselves (not known to Cary) indicate increased activity in the Gallo-Roman period, probably from the 1st century BC onwards.

The Veneti (the Gallic tribe inhabiting the Morbihan region) are known to have been great sea traders. This was the source of their wealth: they struck very good gold staters weighing over 20 Troy grains each.[44] The Veneti are assumed to have had a controlling interest in the cross-Channel Cornish tin trade. Julius Caesar was well aware of their power, and in a battle now known to have been engaged off St Gildas-de-Rhuis, Morbihan, he destroyed the greater part of their fleet in 56 BC. With the Cornish trade thus denied them, or at the very least seriously curtailed, the Veneti may well have initiated an intensification of local tin mining. The possible military presence of Gaulish warriors at Abbaretz has already been noted. Whatever the ultimate fate of Venetic influence and power, some mining continued throughout the Roman period.

Important evidence of post-Roman mining comes from Pont-Nevez, Plélauff, where a 50 m gallery was driven on a lead–zinc lode. Surface pottery implied Roman working, but pit props from the 66 and 70 m levels, gave ^{14}C dates of around AD 460. Nor was tin entirely forgotten. Apart from the Merovingian coins from Abbaretz referred to above, wood from the nearby tin alluvials at La Chênaie and from Le Temple near Limerzel (20 km west of Redon), yielded ^{14}C dates of around AD 650 and AD 1090 respectively. Bronze mediaeval pins of uncertain date come from an alluvial working at Lanmeur north of Morlaix.[45]

Possible documentary corroboration of post-Roman tin mining comes from an indirect source. The 13th century encyclopaedist Bartholomaeus Anglicus quotes Isidore of Seville (AD 570–636) as saying:[46]

> White lead [tin] was first found in the island of the sea *Athlant* in old times, and is now found in many places, for in *Fraunce* and in *Lusitania* in a manner black earth full of gravell, and of small stones, and is washt and blowen [smelted] …

While this passage could refer to the time of Isidore, it could equally refer to an earlier epoch, for Isidore's mineralogy was largely based on other works such as Caius Julius Solinus' *Collectanea rerum memorabilium* written about AD 250.[47]

Résumé of evidence of ancient tin working in France

Massif Central

VAULRY	Undated open-cast pits and slag
MONTÉBRAS	Open-cast workings and spoil heaps of uncertain date
	Two Gallic coins and one early Augustan
	Possible working of amblygonite in bronze age
ÉCHASSIÈRE	Ancient workings including a granite millstone supposedly for preparing the ore

Brittany

PENESTIN	Ancient name of Pénestin, at least as early as 11th century AD
	Possible gold and tin recovered from La Mine d'Or
LA VILLEDER	Open-cast workings and spoil heaps with slag containing specks of metallic tin
	Fibrolite axe of early bronze age

	Middle bronze age palstave		
	Tiles and pottery, perhaps Gallo-Roman	14	DAVIES, 1935
		15	ITC, 1964
	Wooden water conduits	16	DE GOUVENAIN, 1874; DAUBRÉE, 1869
ABBARETZ–NOZAY	Open-cast workings on lodes	17	COLES & HARDING, 1979, 233; DÉCHELETTE, 1924, **2**, 95, footnote 3
	Metallic tin in slag	18	COLES & HARDING, 1979, 233
	Iron age or Roman tools of wood and iron in open-cast workings	19	BRIARD, 1965, Ch. 11, 15–25, is a good account of Breton tin deposits complete with maps
		20	ANON., 1835
	Amphora containing cassiterite	21	BRIARD, 1979, 194, quoting le Comte de Limur (1878)
	Roman coins *c.* 14 BC to *c.* AD 585	22	T. GUIGUES *per* G. D. ENGEL, pers. comm.
		23	CHAMPAUD, 1957, 87–9 deals with Pénestin and La Villeder
LA CHÊNAIE	Wood, ^{14}C dated to *c.* AD 650	24	P. TANGUY *per* O. PADEL, pers. comm.
LE TEMPLE	Wood, ^{14}C dated to *c.* AD 1090	25	ROMÉ, 1980, 6
		26	FAWNS, 1905, 152
ST RENAN	Disc of fibrolite from Bourg-Blanc	27	BAUDOT, 1887; ITC, 1964
	Roman fibulae and slag from the main working at St Renan	28	LACROIX, 1901, **3**, 217–35 deals with tin. He describes La Villeder as having the most important tin veins in Brittany with the largest crystals of cassiterite in France
LANMEUR	Bronze mediaeval pins	29	SIMONIN, 1866, 346–7; SIMONIN, 1868, Figs. 138 and 139

NOTES AND REFERENCES

1   DE LA ROGERIE, 1925, esp. 285–91
2   SCHREIBER, 1794, 72–3 for tin
3   MALLARD, 1866; PATTISON, 1867, is based on Mallard 1866. For Vaulry he quotes Cressac, 1813, *Annales des Mines* (June)
4   DAVIES, 1935
5   GUEDRAS, 1904
6   JONES, 1925, 32
7   ITC, 1964
8   MALLARD, 1866
9   DE LAUNEY, 1901
10 MALLARD, 1866
11 'The Cornish Correspondent', *Mining Journal*, 4 July 1908
12 ITC, 1964
13 DANA, 1951; DE LIMUR, 1893, considered montebrasite to be a blue or green mineral related to turquoise or 'callais' of French writers. He wrongly believed that both montebrasite and turquoise were found in Breton megalithic tombs

30 BAUDOT, 1887
31 CHAMPAUD, 1955
32 ITC, 1964
33 CHAMPAUD, 1957
34 DAVY, 1897
35 ITC, 1964; Carte Géologique de la France, sheet 57, Brest
36 ANON., 1963, *Tin* (Sept.)
37 BRIARD, 1979, 194–5
38 LE ROUX, 1975
39 MASSILLE, 1913
40 BRIARD, 1979
41 Carte Géologique de la France, sheet 57, Brest
42 MASSILLE, 1913
43 CARY, 1924
44 MASSILLE, 1913
45 FLEURIOT & GIOT, 1977, esp. 113–14 for mining
46 WORTH, 1871
47 ADAMS, 1954, 138

15   Tin in Iberia

Tarshish *was* thy merchant by reason of the multitude of all *kind of* riches; with silver, iron, tin, and lead, they traded in thy fairs.

Ezekiel, 27, 12

With these words the prophet described the riches of the port of Tyrus (Tyre) before its downfall at the hands of Nebuchadnezzar, King of Babylon (604–562 BC). The location of Tarshish is disputed. It may or may not have been on the south coast of Spain, though in the time of Ezekiel, Spain's mineral wealth was being exploited, and all the minerals listed are plentiful in Iberia. But the mining of tin has not enjoyed the uninterrupted success of Cornwall's international trade. Even in the recent past Englishmen of letters knew nothing of Spanish tin. The widely travelled John Hawkins wrote on 12 September 1791 to his fellow Cornishman Philip Rashleigh that tin had been:

> lately discovered at Monte Rey [Monterrey] in Galicia. This discovery confirms the assertion of Pliny which before I much doubted.

Within a year 600 t of black tin were raised at this locality south-east of Orense near the Portuguese frontier.[1] Whether Iberian tin was ever entirely forgotten in the tin-bearing regions themselves is difficult to resolve. In the 9th and 10th centuries Iberia north of the Ebro and Douro rivers lay in Christian hands, while the south formed part of the great Moslem empire. Only with the dissolution of the Caliphate of Cordoba in AD 1010 were the Christian kings of León and Castille able to begin their reconquest. Thus, the major tin-mining region of the north-west was largely in Christian hands throughout the period. In the south the Arabs developed the mineral wealth of their territory; Almadén, the most important of European localities for mercury, owes its name to an Arabic word meaning 'the mine'. The Christian north-west remained backward and underdeveloped, though Smith[2] maintained that the tin of the Algarve and Galicia was worked, without giving references for his assertion. Even if tin were not entirely forgotten, in the mediaeval period it can have had little importance beyond a very local level. In the early 1500s, Portugal had close trading links with Anvers

(Antwerp) and maintained a well organized trading 'factory' there. In the list of merchandise exported from Portugal to Flanders in 1495–1521, there are large quantities of copper, iron, mercury, and lead, but no mention of tin. Indeed, the tin traded from Anvers came largely from Cornwall with some, among other minerals, from Germany. It was turned into 'saltcellars, cups, basins, and many other household articles, and exported in great quantity to Portugal and Spain' and elsewhere, usually in amounts impossible to quantify nowadays.[3]

In the 16th century there is confirmation of some Iberian tin production. Agricola says little in *De Re Metallica* other than that 'Some of the Lusitanians melt tin from tin-stone in small furnaces ...' a process which he then describes in some detail. That the tin was won locally is confirmed by Agricola's Spanish contemporary Bartolomé Sagrario de Molina. Perhaps as a counterblast to the superiority of Cornish tin pressed by Biringuccio in his *Pirotechnia* of 1540,[4] Molina sings the praises of Galicia's native wealth:[5]

> Add to Galicia's treasures here adduced,
> That year by year our Realm hath tin produced,
> So plenteous in the Vale of Monte Rey,
> That ores are dressed and smelted day by day;
> The metal white, of quality so rare,
> Not even England's purest can compare,
> With that which holds the mart in Medina's fair.

In spite of Molina's eulogy and the rediscovery of Monterrey in 1791, Iberian tin was of negligible importance until deposits were opened up in the Orense region and elsewhere in the mid-19th century. Even after that, production was small until the 20th century when it doubled in each decade from 1900 to the 1950s. Around 3 000 t in the 1940s, and 2 000 t in the 1950s were the average annual totals of tin-in-concentrates. Mid-20th century production was similar to Cornwall's, though Cornwall has now pulled far ahead.[6]

The Iberian tin deposits are contemporary with, and similar to those of Cornwall, Brittany, and the Erzgebirge. Alluvial, eluvial, and lode tin is currently worked in Iberia, sometimes on a very limited scale described by Fox as 'the scavenging of ore washed to the surface after rain'.[7] The tin belt is large, extending

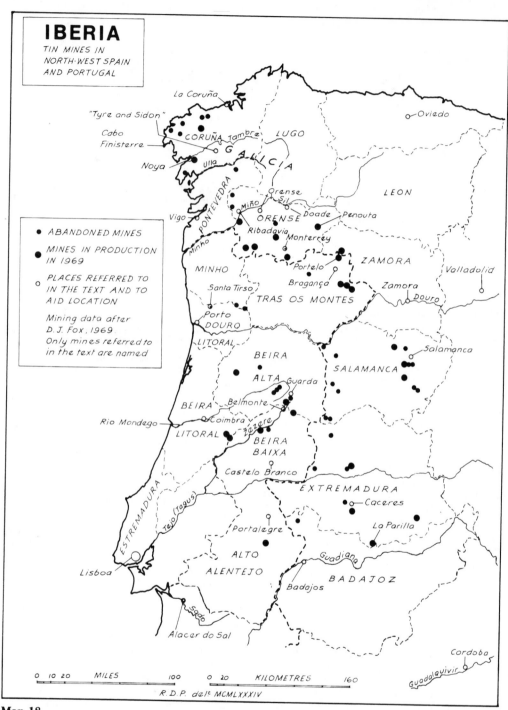

IBERIA

TIN MINES IN
NORTH·WEST SPAIN
AND PORTUGAL

La Coruña

"Tyre and Sidon"

Cabo
Finisterre

Noya

Oviedo

CORUÑA   Tambre   LUGO

GALICIA

Ulla

● ABANDONED MINES

● MINES IN PRODUCTION
   IN 1969

○ PLACES REFERRED TO
   IN THE TEXT AND TO
   AID LOCATION

Mining data after
D. J. Fox, 1969.
Only mines referred to
in the text are named

Vigo

PONTEVEDRA

Miño

Minho

MINHO

Santa Tirso

Porto
DOURO

LITORAL

Orense
Sil

ORENSE   Doade   Penouta

Ribadavia   Monterrey

Portelo

Bragança

LEON

ZAMORA

Zamora

Douro

Valladolid

TRAS OS MONTES

BEIRA
ALTA   Guarda

BEIRA

Belmonte

Rio Mondego   Coimbra   Zezere

LITORAL   BEIRA
            BAIXA

Castelo Branco

SALAMANCA

Salamanca

EXTREMADURA

Caceres

La Parilla

Portalegre

ESTREMADURA

Tejo (Togus)

Lisboa

ALTO
ALENTEJO

Guadiana

Badajos

BADAJOZ

Sado

Alacer do Sal

Cordoba

Guadalquivir

0 10 20   MILES   100   0 20   KILOMETRES   160

R.D.P. delt MCMLXXXIV

Map 18

from Galicia in the north-west along the Spanish–Portuguese frontier into Extremadura, a distance of over 480 km, three times the length of the Cornish-Devonian field. The remoteness and bleakness of many of the mines and the difficulty in obtaining adequate supplies of running water must be remembered when contemplating mining in antiquity. The character of the operations can be judged from the following extracts from Fox's paper.

In 1969 the most productive mine was close to Portelo, Bragança, only 200 m from the frontier, in very bleak country at 1 000 m above sea level where winter temperatures fall to −15°C and snow cover is normal. Many of the miners are land owners in the neighbouring villages to which they return at harvest and other critical periods, a set of circumstances that may have been normal in such inhospitable places in antiquity. The lodes here are nearly vertical, as in Cornwall, varying from about 60 cm to 2 m in width, cutting the Palaeozoic slates.

In Spain the province of Galicia accounts for over half of the country's tin, with Penouta, Orense, being the largest mine. It is also the highest, perched at over 1 400 m in the exposed Montes del Ramilo. It is treeless, parched in summer and snow-bound in winter, and even today accessible only along a gravel track. The hill is cored by a gneissic 'chimney' surrounded by slates, and cut by a stockwork of minute veins 4 or 5 cm wide, varying from near vertical to near horizontal. The whole is deeply kaolinized, hence very friable, allowing open-cast working for most of the 20th century. Alluvial and eluvial tin have also been recovered here. Other mines in the region are of similar character.

The whole of north-west Iberia is rich in abandoned mines, especially in Salamanca where alluvials are common. Howorth described enormous excavations he had seen in Galicia and Tras-os-Montes as 'large enough to contain all the little *ancient* tin-workings of Cornwall'.[8] In Extremadura, Spain, a 'fossil' alluvial deposit is worked at the Santa Maria mine to a depth of

20 m, and payable material may extend down to 70 m. The method of working is very primitive, reminiscent of Cornish tin streaming in the 19th century – mostly pick-and-shovel work. Production is seasonal because the 'bone-dry summer' means a lack of water closing the mill for several months, when the workers help with the agricultural harvest. Ore is prepared for washing during the summer, allowing about 500 kg a month to be prepared in the winter. The problem of water is widespread. In the Parilla Mines, lying on the flanks of low hills impregnated with tin–wolfram veins, alluvials are worked. Water for the mill is 'carefully husbanded during July and August when the area lives up to its name *La Parilla*, the broiler or grill'.[9] Small tin deposits are numerous in Extremadura, usually as veins fringing the granite outcrops. These have been raided in the recent past by bands of six or seven men working by day or night hoping to recover 10 or 12 kg of cassiterite a day, and moving on if three or four days' work yield nothing.

Enough has been said to indicate that alluvial tin and outcropping lodes are abundant. It has not been possible to obtain detailed modern descriptions of the alluvial deposits, but it is clear that the tin is common at or very close to the surface. In this respect they differ from the alluvials of Cornwall where the 'tin ground' was invariably at depth. Almost no tin could be picked off the surface in Cornwall. With Iberian tin so easily available in many places, it is small wonder that almost no prehistoric artifacts can be ascribed to ancient workings. In Cornwall, prehistoric artifacts have been sealed beneath many metres of overburden. The ease of obtaining Iberian tin is well described by Davies who visited workings near Zamora on the Rio Douro in 1890:[10]

the peasants of the district collect the pebbles of black tin from the surface of the newly ploughed fields and smelt them in a primitive furnace in the village ... the peasants dig up the soil and carry it on the backs of mules to a small stream, where, when there is a sufficient supply of water it is washed in a most primitive fashion by women, and from

Tons of tin-in-concentrates from International Tin Council statistics.

	1954	1955	1956	1957	1958	1959	1960	1961	1962	1963	1964
Spain	1 062	880	549	815	879	518	336	378	423	298	92
Portugal	1 297	1 445	1 276	1 127	1 249	1 129	644	729	679	613	687
Total	2 359	2 325	1 825	1 942	2 128	1 647	980	1 107	1 102	911	779
Cornwall	940	1 034	1 044	1 028	1 087	1 252	1 199	1 210	1 181	1 226	1 246

	1965	1966	1967	1968	1969	1970	1971	1972	1973	1974	1975
Spain	112	206	163	142	266	442	402	379	523	650	
Portugal	592	598	631	634	445	435	555	607	533	431	
Total	704	804	794	776	711	877	957	986	1 056	1 081	
Cornwall	1 334	1 292	1 499	1 827	1 648	1 722	1 816	3 327	3 573	3 239	

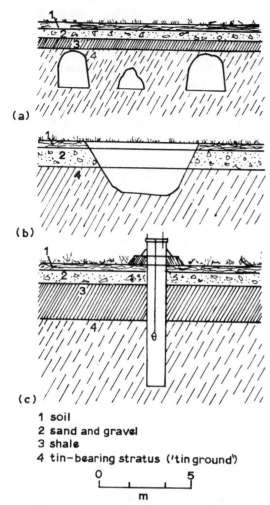

(a)

(b)

(c)

1 **soil**
2 **sand and gravel**
3 **shale**
4 **tin-bearing stratum ('tin ground')**

0 _____ 5
m

a short drivages in the tin-bearing stratum *b* open-cast excavation onto the tin-bearing stratum *c* shaft sunk onto the tin-bearing stratum

40   Idealized sections showing the simple methods of working tin deposits in Orense, Spain (after Garland, 1888)

every 100 lbs of concentrates thus obtained 50 lbs of pure metal are afterwards extracted by the village smith.

In this locality the tin is in a stockwork composed of a network of tiny veins impossible to follow individually. Further along the Douro, near the Portuguese frontier, Davies found that the locals 'spalled' (broke up) the quartz lode, picking out the rich tin and leaving the second-quality ore behind.

Just as the Cornish paid some attention to the tin deposits of Brittany, so they cast more than a sidelong glance at Iberia. Interests included smelting: in 1864, for example, Fox & Co imported about 30 t of 'well dressed black tin' for treatment from the Medina

United Mines near Coruña which had been worked for two years under a Mr Hustler, and yielded 'metal of a superior quality'.[11]

Several Cornishmen have left accounts of visits to Spain, which are very informative about the alluvial deposits and the primitive methods of their working. While it may be ill-advised to draw very close parallels with what may have taken place in prehistory, the reality cannot have been far removed in style of operation. Thomas wrote in 1887 of the tin deposits in a valley 25 km west of Orense:[12]

> Within a few hours' ride on mule back from the town of Ribadavia, in various directions, there are extensive tracts of country covered with a tin-bearing alluvial deposit of from 4 to 12 feet in depth. . . . beneath which are Cornish-like clay-slates or 'killas'. Samples yielded 3 lbs of black tin per ton in the alluvium. Lodes 5 to 8 feet wide existed and cassiterite could be seen in them where they broke the surface.

Twenty years before in the same area he had also seen tin in kaolinized granite, thus emphasizing further the ease of mining.

A much fuller account was published by Garland in 1888,[13] following his visit to the same region, specifically to the village of Doade in Orense, where the unsystematic workings, like those described by Davies, were conducted by agricultural labourers and their families (Fig. 40). Numerous 'shallow pits, trenches' and 'short drivages' littered the deposits which were of great but uncertain extent. Beneath the heather-covered surface the overburden varied between 0·6 and 2·5 m, and averaged 1·2 m, on top of the stanniferous bed which varied between 0·3 and 2·7 m thick, again averaging 1·2 m. Though obviously an alluvial deposit, the 'tin-ground', to use the Cornish expression, was locally called a lode. The tinstones, usually black, varied in size from a pea to that of an egg, with larger lumps only intermixed with quartz and mica gangue. The labourers with their womenfolk and children (Fig. 41):

> with a few simple tools, as pick, shovel, rake, basket, wooden-bowl, and, occasionally when such an important depth as 12 to 18 feet is attained, a windlass of the rudest construction, manage to raise and extract a few pounds of tin-ore per day.

The tin-bearing stratum was not uniformly productive. Of samples taken over a wide area, some would not have been worth working, while others yielded 1·8–6·35 kg of black tin to the tonne, the average being about 2·5 kg. The low specific gravity of the gangue simplified ore dressing, allowing primitive methods reminiscent of those described by Agricola (1556) and Carew (1602):

> Availing themselves of the nearest stream they concentrate the mass by agitating it with a rake in a tiny extemporized sluice, thus getting rid of the clay, the final separation and removal of the gangue being effected by

41 Plate 12 from J. E. Löhneyss, 1616, *Bericht vom Bergwerke*, Clausthal, showing simple methods of washing ore with bowls in tubs (photo courtesy Jiří Majer)

the skilful manipulation of a wooden hand-bowl in a pool of water. The women are especially dextrous in cleaning the ore with the bowl.

In this way four or five pounds of black tin each person per day is often obtained, and for this they obtain from the local buyer 3d. per lb.

Before church service each Sunday morning the tin dealer purchased the ore by the side of the village brook, giving the ore a final washing in 'a wooden strip to remove the last traces of dirt'. After payment, he packed it into sacks on mules' backs for carriage to Vigo for sale. A local buyer in a neighbouring village to Doade showed Garland 'a quantity of small bars of white [smelted] tin, the whole weighing 12 cwt, which he had stored in his bedroom where he also entertained me and my friends very hospitably at luncheon.'

Away from the Spanish–Portuguese frontier, tin is produced in about half-a-dozen small concerns. A mine in the Sierra de Guadarrama, north-west of Madrid, produces tin, wolfram, and a little gold. Another small operation near Puerto de San Vincente, south-west of Toledo, was found at Campillo de la Jara in 1954.[14] More interesting, because it lies within the old Phoenician sphere of influence, is the Sierra Morena north of Cordoba, an area running from Dos Torres (35 km south of Almadén) eastwards to western Jaen in which several small mines occur. The Sierra Morena was, for the Romans, an especially important mining region. 'Ships could reach Cordoba up the Guadalquivir River, and from there boats of hollowed tree trunks could go as far as Mons Argentarius', the Silver

Mountain.[15] For the Romans, Spain was full of mines; *cuniculosa Celtiberia*, as the first century BC poet Catullus put it in jocular form meaning both 'full of rabbits' and 'full of mines'.[16] The coastal strip between Italy and Spain was vital to Rome's trading links and was annexed in 123 BC following the defeat of the Salavii and the founding of the town *Aquae Sextiae*, modern Aix-en-Provence.[17] No better evidence for the importance of the Spanish tin trade in the 1st century AD is needed than the closely datable ingots from the Port Vendre wreck detailed in Chapter 16.

In the tin-producing belt following the Spanish–Portuguese frontier 'there is hardly a stream valley that will not in places show the colours of gold and small amounts of tin'.[18] Unfortunately, evidence for ancient tin-streaming is very scanty, partly for the reasons given above, and either because the evidence has been lost in later workings, or simply by default of recording. Fortunately, in the Zezera valley near Belmonte, Beira Baixa, Allan[19] records that in the working of shallow alluvials by a Portuguese–American Tin Co. from 1913 for 25 years, 10 000 tons of tin metal were produced from an area of 25 km^2, and that the general manager, R. Gruber:

was of the opinion that the valley had been worked over superficially, and a number of Roman coins were turned up by the dredge. Unfortunately these were dispersed without being dated. The dredge was later transferred to the neighbouring valley of the Macainhas river. There further Roman coins were recovered which were dated by the British Museum as all belonging to the first half of the first century AD.

Subsequent abandonment was suggested, perhaps because the area then proved too difficult for the simple recovery methods then being employed. With tin so widely available it would have been pointless to labour to excess in one valley when easier pickings were at hand in the next.

Classical authors say little and much the same thing about Iberian tin, but give sufficient information to show that alluvial deposits were extensively worked. Posidonius (*c.* 135 to *c.* 51 BC) is quoted by Strabo as saying that the tin was obtained in the barbarian country beyond Lusitania, especially by the Artabri of the Cap Finisterre region of Galicia. They dug up the ore using wooden shovels and washed it in sieves 'woven after the fashion of baskets'.[20] Diodorus Siculus (*fl.* 60–30 BC), who may also have derived his information from Posidonius, speaks of 'many mines of tin in the country above [i.e. north of] Lusitania', as well as of the famed Cassiterides off the Iberian coast. He precedes this with the statement that the:

> Tin occurs not found on the surface of the earth, as certain writers continually repeat in their histories, but dug out of the ground and smelted in the same manner as silver and gold.

While this appears to indicate lode mining underground, it is unlikely. Open-cast working of the fine-veined stockworks, as in Brittany, is probable, but Diodorus' description is applicable to alluvial workings, and is closer to the truth in that the tin-ground lay below the surface, if close to it, as seen in Garland's description. The most famous passage is in the *Historia Naturalis* of Pliny the elder (AD 23–79). After mentioning the inevitable 'islands of the Atlantic', he continues:[21]

> It is now known that it is a product of Lusitania and Gallaecia found in the surface-strata of the ground which is sandy and of a black colour. It is only detected by its weight, and also tiny pebbles of it occasionally appear, especially in dry beds of torrents. The miners wash this sand and heat the deposit in furnaces. It is also found in the goldmines called *alutiae* through which a stream of water is passed that washes out black pebbles of tin mottled with small white spots, and of the same weight as gold, and consequently they remain with the gold in the bowls [or 'baskets'] in which it is collected, and afterwards are separated in the furnaces, and fused and melted into white lead [tin].

The use of bowls, or baskets, again reminds us of Garland's description of ore dressing. In Book IV, 20, Pliny mentions the town of Noeia which must be the modern Noya at the mouth of the Ria de Muros y Noya, west of Santiago. The whole area, according to Pliny, was:

> one of the most ancient and notable of the tin districts of Ancient Spain.

The peninsula south-west of Noya contains the San Finx mines in the Montes de Barbanza where Borlase[22] saw:

> very ancient workings, too, consisting of open cuts all along the back [i.e. outcrop] of the lode which is very well defined containing tin of a very high quality.

About 40 km east of Noya near Rio Ulla are similar mines known as Tyre and Sidon with similar open-cast workings along the lode outcrops:

> there forming a pile on either side, just as was the case in the 'old men's workings' as they are called in Cornwall.

Here Borlase traced the worked lode from the outcrops at the river for 900 m in the mountain side.

Well away from the main mineralized belt, tin has been found in the province of Murcia near Cartagena as lenticular deposits in Permian slates.[23] The principal ore of this region is lead, and tin concentrates only yield about 22% compared with 60–65% in Galicia.[24] Murcia was a major mining region in classical times: 40 000 natives worked in silver mines near Cartagena,[25] and a single shaft at Baebelo supplied Hannibal with more than 135 kg a day.[26] The silver came, as most of it does, from the lead sulphide galena. In working these deposits some tin could have been obtained, by chance if not intent, and there are indications in the archaeological record which can be interpreted as proof of Murcian tin exploitation as far back as the early bronze age.

There is a growing awareness that the transition from a non-metal using neolithic to a full-blooded tin-bronze age is an indigenous Spanish development. There is no longer any need to rely on colonists, traders, prospectors, or any other class of immigrants, to account for the emergence of a metal-using culture in south-east Spain.[27] The local sequence develops from a late neolithic (Almeria I) of the 4th millennium BC, through a chalcolithic (Almeria II) in the 3rd millennium, into the bronze age which includes arsenical bronze of the Argar culture which occupied most of the 2nd millennium. The first tin bronze appeared during the Argaric phase, though arsenical alloys predominated. Casting in bipartite moulds was also achieved at this time, and traces of copper working have been recovered from many sites where arsenic–bronze and arsenic–tin–copper alloys were cast in stone moulds.[28]

Harrison[29] has recognized four metal compositions in southern Iberia: 1 pure copper with traces (0·01%) of silver and nickel; 2 copper with 1·5–2·5% arsenic; 3 tin bronze with similar trace elements as above, but with high and very high (over 10%) tin; 4 tin–arsenic-bronze with consistently high levels of arsenic and tin. Metal 3 is known in Pragança, north of the Tagus estuary in southern Portugal. By contrast, the Argaric metalwork of south-east Spain contains very variable amounts of tin, from a non-functional 0·1% to 10% and more. The explanation is simple: Pragança is only 25 km from the Atlantic coast and within easy reach of the principal Iberian tin alluvials, while the south-east of Spain contains only the poorer quality deposits of the Cartagena area. The sporadic tin deposits explain the erratic tin content of Argaric bronzes, 'since the occasional discovery of one of the pockets would provide tin at some times and not at others'.[29] The fact that erratic tin content continued into late Argaric times shows that tin remained hard to obtain; it cannot have been due simply to a protracted period of experiment by metal smiths searching for the ultimate alloy, or the incompetence of smiths.

An excellent account of the nature of the tin in Cartagena was given to Allan[30] by a Mr Cunningham who worked for tin in Almeria from 1927 to 1935:

> Here in a zone two miles long and half a mile wide there were sixteen concessions of varying size along the crest of the hills above La Union. The district was famous for the discovery of rich pockets of tin ore that had made small fortunes for the lucky discoverers. One individual called Pinero living at the time Cunningham was in the district, was reputed to have made over a million pesetas from one of these pockets at a time when the peseta exchange was twenty-eight to the pound sterling. Cunningham operated a mill concentration plant on the Sugunda Aparecida concession, and for some years made twenty to thirty tons per month of 30 per cent tin concentrates which were shipped to England for treatment.

The Fortune mine reached a depth of over 120 m, several others exceeded 90 m, and good gossan outcrops were frequent. Unfortunately, no traces of ancient working were recorded.

Tin would have been difficult to prepare because of the lack of water in the Cartagena area, so that once a ready supply of metal became available from north-west Iberia, local mining might have been abandoned. The south-east is almost a semi-desert with only *c.* 205 mm of rainfall a year at Almeria City, and as little as 120 mm further east. Rainfall is not only sporadic, on no more than about 20 days a year, but it varies considerably annually, so that long-term averages mask periods of serious drought. Irrigation was practised for agriculture, such as a pre-Argaric irrigation ditch at Cerro de la Virgen, 'by no means the driest part of the province',[31] and there must have been similar schemes for the washing and treatment of tin ore.

Early trade with north-west Iberia is possible. Argaric imports are known in Galicia, such as a riveted dagger in a hoard from Ronfeiro in Orense, though bronzes of north-western type are scarce in the south-east and tend to be later, of the middle bronze age.[32] Unfortunately, north-west Iberia suffers from a lack of excavated settlements that could yield ores and evidence of metalworking, while the ancient tin workings themselves are practically undatable.[33] Proof of Roman mining is elusive enough.

When Davies visited the region he could 'find no record of datable objects found in or near the mines'.

Large rock-cut channels for water at Salbe and Ablaneda in Asturias, and at Carvalhal do Estanho in northern Portugal, were only assumed to be Roman.[34] In his later book *Roman Mines in Europe* (1935), Davies can say no more than that tin was probably worked from the beginning of the bronze age because of the abundance of bronze tools in that part of Iberia plus the fact that in the middle bronze age tin percentages in metalwork decrease in proportion to their increasing distance from tin-producing regions. Briard, however, quotes De Serpa Pinto as claiming that stone picks, of presumed bronze age date, had been found in alluvial tin workings at Colineal, Belmonte, and that a bronze age palstave was recovered from a depth of 12 m in a 'mine' at Ouarta-Feira, Sabugal. The latter almost certainly refers to a copper working.[35]

North-west Iberia was not an isolated backwater in the early bronze age. Iberian influence reached Brittany in a pre-tin era, as confirmed by the discovery of the short copper blade known as a Palmella point in the passage grave at Kercadoret, Locmariquer, another dredged up from the Loire at Trentemoult near Nantes and a scattering of others further south. The inference is that contact with Brittany led to an appreciation of the nature and value of tin back in Iberia during the 2nd millenium BC when the northern seaboard of Spain received palstaves, socketed spearheads, rapiers, and dirks of types familiar in north-west Europe including Britain.[36]

To this period can be assigned the opening up of Iberian tin workings, because metalwork begins to take on a local character. This is best exemplified by the manufacture of a palstave with double-loops, very like contemporary French and British examples which may well be prototypes. The double-looped palstave is scarce in Britain but becomes a distinctively Iberian form. There is a heavy concentration of them in the tin-rich regions of Iberia. At Santo Tirso, 18 km northeast of Porto, 36 were found that still showed the casting jets in the shape of an inverted cone. It is to this region that all those with casting jets belong. In all, hundreds of double-looped palstaves have been found in the tin streaming areas of the north-west, many of them made of good quality bronze with about 10% tin (though many others have a high lead content replacing the tin), and all of a characteristic local pattern.[37] With such a concentration of distinctive metalwork, it is pointless to believe anything other than that the raw materials were mined locally.

In 1923 the famous Huelva hoard, probably from a shipwreck, was dredged up in the estuary of the Rio Odiel in the Golfo de Cadiz. It contains bronzes which include the distinctive 'carp's-tongue' swords, a type not out of context even in Britain, as well as axes, daggers, spearheads, helmet fragments, brooches, 'flesh-hooks', and belt-hooks, all variously paralleled in Sardinia, Ireland, Cyprus, northern Iberia, and the

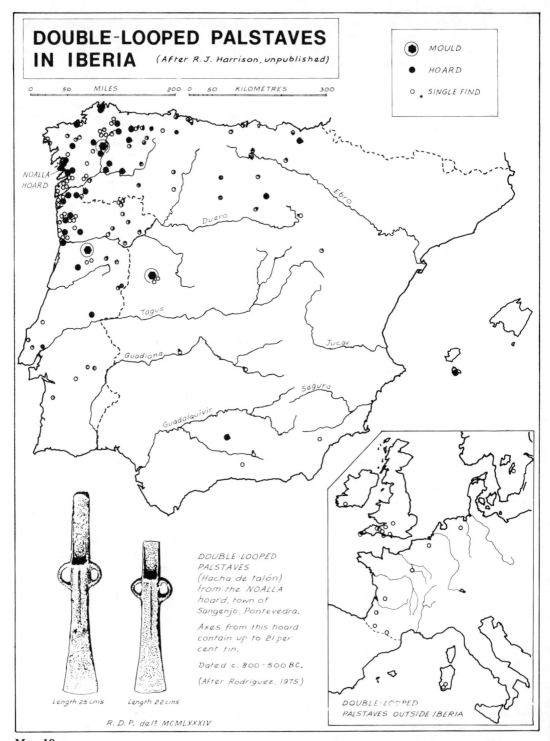

DOUBLE-LOOPED PALSTAVES IN IBERIA   *(After R.J. Harrison, unpublished)*

MILES   KILOMETRES

- MOULD
- HOARD
- SINGLE FIND

NOALLA HOARD

Ebro

Duero

Tagus

Guadiana

Jucar

Segura

Guadalquivir

DOUBLE-LOOPED PALSTAVES (Hacha de talón) from the NOALLA hoard, town of Sangenjo, Pontevedra.

Axes from this hoard contain up to 21 per cent tin.

Dated c. 800-500 BC.

(After Rodriguez, 1975)

Length 25 cms   Length 22 cms

R.D.P. delt MCMLXXXIV

DOUBLE-LOOPED PALSTAVES OUTSIDE IBERIA

Map 19

south of France. Traditionally the hoard has been dated to the 6th and 7th centuries BC, but recent ^{14}C determinations now place it as early as the 10th century.[38] Between 800 BC and the arrival of Roman rule, with the foundation of Phoenician settlements at Cadiz and elsewhere along the southern seaboard, trade was extensive and is well illustrated by the appearance of Celtic objects from northern Iberia in the Mesetas. As Arribas tells us:[39]

> The treasures of the Sierra Morena, a frontier region between [southern] Iberians and [northern] Celts exhibit converging influences of both these people.

In an unpublished thesis Frankenstein deals in detail with the impact of Phoenician expansion in southern Iberia from the 8th century onwards.[40] The Phoenicians stimulated a demand for metals and traded with the tribes to the north. As they had done in western Asia, the Phoenicians treated the local rulers as client kings by giving them 'classical symbols of power and kingship', such as ivory and alabaster goods. To encourage native production they established 'factories' on the south coast, and by 'introducing items such as cloth, dyes, dried fish, oil, unguents, fine wares and metalwork, exchange relations between Phoenician merchants and indigenous rulers could be maintained'. The widespread use of Phoenician luxuries was concentrated in Andalucia and, in Atlantic Iberia, principally in central Portugal and Estremadura, where a plentiful supply of tin was readily available through trade.

Some indication that Iberian tin reached far afield in the 6th century BC comes from the excavations of the Magdalenenberg burials in southern West Germany: an armlet from grave 10 had been repaired with tin, while a metal pin on an armlet from grave 120 was also made of 70% tin with 30% lead. In the words of the excavators:[41]

> It is possible that the Iberian belt-hook from grave 65 indicates the provenance of the metal, the tin-land of Spain.

NOTES AND REFERENCES

1   JOHN HAWKINS, letters of 12 Sept. 1791 and 28 Nov. 1792 (DDR 5757/1/74) at the County Record Office, Truro
2   SMITH, 1967
3   GORIS, 1925, 275
4   SMITH & GNUDI, 1942, 60; Biringuccio wrote of tin, 'the best and the most abundant that is found in the provinces of Europe is that which is mined in England'
5   BORLASE, 1897, 3, quoting in translation from Bartolomé Molina's *Descripción del Reyno de Galicia*, Madrid, 1551
6   ITC, 1964–1974
7   FOX, 1969
8   HOWORTH, 1873, in Discussion, *Proc. Soc. Antiquaries.* **5**, 416
9   FOX, 1969
10 DAVIES, 1901, 188
11 ANON., 1864, 55
12 THOMAS, 1887
13 GARLAND, 1888
14 ITC, 1964
15 ARRIBAS, 1964, 123; GOSSÉ, 1942, has little to say about tin but gives a useful account of the silver–lead mines, notably La Fortuna mine near Mazarron, Murcia, with illustrations of the equipment found there
16 TOYNBEE, 1973, 203
17 WISEMAN & WISEMAN, 1980, 16
18 ALLAN, 1970
19 ALLAN, 1970
20 CLARK, 1965, 194
21 PLINY, *Naturalis Historiae*, Bk XXXIV, 47, in H. Rackham's ed., Loeb Classical Library, 1938–62
22 BORLASE, 1897; Copeland Borlase was a Cornish antiquarian who was also manager of several tin-reservations in Coruña, Zamora, and Tras-os-Montes
23 FAWNS, 1905, 152
24 ITC, 1964
25 RICHARDSON, 1976
26 ARRIBAS, 1964, 122
27 GILMAN, 1976
28 COLES & HARDING, 1979, 225
29 HARRISON, 1974
30 ALLAN, 1970
31 GILMAN, 1976
32 HARRISON, 1974
33 JOLEAUD, 1929; in spite of a lack of evidence, the author assumes Galicia supplied the Chaldeans, the ancient Egyptians, and the Aegean civilizations with tin, referring to an inscription of Sargon I of Akkad and taking 'Anaku-ki' as a geographical location meaning 'land of tin', which he unhesitatingly places in Spain
34 DAVIES, 1933, 44
35 BRIARD, 1965, quoting R. de Serpa Pinto, 1932, 'Explotaciones mineras de la edad de bronce en Portugal', *Comisión de Investigaciones Paleontológicas y Prehistóricas* (1934). I have been unable to find a copy of this paper, but see also Pinto 1933, which lists finds, apparently all from copper mines; no mention is made of pick-axes from alluvial tin workings in Belmonte
36 SAVORY, 1968, 221–2
37 ALLAN, 1970; RODRIGUEZ, 1975
38 SAVORY, *Antiquity*, 1981, 226–7
39 ARRIBAS, 1964, 126
40 FRANKENSTEIN, 1979
41 SPINDLER & SCHWEINGRUBER, 1980, 20

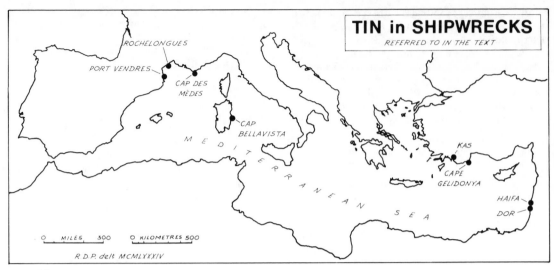

Map 20

16 Tin in shipwrecks

Tin ingots have been recovered from a number of wrecks in the Mediterranean. Best known is that found in 1960 off Cape Gelidonya, Turkey, and excavated by Bass.[1] Brushwood gave a ^{14}C date of 1200 ± 50 BC for a cargo consisting of at least 34 copper ox-hide ingots each weighing about 25 kg neatly stacked with plano-convex 'bun'-shaped copper ingots, the whole bound with matting that subsequently decayed. There was also much scrap metal suggesting that metalworking was carried out on board or at convenient ports. Many bronze tools in the cargo were Cypriot, though it was possible to isolate items belonging to the crew and all these proved to be Syrian.

Thin metal foil folded double and bent into an S-shape contained nearly 53% tin, 19% copper, with lesser amounts of other elements. The high tin content suggests the foil alloy was intended for decoration. Tin foil of similar thickness was used to cover an ivory pyxis found at Mycenae, and good examples are known also from the Athenian agora.[2] Tin-decorated vases are discussed in Chapter 22.

Three piles of a white substance (one pile weighing 8 kg) having the 'consistency of toothpaste' contained 13·84% tin oxide and 71% calcium. Marine concretion preserved the shape of one of the original ingots, a bar 6 cm^2 of unknown length. Its original composition is in doubt because the decomposition of tin to a mobile paste is otherwise unknown, but Bass compared its simple shape to an oblong ingot depicted in the tomb of Nebamūn and Ipuky at Thebes (tomb 181) dated between 1397 and 1353 BC which also shows metal workers and a copper ingot. The oblong ingot was formerly assumed to be of lead, but it is more likely that the Egyptians alloyed tin with copper. For the same reason Bass believes that the blue-grey ingots in the tomb painting of Amen-em-opet (see above, Fig. 4) are also tin. Lead was originally suggested in the mistaken belief that tin ingots did not occur this early.

An even older wreck was excavated by Bass and his colleagues in 1984.[3] It was found by a sponge diver in autumn 1982 lying on a steep rocky shore at a depth of 43–51m and only 100 m off Uluburun near Kas. As in the Gelidonya wreck, only 70 km to the east, the cargo contained neat rows of copper ox-hide ingots with round plano-convex 'bun'-shaped copper ingots be-tween them. Large storage jars or pithoi, terracotta oil lamps or incense burners, a pilgrim flask, and Canaanite amphorae indicate a date probably in the 14th century BC.

Also recovered was a 'greyish brittle material' which proved to be tin metal 99·5% or more pure. How much was found is not stated.

An Eastern origin for all the metal is possible. The report ends by noting that the Kas finds 'remind us of the 14th century scene in the Tomb of Kenamon at Thebes: a fleet of Syrian ships with pithoi on their decks deliver goods to an Egyptian waterfront, one Syrian porter carrying a pilgrim flask and an amphora'.

Analysis of the tin from Kas and Gelidonya is no help in tracing its origin. The conclusion of the Gelidonya report was that the tin was not Cornish, which is probably true, though the reason given is based on unsound evidence. The Gelidonya tin is devoid of any traces of cobalt and germanium. It has become almost a part of folklore that these elements are sufficiently high in Cornish tin to be diagnostic, a notion based on a tentative suggestion by Tylecote,[4] based on analyses in the early 1950s at the Royal School of Mines. Germanium is extremely difficult to detect and the whole idea is regarded as improbable by the staff at Camborne School of Mines (pers. comm.). In any case, cobalt is known from tin elsewhere, as in the cassiterite from El Mueilha and Nuweibi, Egypt (0·01%),[5] while germanium is a trace element together with tin in a copper ore from Kilcoe Mine, Balleydehob, Co. Cork.[6]

The tin from both these Turkish wrecks was doubtless put on board at Syrian ports, to which it may well have been transported a considerable distance, perhaps from Soviet Central Asia. It is tempting to see here proof of the trade referred to in the tablets at Ebla (Tel Mardikh) a millennium earlier. A mould for copper ox-hide ingots has been found in Syria and analysis of the Gelidonya copper indicated significant traces of cobalt, an element absent from Cypriot copper, rare in many Near Eastern ores, but known from Ergani Maden in eastern Turkey.

In the Metropolitan Museum of Art in New York is an ox-hide ingot believed to have come from the Bay of

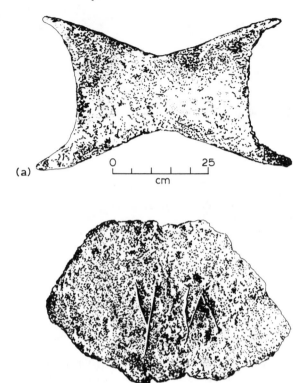

(a)

(b)

a copper ox-hide ingot (16·5 kg) *b* one of five tin ingots, each weighing 4 kg

42   Ingots found underwater south of Haifa in 1982 (redrawn from *Nautical Archaeology*, 12, 178)

Antalya, just east of Cape Gelidonya. When analysed in 1915 it was said to contain 7·5% tin, but re-examination in 1975 proved it to be of pure copper.[7]

Illustrated in Fig. 7, Chapter 2 are two ingots weighing 11·4 and 11·9 kg found in the autumn of 1976 in rather mysterious circumstances south of the Israeli port of Haifa. Dating is tentatively based on the presence of engraved signs likened to the Cypro-Minoan script found on Cyprus and at Ugarit near the Syrian coast between about 1500 and 1100 BC. The ingots contain *c.* 95% tin. They belong to the same general period as the Gelidonya wreck, for similar markings are found on the Gelidonya ox-hide ingots.

In February 1982, two students from Haifa university searched for ingots with a metal-detector south of Haifa underwater. Finds included a large copper ox-hide ingot of 16·5 kg, five irregular tin ingots weighing 4 kg each, and four stone anchors (Fig. 42). The tin ingots might have been Cornish 'Jew's house

tin' were it not for the Cypro-Minoan type signs impressed on them. Samples of the ingots have been sent to Prof. Madden for analysis in Pennsylvania, though whether or not it will be possible to deduce their source is questionable.[8]

The earliest European wreck to contain tin ingots was discovered by divers of the Béziers archaeological society on 21 July 1964 about 500 m off Rochelongues Point, Agde, east of Narbonne.[9] The wreck lay in 7 m of water. First to be found were a bronze socketed axe and a plano-convex copper ingot underneath which were a socketed hammer, an arrowhead, a pin, bracelets, and numerous ingots or fragments. After five seasons' work, the haul totalled 1700 objects, mostly of bronze, as well as 800 kg of very pure copper ingots (over 99%) and weighing nearly 6 kg each. Ornaments were very varied with about 350 bracelets alone. Particularly interesting were belt buckles in the shape of an 'Agde cross', commonly found in burials in Languedoc and Iberia, which dated the wreck to the earliest phase of the iron age. The remaining finds were typically late bronze age. The bulk of the weapons were socketed axes with square cross-section or, less frequently, of round section. Some obsolete wing-hafted axes showed that the cargo consisted partly of old material suitable for recasting.

As well as bun-shaped copper ingots, there were 32 tin ingots, thin leaf-shaped slabs about 14 cm in diameter and weighing 46 g. There was thought to have been granular tin in sacks, but a more recent analysis proved this to be lead.[10]

The whole character of the finds is Iberian. For example, a large fibula decorated with three lozenge-shaped plates is identical to one from tomb 184 at Agullana, a site near the Spanish frontier. The conclusion is that the wreck is of a ship which was working its way up the coast from Spain, halting at suitable sites – river mouths or beaches – where the ship's master expected good trade. The tin must surely be of Spanish origin, as it most certainly is from the next wreck to be discussed.

South-west of Rochelongues and only 10 km from the Spanish frontier lies Port Vendres. In its outer harbour a remarkable wreck was discovered in 1972 and excavated during the following two years.[11] The ship carried a large number of amphorae which probably held oil or salted goods from Spain's southern Roman province of Bacatica. Many amphorae had inscriptions stamped on the handles. Samian ware from the south Gaulish potteries at La Graufesenque north-east of Toulouse bears potters' stamps which narrow the date of the wreck to the '40s (1st century). More important still is a stamp on its metal cargo.

The real interest of this wreck lies in its 14 unusual handled tin ingots (Fig. 43). All are very pure with less than 200 ppm of copper and lead. All are light, the heaviest being only 8·75 kg. In section, each ingot has

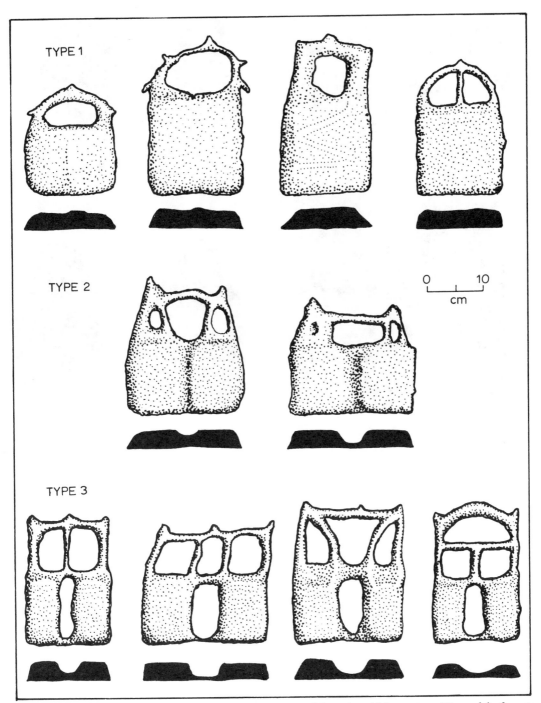

43 Selection of tin ingots showing the variety of forms recovered from the mid-1st century AD wreck in the outer harbour of Port Vendres (redrawn from *Gallia*, 1975, 33, 64)

a upper 23·5 × 18 × 6·3 cm; upper surface (not shown)
stamped MARO *lower* 28 × 32·5 × 5·8 cm
b top right plano-convex 17 × 10·4 × 4·8 cm *handled
ingots* vary in size but *c.* 19·4 × 10·5 × 3 cm

**44   Tin ingots recovered from a wreck south of Capo
Bellavista, Sardinia (photos Soprintendenza alle
antichita, Sassari (No. L 1199 & 1299))**

one flat surface, the opposite one somewhat curved,
and the sides tapering, consistent with casting in open
moulds, probably in sand as no two are exactly alike.
They match sufficiently well to suggest that an already
cast ingot was itself used to form an impression in the
sand for the next ingot. Stamps in relief or intaglio were
impressed on the flat upper surface of the ingots, except for one on a curved underside proving that the
impressions were made when the cold ingots had been
removed from the moulds. Cold stamping is general on
this soft metal to the present day.

There are 74 stamps on the 14 ingots. Decipherment
is sometimes difficult. Decorative motifs, such as an
anchor, dolphin, or *caduceus* (a staff or mace) on the
simply shaped type 1 ingots are also frequent on ingots

or 'pigs' of Spanish lead of the Roman Republic.
Doubtless such devices represent the smelter or exporter. The relief design on one ingot resembles a four-
pronged fork with the handle offset to the left; this may
be the Iberian letter *ti*.

Some stamps are unimportant bits of names and
titles – but not the one reading: L. VALE. AVG. L. A
COM. (L(ucius) VALE(rius) AVG(ustae) L(iber-
atius) A COM(mentariis)). He is identified as the
Lucius Valerius who was made a freedman by Messa-
lina, wife of the Emperor Claudius, and who was entitled to be called Augusta between 41 or 42 and Messa-
lina's execution in 48. *A Commentariis* is a title which
appeared under Claudius for use by freedmen in
charge of day-to-day financial transactions of an official nature.

Colls and his colleagues describe further ingots from
the wreck and comment on a relief in the National
Museum in Naples.[12] This shows the workshop of a
bronze smith or a tin smith. Two objects hanging on a
wall and a third lying on a shelf are tentatively consi-

dered to bear a striking resemblance to the simple type 1 tin ingots from the wreck, but the original description of the objects as handled buckets cannot be dismissed. One of them even has the appearance of wickerwork.

Similar ingots were found in 1954 off the east coast of Sardinia by two fishermen who got their nets caught in the remains of a wreck (Fig. 44).[13] It lay in 7 m of water in a rocky area partially covered with sand south of Capo Bellavista and close to Punta Nera. Prof. Lamboglia visited the site in February 1961 and September 1972, but strong tides and shifting sand made any proper investigation impossible. The hull had almost totally disappeared, but in the apparent hold of the ship were found lumps of agglomerate formed by the oxidization products of the metals cementing together particles of sand. When broken apart, the lumps were found to consist of: 26 iron ingots of *c.* 4 or 5 kg each (total *c.* 65 kg); 2 copper ingots of *c.* 50 kg each (total *c.* 100 kg); 32 tin ingots of *c.* 4 or 5 kg each (total *c.* 160 kg). One of the tin ingots was stamped with the letters MARO with the MA conjoined, but of unknown meaning. The Latin inscriptions and the close resemblance of the ingots to those from Port Vendres suggest the tin is Iberian, a view enhanced by the discovery in the Capo Bellavista wreck of seven lead ingots of Spanish type. Two are inscribed on the ends E D CERDO, with DAGVITIVS G F on the convex surfaces. These date the assemblage to between 30 BC and AD 10.[14]

Tin ingots have been recovered from another Roman wreck at Cap des Mèdes east of Porquerolles, an island south-east of Toulon.[15] The wreck lay at a depth of 29 m about 200 m east of the Cap des Mèdes and was found in 1964 by members of the Hyères Yacht Club. It was examined in February 1966 using a diving bell. The wreck formed a homogenous mass of concretion over 18 m long by about 6 m wide. The cargo consisted of iron objects reduced to scarcely recognizable concretions, though it was possible to make out a tangle of iron bars. Three lead rings and pieces of four amphorae were recovered, the body of one amphora suggesting a wreck of the 1st century BC. Two tin ingots were salvaged. These are oval with a flat upper surface and a rounded and irregular base typical of ingots cast in the ground. They contained 99·5% tin. Both ingots bore two stamps made with the same punch and in roughly the same position on each: above are the letters C V M and below, slightly to the right, M N. The dimensions of the ingots are: ingot C 937: 57 cm long, 21 cm wide, 7 cm thick, weight 32 kg; ingot C 938: 56 cm long, 21 cm wide, 6 cm thick, weight 30·5 kg. It is not certain that the amphorae belonged to the wreck, so dating is far from certain. Nor is the source of the tin known; it may well be Iberian, though the ingots resemble those found in Cornwall.

NOTES AND REFERENCES

1 BASS, 1967; BASS, 1970, 127–31 for a short account
2 IMMERWAHR, 1966
3 BASS, FREY & PULAK, 1984
4 TYLECOTE, 1962, 63; ANON., 1954, 26
5 MUHLY in Franklin *et al.*, 1978, 43–8
6 COGHLAN, BUTLER & PARKER, 1963, 20, Fig. 4
7 WHEELER, MADDIN & MUHLY, 1975
8 GALILI & SHMUELI, 1983
9 BOUSCARAS, 1964; BOUSCARAS, 1971
10 TYLECOTE in Franklin *et al.*, 1978, 52, note 11
11 COLLS *et al.*, 1975; the wreck is numbered 2 to distinguish it from another wreck in the harbour
12 COLLS *et al.*, 1977
13 FULVIO LO SCHIAVO, pers. comm. I am indebted to her for supplying details of the wreck and measured drawings and photographs of the tin ingots; they are illustrated in Gianfrotta & Pomex, 1981, *Archeologia Subacquea*, Milan, 188. At the time of writing two articles by Dr Lo Schiavo are in press; 'Tutela del patrimonio archeologico subacqueo nella Sardegna Centro-Settentrionale', and with A. Boninu, 'Ricerche subacqueo nella Sardegna Settentrionale', read at the 6th International Congress of Underwater Archaeology at Cartagena, 28 March to 4 April 1982
14 TYLECOTE *et al.*, 1983
15 TSCHERINA, 1969, esp. 476–8 and Fig. 25

17 Tin in Wales, Man, Cumbria and Scotland

Wales has recently been added to the long list of countries where tin has been detected. At the Nant mine near Carmarthen, beneath the zone of barite, is a mixed sulphide ore which contains 'appreciable evidence of tin mineralization'.[1] Whatever this may mean for the long-term future of mining in Wales, it is of no interest to the archaeologist.

According to Greg and Lettsom,[2] tin ore was found in the Isle of Man. No doubt as a result of this claim in a respected manual, tin appears on this island in the map of tin locations constructed by Schuiling.[3] Unfortunately, the claim does not stand up to investigation and it is not mentioned in the Geological Survey Memoir.[4] The claim can be traced back to tin and wolfram specimens which were:

> part of a collection that was in the possession of the late Lord Henry Murray, and were obligingly communicated to me [J. F. Berger] by Mr Wm Scott, the Collector of the Custom House at Douglas. Many of them had no labels affixed to them.[5]

Sowerby accepted the specimens in good faith, noting the wolfram 'by favour of His Grace the Duke of Athol and Lord James Murray', and adding 'I understand that Tin has been found there.'[6] This is flimsy evidence indeed for Manx tin. It is far more likely that both the cassiterite and wolfram were Cornish, having found their way north with one of the Cornish miners who were connected with the Manx mines.

There is more certain evidence in Cumbria. In the Carrock Wolfram Mine near Grainsgill, tin had been detected in assays for some years, though not until a few days before the big 1959 conference on mining did Mr J. Hartley report that 'cassiterite had been isolated in specks in the Harding Vein'.[7]

North of the border, tin is said to have been found in a trench at Largs, Strathclyde, in 1927, but this has never been confirmed.[8] Better known is its discovery at the turn of the 20th century near Carn Chuinneage in Easter Ross. Fine-grained cassiterite is distributed irregularly in two bands of streaks and veins of magnetite-rich material 'for a distance of 100 and 250 yards respectively' with 'a breadth of 10 and 15 yards'.[9] A recent reconnaissance of stream sediments revealed uranium, tin, and molybdenum all closely associated with the Carn Chuinneage granite-gneiss dated to 530 ± 10 million years. Some specimens of the magnetite–cassiterite bands assayed at between 300 and 900 ppm of tin, while a spread of 10 to 15 ppm in sediments up to 6 km to the east are thought to have been due in part to glacial and alluvial dispersion.[10] Interesting though this is, there is nothing here to excite the archaeologist.

Two fragments of tin (actually pewter with over 21% lead) were found with a socketed axe, a spearhead, and three penannular rings, all of bronze, in the 19th century in Wester Achtertyre, Grampian,[11] but there is no cause to believe this tin came from anywhere other than Cornwall.

NOTES AND REFERENCES

1 THOMAS, 1972, 383
2 GREG & LETTSOM, 1858, 359
3 SCHUILING, 1967
4 LAMPLUGH, 1900
5 BERGER, 1814
6 SOWERBY, 1804, **1**, 44; 1806, **2**, 125
7 EASTWOOD, in IMM, 1959, discussion 206–7
8 ANON., 1974, 5
9 FAWNS, 1905, 153
10 GALLAGHER et al., 1971, esp. 153–5
11 MACADAM & SMITH, 1872; They describe several early bronze age flat axes with tinned surfaces, illustrating one from Sluie, Grampian

18   Tin and gold in Ireland: bonanza or borrasca?

Cornwall may have been indebted to Ireland for her saints: decidedly not for her tinners.

R. N. Worth, 1872[1]

There has been intensive exploration for minerals in Ireland during the past 30 years without any tin deposits being found; nor is there any evidence that tin was ever worked commercially in the country. In the 9th century Nennius wrote in his *Historia Britonum* that 'copper, lead, iron, and tin' (*stannum*) occurred at Lough Learne in Co. Kerry, a place visited in 1756 by C. Smith, who noted that no tin of 'any purpose' had yet been discovered there, though he hoped it might 'for I have picked up small specimens of ore which contain some tin at no great distance from the lake'.[2] A large number of grooved stone mining mauls have been recovered from the shore of Lough Learne, but these were probably connected with the copper mines of Ross Island and there is no reason to suppose the area ever contained tin in workable quantities.

Several Irish copper ore localities have produced traces of tin. It has recently been confirmed at Allihies, Co. Cork,[3] but its occurrence at Kilcrohane, Sheep Head,[4] in the same county has not been authenticated. Traces of tin have been noted at Kilcoe Mine, Balleydehob, and at Ballycummisk, both in Co. Cork,[5] though other mines in the area gave no detectable tin. The average impurity pattern constructed by the late H. H. Coghlan from 52 samples from various parts of Ireland yielded less than 0·01% tin, insufficient to produce tin bronze.

A little-known tin location in western Ireland is the southern shore of Galway Bay, Co. Clare.[6] The tin is present in ilmenite, perhaps also in epidote, in calcareous nodules derived from a 2 m thick black band at the junction of Carboniferous limestone and granite. The deposit has not attracted attention since 1922 when Beringer of the Camborne School of Mines, Cornwall, analysed specimens and found between 1·3–2·3 kg t^{-1} tin. The tin only appeared in chemical assay and could not be detected by 'vanning' (the simple method of separating cassiterite from gangue minerals on a shovel), so it is highly improbable that any tin was capable of recovery in antiquity.

One of the best-known occurrences of tin is at Dalkey, Co. Dublin, where it was found in a lead–zinc lode, the only lode occurrence known to G. H. Kinahan. Dalkey is one of a number of largely unprofitable lead lodes which outcrop in or near the coast. Hunt,[7] an authority on Cornish mining at pains to record tin occurrences anywhere, quotes an incident in which the mining engineer John Taylor told Thomas Weaver, in or before 1818, that he had discovered 'in the granite, malachite, arsenical pyrites, tinstone, spodumene, and a new mineral killinite' (an alteration product of spodumene) named after Killiney Bay where Taylor found it. Weaver gave more details of the tinstone; it was in a vein cutting a loose block of granite and represented the first discovery of cassiterite in a vein as distinct from particles in a placer deposit. This must be the same location mentioned by Griffiths of cassiterite crystals in the Leinster granite at Malpas Mine, Killiney Hill, Dalkey.[8] Another pebble of Leinster granite containing cassiterite was picked up in the late 19th century out of glacial boulder clay near Greystones, Co. Wicklow: 'In the context of the direction of ice movement, Killiney Hill *could* have been the provenance'.[9] Whether it was or was not, Leinster tin can hardly concern anyone other than the academic mineralogist. The same is true of the only known tin occurrence in Northern Ireland. Seymour noted minute crystals, some of almost microscopical dimensions, in the Tertiary granite of Slieve-na-miskan in the Mourne Mountains, Co. Down.[10] Stream 'concentrations' of around 70 ppm have been mapped here.[11] To put this in perspective, the valueless Scottish tin at Carn Chuinneage gave readings of up to 900 ppm.

The mountains of Mourne, Wicklow, and Blackstairs, lie within the broad belt known as the Southern Caledonides. These include the Southern Uplands of Scotland, the Lake District, the Isle of Man, and north and west Wales. This is a broad trough, a syncline with a central up-warp or geanticline of folded palaeozoic sediments, which has great similarities of mineralization throughout. Wheatly[12] has shown that lead and zinc predominate with copper of economic importance being rare and concentrated close to the geanticline on which lies the famous Parys copper deposit in

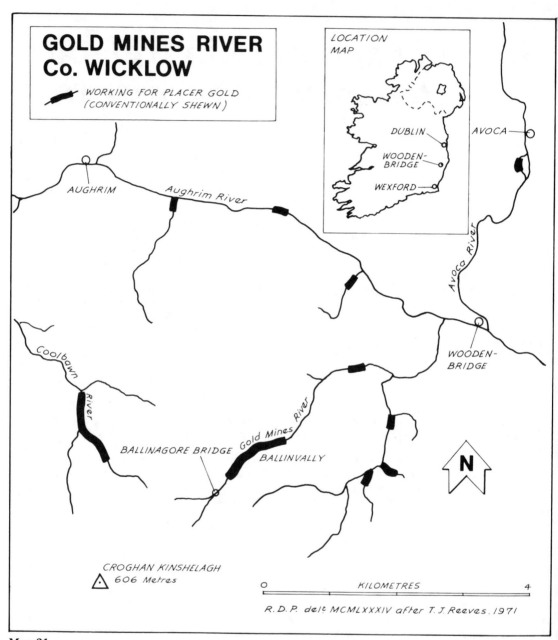

GOLD MINES RIVER Co. WICKLOW

WORKING FOR PLACER GOLD (CONVENTIONALLY SHEWN)

LOCATION MAP

DUBLIN
WOODEN-BRIDGE
WEXFORD
AVOCA

AUGHRIM
Aughrim River
Avoca River
WOODEN-BRIDGE

Coolbawn River
BALLINAGORE BRIDGE
Gold Mines River
BALLINVALLY

CROGHAN KINSHELAGH
606 Metres

KILOMETRES
0   4

R.D.P. delt MCMLXXXIV after T. J. Reeves. 1971

Map 21

Anglesey, and near which lies Avoca, Co. Wicklow, the only mine to work in recent times in the Southern Caledonides. Avoca, which closed in 1982,[13] produced copper and iron with some lead and zinc, but tin has never been more than a footnote there. Greg and Lettsom[14] were aware of the rare copper–iron–tin sulphide stannite in the Cronbane Mine (East Avoca), but it was not a consistent feature of the ore body. Stannite, with major scheelite and other minor sulphides, occurs in dykes on the eastern edge of the Wicklow Mountains (K. F. G. Hosking, pers. comm., 1984).

South-west of Avoca at Ballinvally is what has become known as the Gold Mines River which flows north from Croghan Kinshelagh mountain on the Wicklow–Wexford border. More will be said of the gold later. Whatever the impression created by a book optimistically entitled *The Gold and the Tin in the South East of Ireland*,[15] such a small amount of cassiterite was recovered here that it has been ignored in many standard works dealing with tin locations or simply treated as a curiosity 'of no commercial importance' in others.[16] The first published notice of tin here was in 1801 by Mills, King, and Weaver,[17] who listed all the associated minerals:

> in every instance, where the gold has been found, there have been also found fragments of *magnetic iron ore*, and *quartz* containing *chlorite, iron ochre*, and *martial pyrites*, attended more particularly at the works at Ballinvally, with *specular iron ore, brown and red iron stone, tin stone crystals, wolfram*, and grey ore of *manganese*.

The cassiterite was discovered by Weaver, who did not believe there was much there.[18] Belief that large quantities of tin could have been streamed here in antiquity rests on the statement of William Mallet who, out of a mass of sand 'not exceeding 150 lbs, obtained $3\frac{1}{2}$ lbs of stream tin'. When over 80 years of age Weaver told Warrington Smyth that he (Weaver) did little more than prove the presence of tin, while Smyth (Mineral Surveyor to the Duchy of Cornwall and well versed in the art of streaming) found 'but little of it'. Smyth was surprised at, perhaps suspicious of, Mallet's results, 'a ratio of productiveness which throws into the shade all the richest stream-works ever found in Cornwall or on the Continent'.[19] No one could accept Mallet's result without reservations in a country without a single workable tin lode. The opinion of Kinahan[20] was that Mallet's results might have resulted from an experiment 'made on 150 lbs of the washed sands at the mouth of the buddle in which the tinstone had collected', evidently after the treatment of countless tons of gold-bearing sediments. At all events, Mallet's supposed 3·51 lb (1·59 kg) is all that has ever been recorded from the region. Kinahan hoped that a tin lode would be found in the Woodenbridge area, but trenching and some sporadic boring for a local bedrock source at Ballinvally revealed only barren quartz veins

in the Ordovician sediments and volcanics. Sulphide ores such as iron pyrite ('fool's gold') do contain traces of gold, and the generally accepted theory for the presence of the Ballinvally alluvial deposit is that:[21]

> the small precious metal content of the Avoca sulphides became concentrated at the base of a long developed gossan [the oxidized zone of weathering] to be swept southwards by Quaternary glacial action and re-concentrated in the narrow alluvials of the hillside streams.

A similar explanation accounts for the cassiterite being derived from Avoca, from highly disseminated particles in the granite but not from decomposed stannite, which is no longer considered the parent of wood tin. Stannite breaks down to form varlamoffite. Magnetite is also a possible source of tin, for at Carn Chuinneage in Scotland it yielded 3·22% of tin oxide,[22] but the mineral assemblage at Avoca does not encourage the notion of abundant cassiterite accumulating at Ballinvally.

Several prehistoric objects of tin have been found in Ireland. From Knocknaboul is a bronze age gold object with a tin core, in size and shape similar to the so-called Irish 'ring money'.[23] Three torcs of thin, rather poorly twisted tin were found in or before 1944 at a depth of 2·4 m in a peat bog at Killsallagh, Co. Longford. A broken specimen analysed in 1955 gave the following result (Nessa O'Connor, pers. comm.):

Sn	Cu	Pb	Ag	As	Sb
98·8	·019	·016	<·01	·10	<·01

Similar results were obtained from three tin armlets, part of a hoard from Crannog 61 at Lough Gara, Rathtinann, Co. Sligo.

Sn	Cu	Pb	Ag	As	Sb
98	·01	·12	·01	·05	·01
98·5	·011	·64	·01	·05	·01
98·6	·039	·30	·01	·05	·01

The impurities do not prove the origin of the tin, but none of the elements are inconsistent with a Cornish origin. The present writer must agree with Worth's sentiment quoted at the head of this section. The question now to be asked is whether Cornwall, or Ireland itself, derived any benefit from Irish gold in antiquity. The reason for this apparent digression in a book on tin will become clear.

If Irish tin is as elusive as the proverbial needle in a haystack, few archaeologists have doubted that bronze age Ireland was an important gold producer. The four early bronze age gold torcs from Cornwall, the two from Harlyn Bay accompanied by a flat axe formerly regarded as of Irish manufacture, were seen as Irish gold exports exchanged for Cornish tin.[24] It was George Coffey in 1913 who described Ireland as 'a kind of El Dorado of the Western World',[25] believing that the metal was obtained from Co. Wicklow. Recent analyses have upset this neat picture. Reeves published 12 locations of proven and supposed gold placers in Ireland.[26] These need not be detailed here.

Apart from a minor occurrence 'near the Kilkeel River in the heart of the Mourne Mountains',[27] the only location for which there is any evidence of profitable gold recovery is the Gold Mines River in Co. Wicklow. The earliest prospecting here was about 1770 by a local schoolmaster, while about 1785 a boy, John Byrne, found while fishing a piece of gold weighing 7 g which was sold in Dublin. Between then and 1795 a Dublin jeweller is said to have bought from an unknown person 100–150 g gold each year. The 'gold rush' began about September 1795 when local peasants began working at Ballinvally until the government took possession about six weeks later and guarded it with soldiers. Mining was conducted under the direction of 'some local gentlemen of mining repute' with apparently little profit until a rebellion in 1798 interrupted operations. Reworking continued in 1801 and 1802. The amount recovered was estimated to have been 22 kg by the peasants in the initial strike, 27 kg by the government up to 1802, and 2·4 kg by the Carysfoot Mining Co. from 1857 to 1868.[28]

In 1858, Greg and Lettsom[29] suggested that working might be resumed 'not, it is true, in the hope of reaping an inordinate profit from them, but as a means of affording suitable occupation to the peasantry'. The gold was found in every grade from the minutest grains up to one lump recorded as 0·6 kg, with others of 200, 250 and 500 g. All the gold came from Ballinvally, Ballintemple, and Killahurler, in the same valley. Gold in the neighbouring valleys did not repay the expense of the search.[30]

There is no proof that the gold placers at Ballinvally were worked in antiquity. Even if they were, the recent analyses of ores and of Irish bronze age goldwork do not support the idea that much gold could have been recovered in antiquity. Hartmann[31] found the following impurities in Wicklow gold:

Sn%	Ag%	Cu%	Pb%	Hg%
n.d.	*c.* 8	·064	n.d.	·09
n.d.	*c.* 6	·05	n.d.	·01
n.d.	*c.* 6·5	·05	n.d.	·02
n.d.	*c.* 8	·04	·03	·04
n.d.	*c.* 7	·03	n.d.	·02

(n.d. = not detectable)

No tin was detected in the gold ores, yet the bronze age goldwork usually contained a minute quantity. In a few examples tin has not been detected, otherwise it ranges between 0·27% down to 0·004%. Of 183 analyses of gold bronze age objects from various parts of Britain (but not from Cornwall), only nine contained no detectable tin.[32] Taylor also lists 15 analyses of gold from Wales and Scotland. Apart from two from 'drift'

in Sutherland (Highland) with *c.* 0·000 5 and *c.* 0·000 4% tin, none contained detectable tin. Gold is a minor constituent of Cornish tin streams, but as discussed in Chapter 24, there is good reason to suppose that Cornish streamworks were an important supplier of gold in antiquity.

NOTES AND REFERENCES

1   WORTH, 1872, 10
2   KINAHAN, 1886, 207; JACKSON, 1978. The present writer does not agree with Jackson's conclusion that the Gold Mines River, Co. Wicklow, could have produced enough cassiterite to meet the modest demand of the Irish early bronze age
3   JACKSON, 1978
4   KINAHAN, 1886
5   COGHLAN, 1963; TYLECOTE, 1962, 23
6   MICHAEL COMYN, letter dated 17 Aug. 1922 to Prof. Cole of the Irish Geological Survey (per Dr T. A. Reilly in lit.)
7   HUNT, 1887, 468. John Taylor introduced Cornish mining techniques to Ireland in the 1830s (*see* Burt, 1973, 27)
8   GRIFFITHS, 1828
9   JACKSON, 1978, 122
10 SEYMOUR, 1902
11 BRIGGS, 1976
12 WHEATLEY, 1971
13 THOMAS, 1983. Avoca closed in 1962 'partly because the ore grade did not reach expectation'. It re-opened for copper in 1970 and operated with State assistance until 1982. The only mine still working in Ireland is at Tara, north-west of Dublin
14 GREG & LETTSOM, 1858
15 SIMOENS, 1921
16 PHILLIPS & LOUIS, 1896, 312; FAWNS, 1905, 154
17 MILLS, KING & WEAVER, 1801, 147
18 WEAVER, 1818
19 SMYTH, 1853; the account also appears in Hunt, 1887
20 KINAHAN, 1886, 207
21 O'BRIEN, 'The Future of Non-ferrous Mining in Ireland', in *IMM*, 1959, 15. Note that O'Brien makes no mention of tin at Woodenbridge
22 GALLAGHER *et al.*, 1971, 153–5 for Carn Chuinneag
23 TAYLOR, 1980, plate 35 *d, f*
24 FOX, 1952, 49
25 COFFEY, 1913, 46
26 REEVES, 1971
27 GREG & LETTSOM, 1858
28 REEVES, 1971; HUNT, 1887, 471
29 GREG & LETTSOM, 1858, 239
30 HUNT, 1887, 471
31 HARTMANN, 1970, 24
32 JOAN TAYLOR, 1980

PART 3

SOUTH-WEST ENGLAND
(Cornwall, Devon and the Isles of Scilly)

As for the commodityes of England, and its chiefe richyes looked after by strangers, the chiefe and first is cloth, which maketh all Europe almost England's servant, and weare our liveray. The next is our tin or pewter, which is so excellent in Cornwall that it's only not sylver.

H. Belasyse, *An English Traveller's First Curiosity*, 1657

According to what I have heard from experienced men, the best and the most abundant [tin] that is found in the provinces of Europe is that which is mined in England.

Vannoccio Biringuccio, *Pirotechnia*, 1540

19 Tin in Devonshire

Before turning to Cornwall, the main area of tin production in British prehistory, it is convenient to examine the evidence for early exploitation in the peripheral areas of Devonshire and the Isles of Scilly.

Cassiterite occurs in several areas of Dartmoor and westwards to the Tamar valley. In south Dartmoor, for example, there is a scattering of unimportant tin mines south of the Tavistock–Ashburton road that exhibit the character of the deeper parts of the lodes – in short, the remnant roots of a tin zone steadily eroded away since Permian times. The last tin mine on Dartmoor to work on a commercial scale – Hen Roost near Hexworthy – closed in 1916, though small workings 5

km north of Postbridge carried on into the 1930s.[1] Records of production, though incomplete, amount to little over 300 t of black tin (tin ready for the smelter), but unrecorded alluvial tin must have exceeded that amount.

From a prehistoric point of view Dartmoor is rich in tin. There is no reason why tin streaming, or the opencast working of outcrops, could not have played an important part in the economy of the Moor, intermittently if not consistently, from 2000 BC onwards. The casual discovery of a number of moulds proves the presence of a bronze age metallurgical industry in Devon. Among these are two bipartite

115

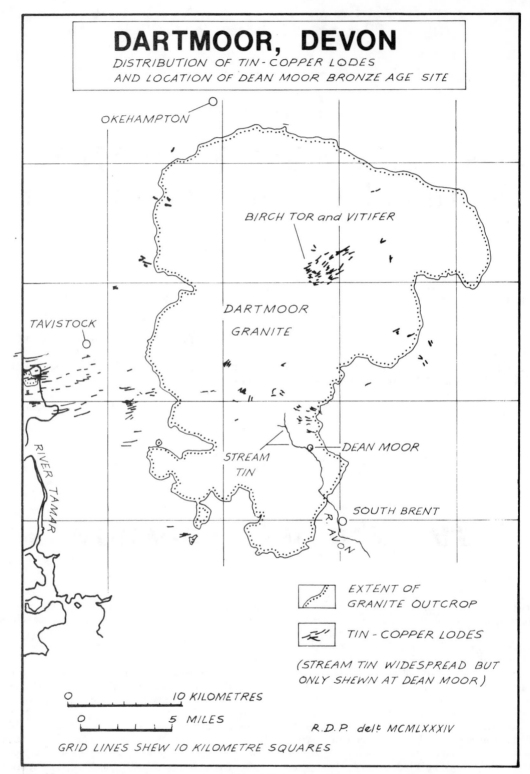

DARTMOOR, DEVON

*DISTRIBUTION OF TIN-COPPER LODES
AND LOCATION OF DEAN MOOR BRONZE AGE SITE*

OKEHAMPTON

BIRCH TOR and VITIFER

DARTMOOR

GRANITE

TAVISTOCK

DEAN MOOR

STREAM
TIN

RIVER TAMAR

SOUTH BRENT

R. AVON

EXTENT OF
GRANITE OUTCROP

TIN-COPPER LODES

*(STREAM TIN WIDESPREAD BUT
ONLY SHEWN AT DEAN MOOR)*

0 _____ 10 KILOMETRES

0 _____ 5 MILES

R.D.P. delt MCMLXXXIV

GRID LINES SHEW 10 KILOMETRE SQUARES

Map 22

45   Tin stream deposits at Huntingdon Cross, Dartmoor (SX 665 662). Mounds of worked-over ground, mostly 1–2 m high, consist of tourmalinized granite and quartz in both angular and rounded fragments; the dark patches are heather tufts growing on top of the mounds (photo Penhallurick, 14 iv 1984)

moulds for casting rapiers found near Chudleigh, one of which also contains the matrix for a ribbed ornamental strip.[2] This and similar finds do not prove local exploitation of tin, but it is hard to believe that local deposits remained untouched throughout pre-history. Six pebbles of stream tin have recently been excavated at the important prehistoric and Romano-British promontory site at Mount Batten, Plymouth. Unfortunately, it is impossible to say if they indicate transport of ore at a mediaeval or earlier period.[3]

It has long been assumed that some of the 'old men's workings' on Dartmoor were prehistoric, as Bray wrote to Robert Southey on 10 April 1832:[4]

On the Moor, also, are several ancient trackways together with stream-works of very high antiquity. Mr Bray is disposed to consider them of the same date with the Druidical remains.

Mediaeval and later workings may well have destroyed prehistoric evidence without trace. Bray says as much in the same letter:

I have heard (though I have never been fortunate enough to see any) that in breaking into old mines in this neighbourhood, heads of axes and other antiquities made of flint have been found...

Support for prehistoric working comes from the Dean Moor excavations in 1954–6.[5] Hut 5B, the eastern unit of a double-hut, was cobbled throughout with small pieces of granite. Among the items trodden into the ground close to the off-centre hearth was a pebble of tinstone measuring about 38 × 44 × 25 mm. All its edges were well rounded, and the specimen had clearly come from a tin-bearing gravel (i.e. a tin-stream deposit). It is rich in tin, consisting largely of granular cassiterite mixed with some quartz and decomposed feldspar, so its specific gravity is only 5·22 compared with *c.* 7 for pure cassiterite. From hut 7 came a small, almost spherical globule of smelted tin metal, about 3·2 mm in diameter, slightly vesicular and typical of smelted tin. Beads of metallic tin known as 'prills' occur inside tin slag and are revealed when the slag is broken open. Worth described such a slag of mediaeval date from Yellowmead, Sheeptor on

Dartmoor.[6] The Dean Moor finds are considered to be later rather than earlier in the bronze age.

Stream tin was plentiful in the Dean Moor area. Old surface workings occur up the Brockhill valley where a late mediaeval 'blowing house' (i.e. smelting house) was excavated shortly before Dean Prior.[7] To the west of the prehistoric settlement, where the Western Wella Brook joins the River Avon at Huntingdon Cross, both valleys show the characteristic hummocks of tin streaming (Fig. 45). These are generally regarded as mediaeval, but they lie less than 2 km from the bronze age settlement, so who is to say that the mediaeval tinners did not work where far earlier generations of 'old men' had streamed? Tin lodes were also worked to a shallow depth in the district, notably at Huntingdon Mine on the site of shallow, open-cast pits. Open-cast workings are also on tin lodes at Wella Brook, and known as the Wella Brook 'Girt' or 'Gert'.

Wedged into the wall of hut 2 of the bronze age settlement at Dean Moor were 23 kg of iron oxide, a mixture of massive haematite and its crystallized variety specularite. Haematite accompanies cassiterite at a number of localities, notably at Birch Tor and Vitifer where alluvials contained up to $3 \cdot 18$ kg t^{-1}, 'and the selected coarse lode material 4 lbs to 11 lbs a ton $[1 \cdot 8 - 5$ kg t$^{-1}]$'.[8] In 1796, Charles Hatchett described in his diary how 'antient channels' had been cut here by tin streamers.[9] The origin of the haematite found at the Dean Moor site is not known, but it may well have been discovered by someone searching for tin in outcropping lodes. Specular haematite is an attractive material, which looks as though something useful could be done with it. Only experiment at Dean Moor would have proved the inability to smelt it with existing technology, though there is no evidence that this was attempted.

A tin ingot of unknown date, but perhaps prehistoric, was reported by Edmund Pearce of Tavistock in a letter to Mrs Bray on 5 October 1835.[10] It was found *c.* 1832 in the ruins of an old smelting site 'near the confluence of the East and West Dart'. He considered it 'the most ancient in existence ... the surface of this block betraying marks of great antiquity'. Unfortunately the ingot does not survive. The only extant Devonshire one was dug up in the grounds of Slade House near Goodamoor in July 1879. Weighing $23 \cdot 35$ kg, it measured 36×20 cm on the upper surface tapering to 28×18 cm in its $7 \cdot 6$ cm depth.[11] Such a regular shape indicates a late date, even up to the 17th or 18th century.

Not far from Slade Hall, at the Shaugh Moor settlement, were excavated seven faience beads. Like other British bronze age beads of this type, the impurities are high in tin and copper. Those from Shaugh Moor also contain traces of lead and zinc. Zinc is uncommon but occurs locally in china-clay samples.

It is thought likely that the beads were made from local china-clay.[12] It is unlikely that china-clay was exploited in the bronze age for its own sake. More probably it was experimented with following its discovery while prospecting for and exploiting tin veins, which are worked with relative ease in shallow open-works where the granite has been decomposed by kaolinization.

Thomas Creber of Plympton St Mary informed John Webster (*Metallographia*, 1672)[13] that:

> Another place they call Armed Pit which holds ore they call *Zill Tin*, which is as small as grit or sand, and neadeth nothing but washing, and is the most easily melted of all sorts of Tin Ore, and lieth in chalk and clay; and this small Ore, because it is rich, they call it fatty Ore.

'Chalk' cannot refer to any Cretaceous rock but describes the softness and colour of kaolin or china clay. Working from this fact, Worth was able to equate Craber's description with an area at the source of the Redlake, a headwater of the River Erme where extensive tin workings are clearly traceable.[14] The excavated gullies (mapped on Worth's plate VII) Worth found by his own experiment to be easily worked to a shallow depth, though at more than 18–22 m heavy timbering was required in the soft ground. As Creber had said, 'The tin occurs in bunches near the lode, in the china clay, in fine to medium grains of great purity ... The soft, thoroughly decomposed rock needs but washing.'

The historic working of Dartmoor tin begins in the second half of the 12th century when for about 50 years known production exceeded that of Cornwall. There is no proof of Saxon working. Local belief that the pennies minted at Lydford are of pure tin is erroneous and based on the misidentification of the very fine silver pennies displayed in the public house next to Lydford Castle. Aethelred II (978–1016) set up a mint in the Castle towards the end of the 10th century.

NOTES AND REFERENCES

1   SHORTER *et al.*, 1969, 166
2   FOX (ed.), 1977, 114 and Plate 14b
3   BROWN & HUGO, 1983
4   BRAY, 1836
5   FOX, 1957
6   WORTH, 1914
7   PARSONS, 1956
8   DINES, 1956, **2**, 724
9   RAISTRICK, 1967, 21–2
10 BRAY, 1836
11 ROWE, 1896, 175
12 PEEK, R. A. & WARREN, S. E., 1979, 26–7
13 BURNARD, 1888–9
14 WORTH, 1914

20 Tin in the Isles of Scilly

> Seeing they have veins of Tinne, as no other Iland hath beside them in this tract, and considering that two of the lesser sort, to wit *Min an Witham* [Menawethan] and *Minuisisand* [Illiswilgig?] may seem to have taken their names of *Mines*, I would rather think these to be CASSITERIDES...
>
> Camden, *Britannia* (1636, 227)

The Revd Richard Polwhele augmented Camden's view so enthusiastically in his *History of Cornwall*[1] that he regarded the Isles of Scilly as the principal source of tin throughout antiquity to the exclusion of the Cornish mainland. The Greeks and the Phoenicians shared in the islands' wealth – of lead as well as tin – while Julius Caesar's legate Publius Crassus crossed to the islands and taught the inhabitants to sink perpendicular shafts, thus bringing lode mining 'for the first time' to Britain. In a footnote he poured scorn on William Pryce's wholly sensible assumption that the earliest tin workings were alluvial and in Cornwall, countering with his own view that the '*Koffens*' (coffins) on Tresco proved that working for tin in lodes had 'begun in Sulley, *about the very period of the Incarnation itself*'. Had Polwhele directed his fertile mind to the geology of Scilly instead of misinterpreting the texts of classical and other authors, his conclusions would have been considerably less fanciful.

A study of Scillonian place-names by Thomas[2] has established on a firmer footing the theory that all the islands except St Agnes – or just Agnes (from old Cornish *ek enes*, 'off island'), as he insists it should be – comprised a single island, not just throughout prehistory but as late as mediaeval times. The suggestion is supported by various classical writers. Sulpicius Severus writes of *Sylina Insula* as the island to which the heretic Instantius was banished in 387. Similarly in Solinus (*fl.* AD 200) there is the following description of Scilly in the singular, which is also instructive for its total lack of reference to minerals:[3]

> A rough strait also separates the island of *Silura* from the shore which the British tribe of the *Dumnonii* occupy. The inhabitants of this island preserve the ancient customs; they refuse money, give and accept things, obtain their necessities by exchange rather than by purchase, are zealous in their worship of the gods, and both men and women display a knowledge of the future.

It has been claimed, therefore, that tin lodes exist and are now hidden beneath the waters of St Mary's Sound. Such a view is naïve and not supported by any evidence in the exposed granite islands and hundreds of rocks that comprise the archipelago. Cassiterite has been found in the Isles of Scilly; it would be surprising if it had not, but the quantities are so meagre that it cannot be assumed that tin played any significant part in the history of the islands at any time. Hope springs eternal, and several attempts have been made to work tin. An early and particularly informative account is provided by Turner in a MS dated 1695.[4] He was informed by 'a person of integrity' that:

> There is not any Tin gott in these Islands, nor can he learne that there ever was any gott heretofore. He hath carefully view'd and searched all these Islands but found no Pits, nor other signs of mineing, besides only one Pitt, which was sunk 40 yeares agoe by command of Gr. Fr. Godolphin, who finding only one inconsiderable vein of Tin, he desisted. These islands therefore yielding, nor as far as appears ever having yielded any of the mettall, they cannot be the Insulae Cassiterides of the Ancients, tho Dr Haylin & others have been of the opinion that they were, these islands borrowing their name from CASSITEGOS, by which name the Greeks intended Tin, some whereof is found here.

Turner does not say where Godolphin's tin pit was, but it may well have been one of those seen in the following century by the Revd Borlase on Tresco. It also suggests that the other pits in the same area of Tresco were sunk between 1695 and 1754. In 1750, Heath[5] writes of the interest in Scillonian tin and other metals, having been told the islands afforded lead and copper as well. Heath was a gullible victim of miners' tales which are no wit less extraordinary than those of fishermen. According to Heath:

> The Tin is discoverable by the Banks next the Sea, where the Marks of the Ore, in some Places, are visible upon the Surface: this I was assured by some very considerable *Cornish* tinners, in the Year 1744, who desired me to make Representation thereof to the present Proprietor, for obtaining his Lordship's Consent for the working of Tin and other Metals in *Scilly*, wherein they propos'd a certain Share to his Lordship free from Expenses; but I did not then succeed.

46   Tinners' trial trench at the north end of Tresco, Isles of Scilly, probably dug in the late 17th or early 18th century (photo Gibson, late 19th century)

Evidence for tin in Hughtown, St Mary's, is given by Heath[5] where he discusses the town well, 'opposite to the landing place':

> When the Rubbish with which it was filled up was removed, the Miner discovered a rich Vein of Tin Ore, which promised Encouragement for working it as a Tin-Work; but there being none to undertake it, the Well was cased up with Pieces of Rock-stone.

When the Revd William Borlase made his historic visit to the islands four years after the publication of Heath's book, no one knew anything about the tin lode at the town well, so it could hardly have been a memorable find.

Borlase found evidence of a search for tin at the north end of Tresco (Fig. 46),[6] a well known tin locality.[7] In the summer of 1979, Charles Smith, geologist at the Geevor tin mine in St Just, found minute specks of cassiterite in a vein at Piper's Hole. This part of the island contains numerous east–west quartz veinlets rich in chlorite and tourmaline,

minerals known to the Cornish miner as 'peach' and recognized as good indicators of possible metallic mineralization. Here Borlase came across:

> a row of shallow Tin-pits, none appearing to be more than four fathoms deep, most of them no deeper than what the Tinners call *Costean* shafts, which are only six or eight feet perpendicular ... This course of Tin bears East and West nearly, as our Loads, or Tin Veins, do in *Cornwall*. These are the only Tin Pits which we saw, or are anywhere to be seen, as we were informed in these islands.

While it is true that some of the Tresco veinlets contain tin, it is clear from the very shallow nature of the 'Costean shafts' or trial pits that they contained nothing of any importance.

Perhaps as a result of Borlase's account in a book that commanded considerable respect in his own day, or perhaps from a rereading of Heath's story, further exploration for tin took place on St Mary's in the 1790s. At all events, Troutbeck records[8] that on the outer defence slope, or 'glacis', of the Garrison:

is an old tin pit wherein some miners were lately employed, but as they could not raise ore, of a quality and a quantity sufficient to defray the expense, they are discharged. This pit is left open and is very convenient for catching a drunken soldier reeling from a gin shop to his barrack in a dark night if he should stagger down the hill from the road. Close by this tin pit is a hill called Mount Hollis.

This enterprise near the Garrison entrance was named 'Wheal Hollis Tyne Work' (after the hillock) in an account in the Leases of the Duke of Leeds preserved at the County Record Office at Truro. Tangye gave an excellent summary of this abortive attempt at mining in 1791.[9] In that year, two miners, one a Cornish mine captain named Trestrail, made a thorough search for tin over a period of a year. Mount Hollis was the largest undertaking, being 11 m long and 9 m deep, and involved considerable effort.

In 1791, Andrew Harris, a smith, was paid to sharpen 'The Tynner Tools at Wheal Holly Tyne Work', and also to make 'iron work to a new kibble' for raising ore. The 'Account of Expenses Trying the Tin load at Mount Hollis, in Scilly by order of Geor. Brooks Esq., between 10th Sept., 1791 and 13th Jan., 1792', show the purchase of 'Cordage, candles, pick hilves, powder, gadds'. From 26 September to 23 October 1791, were purchased '10 lbs of gunpowder at 1/4 for the tin work', which indicates the intractable nature of the ground being worked.

For two mens eating 3wks. by order of Mr Carn	£2.2.9.
For two mens Beer 3 weeks by order of Mr Carn	£0.7.0.
For two mens Rum for 3 weeks by order of Mr Carn	£0.1.6.

A specimen of tin was eventually obtained and conveyed to St Michael's Mount.

Paid to men taking out the Tin	£0.4.0.
Freight of Tin – Sam¹ Tregarthen	£1.1.0.
To paid shipping the stuff on board the Sloop Friendship. Keeleg to the Mount and paid the Pilot expenses	£0.5.0.
Feb. 4th, 1792. A horse to Thomas to Germoe to acquaint Capt. Phillips of the arrival of the Tin	£0.1.6.

A full report of the venture survives in a letter of 6 May 1792 from John Vivian of Truro to Thomas Williams Esq:

> Capt. Trestrail and his companion are returned from Scilly. The place from whence the stone of Tin which was sent to you, was broken, is near the Town of St Mary's – the ground has been tried to the satisfaction of miners heretofore:
>
> It has been opened about 6 fathoms in length, and 5 in depth, there is a regular load spotted with Tin but in so small a degree that they could not break a single stone of tin to bring home with them.
>
> Capt. Trestrail says he is confident that according to the present appearance, there is not Tin enough in the load to pay for breaking.

The report is full of good sense; it mentions that no lode of promise had been found anywhere in the islands. If such a lode had been found, it would not have been practical to extract it. No stream of water exists for the working of stamps to crush the ore, the granite was extremely hard and the expense and distance involved added to the disadvantages. It was decided that there was 'no encouragement to lay out a shilling in mining in Scilly'. Such a conclusion is the only possible one for would-be prehistoric tinners.

Gibson drew attention[10] to the curious millstone cut out of an exposed outcrop of granite on the hilltop between Old Grimsby and the Abbey Pool, Tresco. It still survives, though in a more ruinous state than when Gibson photographed it *c.* 1885. He thought it might have been for pulverizing tin, even though the islanders called it, as they still do, a cider mill. Worth[11] had no hesitation in calling it a cider mill, but his assumption that it was old enough to be connected with Tresco Abbey cannot be proved. The mill's edge-runner stone survives, built into a building wall lower down the hill.

Those who have sought to equate the Cassiterides with the Isles of Scilly clutch at straws, the most substantial of which is a small tin ingot. In 1948, below the high-tide line on Par Beach, Highertown Bay, St Martin's, Lewis found the remains of a circular hut that contained 'Roman pottery of the third or fourth centuries AD, and pieces of cassiterite or tin ore.'[12] What survives at the County Museum, Truro, is an ingot of smelted tin suffering badly from corrosion (Fig. 47), accompanied by two joining fragments of a pot with everted rim typical of the Cornish iron age and Romano-British period. Also found with it was a small pebble typical of those found in Cornish tin streams, though recent analysis has shown that it contained only 0·001% tin. Not all pebbles from the 'tin ground' in Cornish tin stream deposits are rich in tin.

The Par Beach site lies only a kilometre from the tiny island of Nornour, where a site has yielded nearly 300 brooches, 30 finger rings, some half-a-dozen bracelets, and a variety of other ornamental bronze items, all of Roman date.[13] All the bronzes were assumed to have been the products of a local workshop, but recent research indicates that they were all imported to satisfy the needs of a Roman shrine established at a safe anchorage in what was, at that time, a sheltered bay between the present islands of Nornour and St Martin's.[14] It is not possible to say whether or not the Par Beach tin ingot was connected with some industry established at this safe anchorage; what is clear is that the tin must have been imported from Cornwall.

The Isles of Scilly can have played no part in the prehistoric tin trade other than acting as an anchorage for trading ships, either by design or to escape the

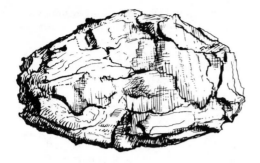

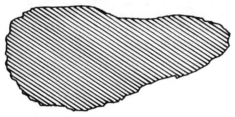

47   Much corroded tin ingot (max. length 7·8 cm) from St Martin's

worst of the Atlantic weather. This can change in mood with great rapidity in the dangerous waters of the western approaches to the English and Bristol Channels.

NOTES AND REFERENCES

1   POLWHELE, 1803, **3** (supplement), 50–9
2   FOWLER & THOMAS, 1979
3   RIVET & SMITH, 1979, 85
4   ANON., 1964; for an earlier speculative attempt at working tin on Scilly, note an indenture dated 1563 between Martin Dare, John Elliot, and Roger Carew, referred to by Bowley, 1957 (4th ed.), 29
5   HEATH, 1750, 25, 78
6   BORLASE, 1756, 45, 71
7   BARROW & TEALL, 1906, 10–11
8   TROUTBECK, 1796, 53
9   TANGYE, 1970, Appendix 67–8
10   GIBSON, 1885
11   WORTH, 1953, 382
12   O'NEIL, 1949
13   DUDLEY, 1967
14   THOMAS, 1965, esp. 163–72

21   The Phoenician myth

To the astonishment of Cornishmen, especially of those best acquainted with the subject, it was, at the Truro meeting of the Cambrian Archaeological Association, in August last, gravely questioned whether the Phoenicians ever visited Cornwall.

Edmonds, *J. Royal Inst. Cornwall*, May 1863

Two myths concerning British prehistory tenaciously maintain their hold on the popular imagination. The first is that Stonehenge was built by the Druids. The second is that the Phoenicians came to Cornwall for tin, which they supplied to all the ancient civilizations of the Mediterranean and the Near East (Fig. 48). The story is so firmly embedded in the south-west that it is commonly set down as a general introductory remark to prove the antiquity of Cornish mining in otherwise authoritative books dealing with recent aspects of Cornish economic history. The myth has also found its way to Brittany,[1] doubtless as a recent Cornish export.

In the past no source of tin was known other than Cornish tin, so William Pryce can be forgiven for writing in the Introduction to his *Mineralogia Cornubiensis* (1778):

Tin in its Mineral state, being totally unknown to all other countries but our own, affords ample reason to assert, that we supplied all the markets of Europe and Asia with that commodity in early ages.

For Pryce the 'few Tin Mines in Germany' were not discovered until the time of King John. Even as late as 1848, when French and Spanish tin deposits were well known, Phillips could entertain the same idea – that Cornwall and 'the Asiatic Isles' were 'almost the only sources of Tin of the antient world'.[2]

That no Phoenician object has ever been found in Britain does not deter the supporters of the voyagers from the Levantine coast. Belief does not stop at accepting Tartessos in southern Spain (perhaps the Biblical Tarshish) as the end of the trade route from Cornwall, but continues to the very ports of Tyre and Sidon. Popular accounts are poetic in their enthusiasm:[3]

A Day on St Michael's Mount . . . here we may get visions of an all but forgotten world. We may gain peeps into shadowy Tyre and hazy Phoenicia. From this legendary resting place of the Archangel Michael we may again

watch the Trojan galley-fleets, as they skim, with swan-like grace, the blue expanse called Mount's Bay, and drop anchor one by one at Iktin, 'the hill-tin-port' of the Old World.

It was nothing for the Cornish, indeed for the British, to accept as gospel truth that their precious tin must have been used in the manufacture of Achilles' shield and helmet, in the construction of the Tabernacle of the Israelites, and above all in the temple of Solomon. Two mines in particular are credited with sending tin to Solomon: Ding Dong in the middle of the Land's End peninsula, and Great Work near Godolphin.[4] The following whimsical passage must surely have been written with tongue in cheek, but belief in the Phoenician presence has been so strong that it is no more than a quaint elaboration of the supposed truth:[5]

When St Ives first became a resort for Artists is not known. Probably the Phoenicians, who seem to have been at the bottom of everything in Cornwall, founded an art colony there, and took back with them, with their tin and copper, impressionist sketches of St Ives Bay and the surrounding country.

Perhaps it was Cornwall's strong attachment to Methodism which made its inhabitants yearn for some direct link with the biblical lands. And who better to seek help from than Solomon and Christ? The Phoenician ruler Hiram I of Tyre (970–936 BC) made a treaty with Solomon allowing Phoenicians to use the port of Ezion Geber on the Red Sea, and permitting his craftsmen to work in the temple of Jerusalem. As a result, the temple exhibited Phoenician motifs and used bronze, the tin for which naturally must have come from Cornwall. That Phoenician involvement beyond the Pillars of Hercules cannot have existed much before 800 BC counts for nought.

Joseph of Arimathea is described in the Bible as a wealthy man and a councillor. He is popularly believed to have been a merchant who traded in Cornish tin, making his journeys in Phoenician ships. The Jews were supposedly regular voyagers to Britain with the Phoenicians from ports such as Tyre and Sidon. Joseph, it is conjectured, brought the young Jesus with him on one of his voyages, at a time when

48   'Phoenicians bartering with Britons' by Sir Frederick Leighton, 1894 (photo *The Guildhall Library*)

the Bible is silent about the events of his boyhood. The adventurous young Jesus did not confine his peregrinations to Cornwall, but made his way to Glastonbury.[6] Baring-Gould gave a possible explanation of the legend:[7]

> Another Cornish story is to the effect that Joseph of Arimathea came in a boat to Cornwall, and brought the child Jesus with him, and the latter taught him how to extract the tin and purge it of its wolfram. This story possibly grew out of the fact that the Jews under the Angevin kings farmed the tin of Cornwall. When the tin is flashed [in smelting], then the tinner shouts 'Joseph was in the tin trade', which is possibly a corruption of 'St Joseph to the tinner's aid'.

The Jewish connection with tin has long been part of Cornish folklore. Richard Carew, who seems to have been ignorant of the Phoenician links forged by some of his contemporaries, faithfully recorded in his *Survey of Cornwall* (1602) that the Elizabethan tin streamer:

> maintains these works to have been very ancient and first wrought by Jews with pickaxes of holm, box, and hartshorn; they prove this by the name of those places yet enduring; to wit, Attall Sarazin, in English, the Jews' Offcast, and by those tools daily found amongst the rubble of such works … There are also taken up in such works certain little tools' heads of brass which some term thunder axes, but they make small show of any profitable use. Neither were the Romans ignorant of this trade, as may appear by a brass coin of Domitian's found in one of these works and fallen into my hands; and perhaps under one of those Flavians the Jewish workmen made here their first arrival.

Almost the same account appears in John Norden's *Speculi Britanniae* (*c.* 1584), as he obtained much of his Cornish information from Carew. Carew's 'attall' is Cornish *atal* (rubbish or mine waste). Until quite recently tin ingots dug up in the county were invariably referred to as 'Jews' House tin'. The Jewish link is based on the historical fact that the Jews were involved in mining and smelting, especially following the Norman invasion and until their expulsion from England in 1290.

In 1853 a seated figurine made of tin with a little zinc was dug up 3 m below the surface of Bodwen Moor, Lanlivery, near the site of an old smelting house or 'Jews' House' (Fig. 49). The image is difficult to date but has been ascribed to the period of Richard Earl of Cornwall and King of the Romans (AD 1209–72). It certainly resembles the seated king in the chess set from Uig, Isle of Lewis, found in 1831 and dated to the mid or late 12th century, so could well belong to the 12th or 13th centuries. Richard was granted the wealthy earldom of Cornwall on 13 February 1225, and in November he obtained control of the tin mines held by his mother Queen Isabella. Richard controlled the coinage of tin very tightly in order to finance his high-sounding title.

When found, the tin figure wore a simple crown which has since become detached and lost, but it still bears Hebrew characters of uncertain meaning, which have been interpreted as signifying 'Rapacious Eagle, and 'Jehovah is our King'. The eagle, derived from Roman standards, is an emblem associated with Richard. It is conceivable that some mediaeval Jewish merchant, groaning under the latest of Richard's impositions, made this caricature of the 'Rapacious Eagle' as a final gesture of defiance, asserting that Jehovah, not Richard, was his true King.

Unlike the Jews, the Phoenicians are not part of Cornish folklore. They were simply one of several peoples used before archaeology developed to fill the empty void in British history between the Creation of the World (4004 BC in Archbishop Ussher's chronology of 1650) and the arrival of Julius Caesar in 55 BC.

In his 9th century *Historia Brittonum*, Nennius places Brutus, Trojan grandson of Aeneas, as the first settler in Britain. This idea was enhanced by Geoffrey of Monmouth in his *Historia Regnum Britanniae* (*c.* 1147) when he makes Brutus land at Totnes in Devon in 1170 BC. All sorts of people known from biblical and classical sources were invoked to fill our short prehistory. Those who had no time for Trojans, or the lost tribes of Israel, preferred the Druids or the Phoenicians. The Phoenician link was first proposed by John Twynne (1501?–1581), master of the free grammar school at Canterbury. In 1590 his *De Rebus Albionicis Britannicis* was printed by his son Thomas. In it, as Daniel tells us:[8]

> He argued that the Welsh word *caer* was Phoenician, that Welsh coracles were Phoenician; so too was the dark blood of the Silures – even the dress of the Welsh women in the sixteenth century was … a survival of an old Phoenician form of dress – even though it was of course a provincial survival of English court dress.

The same line of thought was developed by Samuel Bochart in his *Geographica Sacra* (1646); in his opinion the name Britannia was first applied by the Phoenicians, in whose language *Bartanac* signified 'the land of tin'. The pre-Roman inhabitants of Britain were certainly not unified enough to call themselves British or their country Britain. Bochart was astute in seeing that the name was most likely given by foreigners. His suggestion was followed up by Edmonds in an article published in 1871 entitled 'The name of Britain and the Phoenicians'.[9]

A truly splendid monument of Phoenician fancy was erected in the 19th century by Joachimus Laurentius Villaneuva. His Latin text was painstakingly translated by O'Brien in 1833, for whom everything Irish was Phoenician.[10] Even the Dumnonii of south-west England were, like the Damnii of Ireland, 'Spanish Phoenicians'. The Scots, to break the monotony, were 'Scythians, a people of northern Asia'.

49   **Hollow tin figurine; height 14·2 cm (courtesy County Museum, Truro)**

His antediluvian views were summed up on pages 47–8. The terms 'Phoenician' and 'natives' were, as regards Ireland, 'convertible terms', and in a footnote he states:

> It is more than probable that Ireland remained desert and uninhabited from the creation to the deluge. No history, not even that of Moses, offers anything which can lead us to suppose, that before the universal deluge, men had discovered the secret of passing from one country to another that was separated by water. The ark, which was constructed by orders of God himself, and which served to preserve man on the watery element, is the first vessel of which we have any knowledge.

Elsewhere he derives the name 'Druids' from a Phoenician word 'dor-ida . . . a progeny of wise men or benefactors'. The abundance of tin in Britain gave rise to the name of the country, 'being compounded of Bruit, "tin", and Tan, "country", corresponding to "Cassiterides", the mercantile name given by the Phoenicians to both Ireland and England'. Later he alters his etymology to 'Bara anac, Phoenician for *land of tin*, which gave rise to the name Britannia.'

Etymology is a quagmire non-philologists enter at their peril. The name 'Britain' is not certainly understood, and I can do no more than paraphrase the observations of modern authorities. An original pre-Roman Celtic name *Pretani* is envisaged, probably used by the Gauls when speaking of the islanders to Greek traders from Massalia (Marseille), some of whom were certainly interested in the tin trade. Greek πρετται had its initial P changed to B when transliterating into the Roman alphabet. The currently accepted meaning is that *Pretani* meant 'figured folk' or 'tattooed folk', in deference to their practice of decorating themselves with woad.[11] John Twynne's Phoenician connection was elaborated by Aylatt Sammes whose *Britannia Antiqua Illustrata* (1676) is subtitled 'Antiquities of Ancient Britain derived from the Phoenicians'. Sammes took the names 'Erth' in 'Meneg' (i.e. the Meneage district of the Lizard peninsula, though the location of his Erth is uncertain) as Phoenician 'to favour his Hypothesis that this Part of *England* was peopled by the *Phoenicians* who traded hither, but these are uncertain Conjectures not to be depended on', as Cox wisely observed.[12] The first scholar to link the Phoenicians specifically with the tin trade was William Camden, whose *Britannia* was first published in 1586 and went into five further Latin editions before the English version appeared in 1610. Camden identified the Isles of Scilly with the Hesperides and the Cassiterides of ancient Greek writers, and translated a poem of Dionysius Periegetes, the *Orbis Descriptio* of *c.* AD 125, as:

> The Islands nam'd Hesperides do lie,
> And those well stor'd with Tin, a rich metall.
> But would ye know the people? then note well,
> The glorious wealthy Spaniards therein dwell.

However, as the last line tells us, the poem has nothing to do with the Isles of Scilly. *Hesperides* means the 'western land' (Greek ἕσπερος) and the name was given by the Greek poets of southern Italy to Spain, sometimes called *Ultima Hesperia* to distinguish it from *Hesperia Magna* or Italy.

The Phoenicians were not bad candidates to select as initiators of the Cornish tin trade; they were noted traders early in the 1st millennium BC. Homer described them in derogatory terms through the lips of Telemachus in *The Odyssey*:[13]

> One day the island was visited by a party of those notorious Phoenician sailors, greedy rogues, with a whole cargo of gew-gaws in their black ship.

In the reign of the Egyptian pharaoh Necho II (609–593 BC) and at his command, a Phoenician ship circumnavigated Africa. The story is related by Herodotus, who did not believe it for the very reason that modern scholars accept it: the sailors reported that as they rounded the southern end of Libya (South Africa), they 'had the sun on their right, to the north of them'.[14] Herodotus then says that the next people to make a similar journey were the Carthaginians, although Sataspes who was in command sailed in the opposite (anti-clockwise) direction and only reached as far as the region inhabited by pygmies. The publicity enjoyed by these momentous voyages shows they were exceptional and not commonplace wanderings.

Two other Phoenician explorers, Hanno and Himilco, are not mentioned by Herodotus who died in 425 BC, so their journeys may have been after that date. However, others place them earlier – Hanno not long before 517 BC, and Himilco 'within the time of Carthage's greatest flourishing, so before 480'.[15] Hanno turned his attention to west Africa and reached the mouth of the Niger, and perhaps the coast of Gabon.

Himilco is mentioned by Pliny, but what little survives of his journey is in a geographical work in verse, the *Ora Maritima* or 'Seacoasts', by the late 4th century AD Latin poet Rufus Festus Avienus.[16] The journey described is to Massalia (Marseille) from some point in the Atlantic beyond the Pillars of Hercules. The traditional interpretation of the Atlantic voyage is up the coast of Iberia, across the Bay of Biscay to the Oestrymnides (identified as islands off southern Brittany), past a 'storm-washed crag' (Ushant), arriving two days later at a holy island called Hierni, taken to be Ireland (Hibernia), near which dwelt the Albiones – presumably in Albion (the earliest name for England). This interpretation first proposed by Adolf Schulten seems watertight. Unfortunately for Schulten and later followers, insufficient attention was paid to the sailing times for the skin-covered coracles described in the *Ora*

Maritima. Recently Hawkes has scuttled this interpretation in a brilliant piece of reasoning.[17]

The *Ora Maritima* draws some of its content from the lost geography of Ephorus of Cynae (*fl.* 340 BC), but is largely based on material from before 500 BC when the Carthaginians blocked the Straits of Gibraltar to all but their own shipping. Ephorus, writing after the Carthaginian blockade and well aware of Cornish tin passing to Massalia across France, mistook Hierni for Ireland and the Oestrymnides for Breton Isles. In reality the *Ora Maritima* deals with a much shorter Atlantic coastline, no more than seven days' sailing in a skin-covered boat from the Mediterranean or 'Sardinian Sea'. All the places mentioned need not be placed further north than the Ophiussae peninsula, agreed by all interpreters of Avienus as being Cape Roco at the mouth of the river Tagus. Oestrymnis becomes the now silted-up mouth of the Guadalquivir – Avienus' description of it as 'great' being 'due to Ephorus' error in supposing it the whole Bay of Biscay'.

Pliny described a people called *Albiones* living on the Biscayan shore of Spain, and there is no reason why others of the same name did not live in southern Spain or Portugal, just as the *Cempsi* of the Pyrenees bear the same name as inhabitants near Cape Roco. The sacred isle *Hierni* must have been in the south, for Avienus placed it only two days' voyage from the Oestrymnian gulf – *soles* implying two daylight journeys broken by a night on land. Brittany to Ireland with bed-and-breakfast in west Cornwall is out of the question. The sailing days of the periplus 'in days, need fit nothing north of Cape Roco and the Tagus', an interpretation which very well fits the archaeological record because 'it is there that the distributions of Mediterranean finds begin to peter out'.

Nothing Phoenician has yet been found in northern Iberia, let alone in Cornwall or Ireland. About the most northerly Punic finds have been at the native settlement of Santo Olayo at the mouth of the Mondego river, about 110 km south of Porto, while at Alcácer do Sol, near the mouth of the river Sado 80 km south-east of Lisbon occurred pottery 'from some ill-documented graves' suggesting Phoenician burials there in the 7th century BC.[18]

The classical writer to whom supporters of the Phoenician–Cornish link have invariably turned is Strabo (*c.* 54 BC to *c.* AD 24), a native of Amasia, capital of Pontus in Asia Minor. He was a much travelled man whose great work on geography in 17 books survives almost intact. He refers several times to the Cassiterides, or Cattiterides (double S and double T being interchangeable in Greek), and the island of Bretannika, often spelled with an initial P. The two locations may be referred to in the same passages.

Until the end of the 18th century when the reality of Spanish tin became known to English scholars, and the early 19th century when the French tin deposits first attracted modern attention, the only tin source in western Europe was known to be in Devon and Cornwall. Therefore it was natural to assume that the Cassiterides lay near Britain, the Isles of Scilly, or referred to Cornwall itself. Doubts as to their identity could be interpreted as Strabo's hazy knowledge of the western shore of Europe, understandable for someone brought up on the shore of the Black Sea. His geography of the Atlantic was certainly imprecise. The Cassiterides cannot exist as tin-bearing islands, for there are none off Britain or off the Continent. They are as mythical as the Amber Isles (*Electrides Insulae*) drawn on a map by the 12th century Canon Heinrich of Mainz; these were not islands but the Baltic coast.

In the light of present knowledge, the most that can be said is that the Cassiterides refer to the tin-bearing regions of western Europe and Britain – a more precise source meaning different things to different people. For Strabo the Cassiterides were distinct from Britain, lying 'in the open sea opposite the Artabri [of Galicia], set approximately in the Brettanic latitude'. In book III he describes the islands in more detail in this famous passage:[19]

> The *Cattiterides* are ten in number and they lie close to one another in the open sea to the north of the harbour of the Artabri. One of them is uninhabited, but the rest are occupied by people in black cloaks, who are clad in tunics reaching to their feet, wear belts around their breasts and walk about with canes, like the Poenas in tragedies. They live off their herds, pastorally for the most part. Having mines of tin and lead, they exchange these and hides with merchants for pottery, salt, and bronze vessels. In former times only the Phoenicians carried on this trade, from Gades, concealing the route from everyone else. Once when the Romans were following a certain ship master so that they too could discover the markets, out of spite the ship master deliberately drove his vessel onto a shoal, bringing his pursuers to the same disaster and, having escaped on a piece of wreckage, he recovered from the State the value of the cargo which he had lost. But by trying many times the Romans learned all about the route, and when Publius Crassus had crossed over to them and had learnt that the metals were being dug from only a shallow depth, and that the people were peaceful, he provided plentiful information to those who wished to trade over this sea, although it is wider than that which separates *Prettanike* [Britain]. So much for Iberia and the islands lying near it.

Strabo always speaks of the Cassiterides in a Spanish context, and the story of the sea captain, whether true or not, refers to the Phoenicians' attempt to keep to themselves knowledge of the Iberian tin suppliers. Publius Crassus has been identified as the legate of Julius Caesar who visited Britain, but Strabo's story surely refers to the legate's grandfather with the same name who was consul in Spain in 97 BC, won a victory over the Lusitanians in 93, and then must have visited

50 Bronze bull from St Just-in-Penwith (sketch actual size)

the tin-rich regions of the Artabri in the north-west. That he is not Caesar's legate is supported by Caesar's own ignorance of British tin. In reporting its occurrence in his *De Bello Gallico* (Book V, 12), he assumed it was in the interior of the island, a statement he could not have made had one of his own legates been sent to find out more about it.

A Spanish location is also inferred by Pomponius Mela, a native of Spain in the 1st century AD, who in his own *Geography* places the Cassiterides before his description of the Breton Île de Sein (*Sena*), hence south of Brittany. Once the geography of western Europe became better understood, the Cassiterides disappeared from the literature. Further conjecture is unprofitable.

Many archaeological finds in Britain have been attributed to Phoenician traders. Those who followed in the footsteps of Aylett Sammes knew no bounds:[20]

> The bronze swords, daggers, and spear-heads which have neither a Greek nor a Roman type were probably first introduced by the Phoenician traders.

In Cornwall the mediaeval tin figure from Bodwen (Fig. 49, *above*) has been called Phoenician, but the one single item which has attracted most attention is the small bronze bull from St Just-in-Penwith (Fig. 50). This was described by Poole[21] as 'precisely what a Phoenician trader would have carried'. Buller[22] relates that it was discovered in January 1832 by John Lawry, a workman demolishing a stone or Cornish 'hedge' and 'trenching ground for planting' at the vicarage. He uncovered the foundations of an old building 'and from the quantity of ashes remaining, it was conjectured that the premises had been burnt; near this place he found a bronze figure of a bull . . . It was shown to some of the most learned antiquaries in London who pronounced it to be Phoenician'.

Buller was sceptical of this identification and thought it might be Greek. Taylor[23] also took a different view suggesting it belonged to a worshipper of Mithras who 'fixed his temporary abode in what is now St Just vicarage garden at no great distance from the tin streams which have been worked from time immemorial'. After consultation with Hall of the Egyptian department, Smith of the British Museum wrote to Truro Museum on 21 September 1929 stating that the bull was Egyptian of the Roman period, an 'Apis Bull, which could easily have reached Cornwall in Roman days in connection with the cult of Isis.'

More recent opinion is not enthusiastic about Hall's suggestion, according to letters in August 1981 from Andrews in the Department of Egyptian Antiquities. The Apis bull was not much worshipped outside Egypt, though it became confused with such cults as that of Mithras. Bronze Apis bulls, when not lying down, are usually sculpted striding out, and in Roeder's *Ägyptische Bronzefiguren* (1956) only one bull resembles the St Just one. It also stands with feet together and possesses a curiously thick and upright neck (Pl. 49, Fig. f), which is quite untypical and shows 'Greek influence'. Unfortunately, this bronze is without provenance, and Greek influence will only date it to 300 BC or later. At present it is not possible to say whether the St Just bull may have come from Egypt or elsewhere in the eastern Mediterranean, or whether, as Andrews suspects, it is of local, perhaps Cornish, manufacture. What is certain is that it is not Phoenician.

It is strange that George Smith failed to mention the bronze bull in his book *The Cassiterides* (1863). Like others who have envisaged Mount's Bay thronged with the ships of Tyre and Sidon, Smith relied on less tangible evidence. At least his contemporary Poole thought that by studying the weight of Cornish tin ingots (two of them known to be Roman) he could prove to his own satisfaction that the Phoenicians came to Cornwall. He did. Even so, the system of weights he deduced was Greek, the St Mawes ingot of 71·6 kg being two 'Attic Commercial talents', with other ingots proportionately less. Greek weights were explained away by the Phoenician use of weights 'most useful in the markets of the Mediterranean'.[24]

The search for Phoenician traces in names has been quite ingenious at times. The 19th century enthusiasts would have done well to heed the words of William Borlase in his *Antiquities of Cornwall* (1754):

> That the western parts of this island were first discovered by the Phoenicians, and by them inhabited, has no other foundation than that the names of places in these parts may be derived from Phoenician words, which is too deceitful a ground to build on, especially considering they may all be found in the British tongue . . .

Edmonds[25] was convinced that the word 'tin' was Phoenician, along with all the modern European words for the metal derived from it. In fact the origin is unknown. An explanation as good as any is that the word is British, surviving in Cornish as *sten*, passing into Latin as *stannum* which superceded the earlier 'white lead' *plumbum album*, sometimes *plumbum candidum*. Pliny confirmed the Latin meaning by giving it the Greek equivalent κασσίτερος, for which the origin is also uncertain. Whatever the Phoenicians of southern Iberia called tin, the basic Hebrew word was *bedil*.[26]

Edmonds then proposed that *Iktis* – or better *Iktin*, as it could have been – the name generally accepted as referring to St Michael's Mount, 'indicates the very port in our island from which the Phoenicians exported their tin'. Then taking the Cornish word *bre* (hill) he reconstructed an alternative name for St Michael's Mount – *Bretin*. Just as the English name has given its title to Mount's Bay, so he proposed the Cornish name became applied to the whole of Britain, which 'is nothing more than we should naturally have expected; for the Mount is the most strikingly beautiful object in Mount's Bay, and was better known to the Phoenician tin-traders than any other place in Britain'.

Finally, Edmonds suggested that *Belerion*, the name Diodorus Siculus used to describe Cornwall or, perhaps, just the Land's End peninsula, must have been given to the region first by the Phoenicians. This was in honour of the god Bel or Baal, 'identical with Apollo or the sun', the worship of whom 'survives to this day in the district of which Penzance is the capital' and where 'solistical festivals and midsummer fires' also survive.

Other claims have been delightfully wild and devoid of any explanation. A correspondent to the *Western Antiquary* in December 1882 had 'heard a vague legend about the Ninnis clan being descended from a Phoenician princess', while Bawden in the same journal of April 1883, noted that all British surnames ending in 'is' – Davis, Ellis, Morris, Harris, etc. – 'are believed to be Phoenician. Even those most famous of all Cornish names, *Tre* (dwelling), *Pol* (pool), and *Pen* (head), which prefix a great many Cornish surnames, have not escaped unscathed in the popular imagination. The explanatory text of a postcard of Phoenix United Mines near Liskeard, produced about 1905, proclaimed that the 'names of families commencing with Tre, Pol, and Pen, are evidence of Phoenicians visiting this neighbourhood.[27]

Etymological pitfalls are legion in Cornwall. Apparently straightforward English may have a Cornish pedigree; Camborne miners in the 19th century used to put a small clay image over the first set of timbers placed in a new level while uttering the words 'send for the merry curse and the priest'. This nonsensical formula only becomes intelligible if Morton Nance's suggestion is accepted, that it is a corruption of the Cornish '*Synt Meryasek ny a'th pys*' (Saint Maryasek, we pray thee). This is a plausible invocation for 16th or 17th century Cornish miners

speaking their native tongue when deep mining was in its infancy.[28]

Cornish place-names readily lend themselves to the most outlandish interpretation. Buller[29] supplied 'evidence' of Jewish traders in Phoenician ships with the following names: Bojewen in St Just-in-Penwith, Trejewas, and Marazion. Far from meaning 'the Jew's dwelling', Bojewan was Bosywen in 1302 and derives from Cornish *Bos*, dwelling, plus a personal name comparable to the Welsh *Yuein* from the Latin *Eugenius*. Trejewas, a name not identifiable in modern gazeteers, at least bears some resemblance to Tregew found in several parishes and meaning 'farm in the hollow'. The 'zion' of Marazion has nothing to do with Zion, Jerusalem. Originally, Marazion was *Marghasbigan* (*c.* AD 1200), which has the prosaic meaning of 'little' (*bihan*) 'market' (*marghs*), distinguishing it from Market Jew which was more or less contiguous and survives in the town of Penzance as Market Jew Street. Around AD 1200 it was spelled *Marachadyou*, usually accepted as meaning 'Thursday' (*De Yew*) 'market' (*Marghs*), or just possibly *Marghs Deheu*, 'South Market'.[30]

Even those justly famous Cornish delicacies, clotted cream and saffron cake, have been credited with a Levantine origin. To quote the Rt Hon. Lord Halisbury in his presidential address to the Cambrian Archaeological Society at their Launceston meeting in 1895:

What has been called Devonshire Cream, greatly, I believe, to the indignation of the more western part of our island once called Damnonia, I am told is made by a process quite familiar on the Syrian coast; and if the conjecture as to the *Cassiterides* or Tin Islands be correct, the Phoenicians were in Cornwall, and, undoubtedly, were great builders of Castles, but for all that I cannot insist upon the theory with which, at an earlier period, I was enamoured.

NOTES AND REFERENCES

1   ROMÉ, 1980, contains a short section on 'The Phoenicians and Tin'
2   PHILLIPS, 1847, 83
3   HAWKINS, 1896, 29
4   RUNDLE, 1900, 78
5   BADCOCK, 1896, 61
6   DOBSON, 1936; this curious work went into 20 editions and reprints up to 1974
7   BARING-GOULD (ed.), 1912, 57
8   DANIEL, 1964, 21
9   EDMONDS, 1871
10   O'BRIEN, 1833
11   RIVET & SMITH, 1979
12   COX, 1720–33, 310
12   RIEU, 1970, 418–22
14   SÉLINCOURT, 1976, 284
15   HAWKES, 1975

16 CARPENTER, 1966, Ch. 6 'Avienus', containing a translation of the surviving 714 lines of the *Ora Maritima*. Carpenter follows the traditional geographical interpretation, *contra* Hawkes 1975

17 HAWKES, 1975

18 CULICAN, 1966, 117–18

19 RIVET & SMITH, 1979

20 HUNT, 1887, 15

21 POOLE, 1865

22 BULLER, 1842, 6

23 TAYLOR, 1925

24 POOLE, 1865

25 EDMONDS, 1871

26 MUHLY, 1973, Ch. 3 'The Words for Tin'

27 ALLEN, 1856

28 THOMAS, 1950, 11

29 BULLER, 1842

30 GOVER, 1948

22　The Mycenaeans and the tin trade

The 5th century BC historian Herodotus was well aware of the Cassiterides, the half-legendary tin islands. However, though he collected material for his *History* during the greater part of his life he could say nothing of the islands, not even precisely where they were.[1] Nevertheless, they were for him the place 'whence we get our tin' (Bk. 3, 115), whether the source was Iberia, France, Cornwall, or any combination of the three.

After about 800 BC the Greeks could have obtained tin from Iberia through the Phoenicians. Kyme, a commercial port of Asia Minor by the 8th century, may have been the birthplace of a sea captain called Midacritus who, Pliny tells us,[2] was the first man to bring tin from an island called Cassiteris (Bk. 34, Ch. 47). The story may be untrue. If it hides a grain of truth it may illustrate an early Greek attempt to obtain tin from a western source without relying on Phoenician merchants. If it is considered necessary to place Midacritus' tin island on a map there is no need to look further than the trading centre of Gadir (Cadiz), built on an island as a port for the transhipment of Iberian tin. Alternatively, Sardinia was an important metallurgical centre trading its own copper and Iberian tin. Whatever route was chosen later, it is known from Pliny in the 1st century AD that the Greeks imported Iberian tin from Galicia and Lusitania (Bk. 7, Ch. 57).

An early Greek venture west of the Pillars of Heracles is recorded by Herodotus (Bk. 4, 154). Colaeus, a merchant of Samos, was driven by contrary winds and 'a piece of more than human luck' as far west as Tartessos, a place 'not at that period exploited' (*c.* 638 BC). On his return the merchants 'made a greater profit on their cargo than any Greeks of whom we have precise knowledge'. One-tenth – six talents – they spent on making a bronze vessel 3·5 m high, which they placed as an offering in the temple of Hera. It is assumed their cargo was metal, though whether of silver ore or of tin and copper Herodotus does not say. However, it is tempting to assume that the bronze vessel was fashioned from part of their cargo.

Any Greek trade with Tartessos, or elsewhere beyond the Pillars of Heracles, must have ceased soon after 500 BC, for it was in the first decades of the century that Phoenician Carthage helped its ally Persia in its war with Greece by cutting off all Greek trade with Tartessos by fortifying Gadir – which in Phoenician means a stronghold. If this severed a vital supply of tin for bronze armaments it must have stimulated the Greeks of Massalia to increase their trade with the 'barbarians' of northern Europe and to develop their own route to the Cassiterides, which the Greeks believed were ten in number lying somewhere north of Iberia.

It is doubtful whether the Greeks of Massalia, or anywhere else at this time, knew of the existence of tin in central Europe. By the time of Herodotus the 'Ore Mountains' may have passed their prehistoric importance as a source of tin, but it is not improbable that Erzgebirge tin reached the Mediterranean world at an earlier period.

The Mycenaean love of amber is well known. In spite of Rottländer's conclusion that 'the amber routes appear to be fictitious',[3] many archaeologists accept that Baltic amber was traded to the Mediterranean, a trade well attested by Pliny (Bk. 37). There is amber in Romania and Sicily, while in the 19th century it is said to have been plentiful enough in the Basses-Alpes of France for the peasants to use it as a fuel.[4] Studies by Beck and others showed that Baltic amber possessed an absorption pattern not seen in amber from elsewhere: 28 beads from an Etruscan necklace yielded typical Baltic amber spectra while only three beads might have been Sicilian. The bulk of Mycenaean amber was Baltic, though quantities were neither regular nor extensive. Trade occurred *c.* 1600 BC, possibly *c.* 1500 BC and again *c.* 1200 BC. The distribution of amber finds suggests close contact with Italy and a trade route down the Adriatic.[5] If amber crossed the Alpine barrier, then why not tin?

It cannot be denied that obtaining a regular supply of tin must have been a problem to Mycenaean metal smiths. Yet it was obtained in sufficient quantity to be

51 Stemmed cup or kylix with remnants of tin incrustation; Mycenaean from Ialysos, Rhodes; height 14·5 cm (photo British Museum)

used not only in the manufacture of bronze, but also unalloyed to decorate objects, even though surviving examples of these are rare. In 1905 Evans read a paper in which he described a handled cup from Zafer Papoura, Mycenae, 'covered with a kind of black varnish'; he concluded that the coating 'may have been intended to produce an illusion of metal work for funeral show'.[6] Thanks to the research of Immerwahr,[7] initiated by a study of 12 vases from the Tomb of the Bronzes at Mycenae, it is now known that Evans' vessel belongs to an unusual class of funerary vases covered with a layer of tin foil. Since many vases were cleaned after excavation with dilute hydrochloric acid, in which tin metal dissolves, the coating of many vases may have been destroyed. Even so, tin-incrusted vessels have been recognized from well equipped tombs at Salamis on Cyprus,[8] Knossos, Ialysos on Rhodes, Mycenae, and Dendra near Mycenae.

Immerwahr listed 39 of seven basic forms, the kylix or stemmed cup comprising 25 of them.

Figure 51 is a vessel from Ialysos, a site which has also produced amber. It is a kylix from tomb 5, presented to the British Museum in 1870. The entire surface of light brown clay, except the inside of the foot, is covered with a red-black varnish; this was coated with tin, most of which has flaked away but shows white under the handles.[9]

Next to the Tomb of the Bronzes at Mycenae is the Tomb of the Ivory Pyxides. A large ivory pyxis from here was found to be lined with tin. All these objects belong to the late 15th or first half of the 14th centuries, a period when the major European tin-fields were in production. As will be detailed later, Cornish production was well underway, though it is disputed whether Cornish tin reached Mycenae. This is also the period of the 'tin' pilgrim bottle from Egypt, XVIII

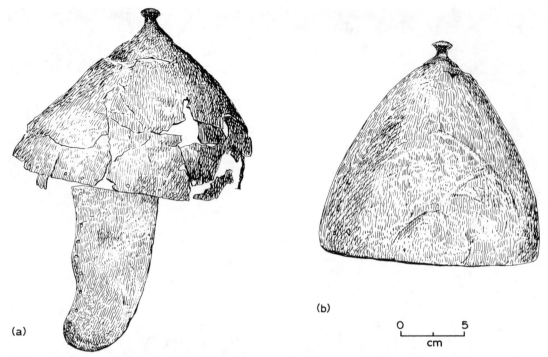

(b)

(a)

0   5
cm

**52   Bronze helmets from *a* Knossos, found in 1951, and *b* Beitzsch hoard, Brandenburg, in 1847
(redrawn from photos in *PPS*, 1952)**

Dynasty (*see* Chapter 1, Fig. 2). The vessel may have Aegean parallels in shape, if not in material, so that the tin for all of them could have come from the same mining region. Immerwahr notes several north European items in Mycenaean graves: amber, boar's tusks, menhirs and stelae, as well as 'a barbaric abundance of gold'.

A tin trade with the Erzgebirge is thus not out of the question, a view strengthened by the fact that the Mycenaean type C sword (a long-horned rapier with midrib and rivetted hilt) occurring at most sites that have produced tin-decorated pottery, was diffused in a less luxurious form as far north as the Danube. Thus the Mycenaeans were brought within reach of a well worn trading route to central Europe.

Vladar sees in what are now parts of Czechoslovakia and Hungary similarities in artifacts and their decoration, which he attributes to Mycenaean influence.[10] Bouzek has also drawn attention to bronze figurines from three north European localities that he regards as Aegean or Near Eastern imports, or local imitations of them.[11] The two male figurines (now lost) from Stockhult, Skåne, Sweden, were found with metal axes, spears, and various ornaments dated to around 1500 BC or not much later. Torbrügge has compared them to four wooden figurines in a model boat from Roos Carr, Yorkshire.[12] So not everyone accepts

Bouzek s view that the Stockhult figures' conical hats and 'loin cloths' are of east Mediterranean inspiration, though the possibility needs consideration. Bouzek also notes a figure from Sernoi, on the Lithuanian coast, which resembles one from the Greek island of Delos found in a Mycenaean hoard dated to the 14th or 13th century BC. Some link with the Baltic amber trade is possible here, while the figure of a nude female charioteer from Kourim in central Bohemia is close enough to the Erzgebirge to make tin trading a possibility.

Better evidence comes in the form of a bronze helmet from Beitzsch in East Germany, which closely resembles one from Knossos (Fig. 52).[13] Similar helmets are known from Oranienberg and Kreis Guben in Germany and from Lúcky in Czechoslovakia. Cheek-guards on the Knossos helmet, lacking on the one from Beitzsch, are paralleled in a find from Weissig in Saxony. Branigan accepts the Beitzsch helmet, and a sword from Ørskovhede excavated in 1906, as particularly important 'since both items have close Aegean parallels' (Fig. 53).[14] The sword contains 9% tin. Randsborg[15] noted a similar example from Lyons, suggesting the route by which the Ørskovhede sword reached Denmark. Randsborg favoured an Aegean origin but admitted that it could be a European copy.

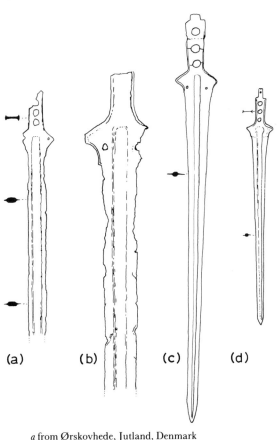

a from Ørskovhede, Jutland, Denmark
b from Lyons, France
c from chamber-tomb 78 at Mycenae, Argolis
d from a chamber-tomb at Ayios Ioannis near Knossos, Crete

53　Mycenaean swords (after Klavs Randsborg, 1967)

Trade between coastal sites in southern Italy, Sicily, and Mycenae is well documented,[16] and there is growing evidence of Mycenaean pottery of a late period (IIIb and IIIc) from the lower Po valley and elsewhere in Italy (Macnamara, *in litt.* 1981). Later contacts between Italy and northern Europe have been discussed by Sestieri.[17] As an example she describes from the Poggio Berni hoard in the north of peninsular Italy a spearhead comparable to north European types, a bracelet of twisted bronze and a knife of types found in the Tyrol and Germany in the earliest Hallstatt phase dated to the 11th century BC. Further examples need not be cited here. What is apparent is that while some archaeologists dispute early Mycenaean contacts with northern Europe, *c.* 1500 BC, later Greek contacts from around 1200 BC are increasingly firmly established.

It is premature to suppose that the 'New Archaeology' with its condemnation of old diffusionist ideas

has finally severed contacts between barbarian Europe and the Mediterranean world of the mid-2nd millennium BC. As Coles has recently pointed out, 'One major concern of British Bronze Age studies is the relationship, if any, of the "Wessex Culture" to Mycenae and the Aegean, and to central Europe ... here radiocarbon dates have proved equivocal.'[18] There may yet be a return to the heady days of the 1950s when no one seriously challenged Childe's view that:[19]

> In the Early Bronze Age peninsula Italy, central Europe, and the west Baltic coastlands, and the British Isles were united by a single system for the distribution of metalware, rooted in the Aegean market.

The segmented beads of faience well known from early bronze age 'Wessex' culture graves in southern England; the beautiful ribbed gold cup found in a barrow at Rillaton in Linkinhorne parish, Cornwall in 1857; the remains of a dagger from a barrow at Pelynt, also in Cornwall; the rather similar dagger carved on stone 53 of Stonehenge's sarsen stone trilithon; even the 'entasis' or tapering shape of the trilithons themselves; and a small number of 'double-axes' from widely spread British localities – all these were taken as firm evidence for direct contact between Britain and Mycenaean Greece or, in the case of faience beads, more particularly with Egypt where faience beads were made notably at Tel-el-Amarna *c.* 1450 BC.

A Mycenaean thirst for Cornish tin traded by the rich Wessex overlords was seen as the important initiator of links between Britain and the eastern Mediterranean. Archaeologists were just beginning to wonder how long it would be before a Mycenaean trading post was uncovered on the south coast of England, when the ^{14}C revolution apparently overthrew this cosy picture. Suspicion grew that the Wessex bronze age was earlier than the rise of Mycenaean Greece, so that any similarity between the Rillaton gold cup and a ribbed example from shaft-grave III at Mycenae was fortuitous (Fig. 54).[20] Native British parallels were invoked, such as the gold armlet from Cuxwold in Lincolnshire,[21] and a handled pottery vessel from Balmuisk in Perthshire.[22] Accounts of a recently rediscovered gold cup from Eschenz, Canton Thurgau, Switzerland, and a silver cup from Saint Adrien, Côtes-du-Nord, Brittany,[23] do not even consider the possibility of Mycenaean influence. Both were described as part of the increasing corpus of good quality north European metalwork owing no allegiance to southern forms.

However, an increasing number of ^{14}C dates for the Wessex bronze age confirms that the traditionally late phase of the Wessex culture flourished around 1400 BC. Faience beads have proved on analysis of their blue glazes to be less like their supposed Egyptian prototypes than was formerly envisaged (Fig. 55), and there can be little doubt that both British and Czecho-

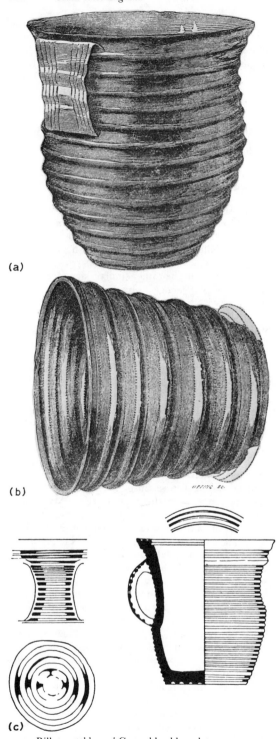

(a)

(b)

(c)

a Rillaton gold cup *b* Cuxwold gold armlet
c handled food vessel from Balmuick
54   Finds from 'Wessex culture' graves

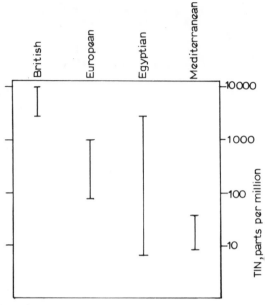

**55   Tin content of faience beads from different areas
(after Aspinall *et al.*, 1972)**

slovakian beads were manufactured locally.[24] Even so, this does not prove that north European and Mediterranean beads were made without any awareness of each other's existence, no matter which beads were made first.

The late phase of the Wessex culture, outlined by Burgess, matches 'exactly the horizon for the closest parallels for British faience'.[25] A Mycenaean link is again feasible. To return to amber for a moment, Harding has pointed out that a diagnostic bead type used in Mycenaean shaft-graves between 1600 and 1400 BC is the so-called spacer-plate, very similar to spacer-plates found in Wessex graves as well as in southern Germany.[26]

Some supposed Mycenaean links are justifiably discredited. A characteristic artifact of Mycenaean Greece is the double-axe, a two-bladed axe-head with a central hole for the attachment of the handle. Double-axes occur in Britain, including a copper one from Topsham in south Devon. However, there is a respectable tradition in northern Europe for earlier double-axes made of stone. Following the discovery that a bronze one in the Royal Scottish Museum had been bought in Attica, all other British examples were scrutinized. The one from Topsham could well have been brought to Devon in recent times by a sailor, and the find-circumstances of all British examples are unsatisfactory.[27] This revelation does not invalidate the existence of links, however tenuous, between Britain and Mycenae.

One object in particular is recognized as a

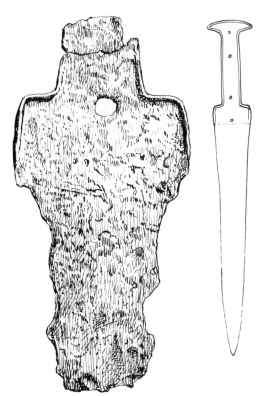

56 **Pelynt dagger (length 11·2 cm); outlined is Furmark's type *b* dagger from Mycenae for comparison**

Mycenaean export: the Pelynt dagger (Fig. 56), especially important because of its discovery in Cornwall and the implications for the tin trade. Unfortunately the circumstances of its discovery are shrouded in mystery. The earliest reference to it appears in an album of water-colour drawings compiled by Borlase and dated March 1871.[28] Above the Pelynt dagger is another water-colour of a 'brasen spear head', in fact a Wessex dagger, 'found in a barrow in the parish of Pelynt, presented by Mr J. Couch'. Both objects are preserved in Truro Museum.

In his report to the Royal Institution of Cornwall for 1845,[29] Jonathan Couch, a well known naturalist and antiquarian from Polperro, described the discovery of the 'brasen spear head' late in 1834 in a barrow south of Pelynt church. The only other metal artifact he mentions is a 'celt', now lost, which was a corroded middle bronze age palstave. Couch presented sketches of the palstave and Wessex dagger to the Royal Institution of Cornwall's museum at Truro. However, he did not sketch nor describe anywhere the Mycenaean dagger. When Childe described the Mycenaean dagger in 1951,[30] he misunderstood the account of the Pelynt barrows published by Borlase in

1872[31] and wrongly assumed the lost palstave and the Mycenaean dagger were one and the same. There is no accession in the museum's records that fits the Mycenaean dagger. It is uncharacteristic of Couch's meticulous nature to have made no reference to it had he found it. Couch invariably stuck labels on the objects he presented to Truro museum, which may have existed on the Mycenaean dagger before it was over-cleaned about the time Childe examined it. Certainly, when the Curator George Penrose made outline drawings of all the museum's prehistoric metalwork, he unhesitatingly assumed the Mycenaean dagger was from Pelynt.

In 1873, three years after Couch's death, Nicholas Whitley exhibited at the Society of Antiquaries in London 'two daggers (from Pelynt)'.[32] These must surely have been the Wessex and Mycenaean daggers. The evidence that the Mycenaean dagger came from Pelynt must rest on Borlase's caption to his water-colour of 1871: 'Another, as last', implying that it was found, like the 'brasen spear head' 'in a barrow in the parish of Pelynt'. This is not very satisfactory evidence for such an important piece, as Macnamara has pointed out,[33] but its Cornish provenance deserves the benefit of the doubt even if it is uncertain when or from what barrow it was obtained. It is not the sort of object likely to have been picked up as a souvenir in the Aegean.

There is no question that the dagger is Mycenaean, and if it is accepted that it was found in a Cornish barrow its dating becomes important. According to Macnamara it shows striking parallels with daggers or short swords made *c.* 1300–1230 BC (Mycenaean III*b*) when 'there was a considerable contact between Greece, along the great corridor of the Adriatic and northwards into central Europe'. The Pelynt dagger resembles a sword from Surbo in Apulia, southern Italy. The group of Surbo bronzes includes, among other items, a 'winged-axe' not unlike a mould found in the house of the oil merchant at Mycenae and dated to the 13th or 12th centuries BC.[34] It has a central European ancestry comparable to finds from Czechoslovakia.[35] According to Macnamara the evidence for Mediterranean contact with central Europe was at its peak at just the time when the Pelynt dagger is presumed to have found its way to Cornwall.

Branigan[36] has recently published a Cypriot 'hook-tang weapon' (Fig. 57), *c.* 1650–1400 BC, ploughed up about 1950 near Torrington, Devon. Five were already known from Sidmouth, Devon, one from Egton Moor, Yorkshire, and seven from Plouguerneau in northern Finistère, Brittany. Whether all are 'archaeological' finds rather than lost relics of recent travellers is questionable, but the Torrington example strengthens the argument in favour of contact between Britain and the Mediterranean civilizations in the middle of the 2nd millennium BC. Yet even accepting this, the Pelynt

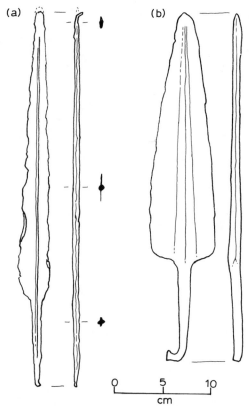

(a)   (b)

0   5   10
cm

57 Cypriot 'hook-tang weapons' from *a* Torrington and *b* Sidmouth (after Branigan, 1983)

dagger and approximately contemporary finds are slim proof of an extensive tin trade flourishing in the ensuing centuries. Contact between Cornwall and the Mediterranean must have remained tenuous, otherwise, Herodotus ought to have discovered rather more about the shadowy edge of Europe 'whence we get our tin'.

NOTES AND REFERENCES

1   SÉLINCOURT, 1972
2   RACKHAM *et al.*, 1938–62
3   ROTTLÄNDER, 1970
4   BECK *et al.*, 1965
5   HARDING & BROCK, 1974
6   EVANS, 1906
7   IMMERWAHR, 1966
8   KARAGEORGHIS, 1969, 32, 54, 97, 127
9   FORSDYKE, 1925, **1**, (1), 153, item A860; the tin is described in error as 'thick unfired grey pigment'
10   VLADAR, 1974, quoted in Phillips 1981, 220
11   BOUZEK, 1972
12   TORBRÜGGE, 1968, 138, 149
13   HENCKEN, 1952
14   BRANIGAN, 1972
15   RANDSBORG, 1967
16   HARDING, 1975
17   SESTIERI, 1973
18   COLES, 1982, esp. 271
19   CHILDE, 1958, 166
20   RENFREW, 1968
21   TROLLOPE, 1857, *Arch. J.*, 92; there is also a fragment of a gold bracelet from Mountfield, Sussex, illustrated in the British Museum *Guide to the Bronze Age Antiquities*, 1920, 52, Fig. 39. For similar gold and bronze bracelets from Denmark see Torbrügge 1968, 94, for bronze examples from Røgerup, Zealand
22   CLARKE, 1970, **2**, 417, item 1081; the Rillaton gold cup is the full-colour frontispiece to **2**
23   ASHBEE, 1977; ASHBEE, 1979
24   ASPINALL *et al.*, 1972
25   BURGESS, 1974, esp. 188–9
26   HARDING, 1980
27   BRIGGS, 1973 and 1975; HAWKES, 1974
28   BORLASE, 1871
29   COUCH, 1845
30   CHILDE, 1951
31   BORLASE, 1872, 189
32   WHITLEY, 1873; he gives no description of the Cornish bronzes
33   MACNAMARA, 1973
34   MACNAMARA, 1970, 245 for the Pelynt dagger
35   GIMBUTAS, 1965, 114 and Plate 19 No. 3
36   BRANIGAN, 1983

23 Massalia, Pytheas, Ictis and Belerion

The Greek cities of the western Mediterranean were prosperous – *nouveaux riches*, as Boardman aptly tags them.[1] None were more so than Massalia (modern Marseilles), whose inhabitants flaunted their wealth by building a marble treasury or thesaurus at Delphi in the 6th century BC.[2] Trade was the reason for the foundation of the city, and colonists from there soon founded the appropriately named Emporion (modern Ampurias) in northern Spain near the French frontier. The earliest authenticated Greek pottery, Corinthian ware of the late 7th century, has been found at Saint-Blaise and La Couronne close to Marseilles, and Massalia itself was settled by Phocaean Greeks from Asia Minor *c.* 600 BC. The city possessed an ideal harbour 0·8 km long by 0·4 km wide, and also commanded the important trade-route up the Rhône valley.

By this route travelled the magnificent bronze krater, or handled-bowl, probably made at Sparta *c.* 500 BC, which came to rest in a princely burial at Vix near Châtillon-sur-Seine, close to the source of the Seine.[3] It is the largest Greek krater that survives and resembles another, a Spartan gift to Croesus on the occasion of an alliance with Lydia around 546 BC. The Vix krater, over 1·6 m tall and weighing 208·7 kg, may well have been occasioned by some trading treaty. The tin used in its manufacture could have come from France or Cornwall.

It is hardly surprising that the majority of Greek coins found in France are Massaliote, 228 of bronze and 21 of silver compared with only five from Minorca and one from Syria.[4] Examples are two Massaliote coins from tin-producing Brittany: a silver drachma bearing the head of Artemis and on the reverse a lion above the name ΜΑΣΣΑΛΙΗΤΩΝ, of the 2nd century BC found by a farm worker at Château de Lesmel, Plouguerneau in north Finistère. A bronze sestos of similar date was found 'near the edge of the sea' at Jard, Vendée.[5] There is also a gold stater of Cyrene (*c.* 322–313 BC), an important Greek settlement on the north African coast between Alexandria and Carthage. This was found on the beach at Ploudal-mézeau, north-west Finistère, where it had been washed up embedded in deep-growing seaweed.[6]

Greek coins are more numerous in Britain than is popularly believed. Some may be modern losses, but the coins are too many for all to be attributed to such carelessness. Some must have arrived during the Roman occupation, but their distribution is essentially southern reflecting pre-Roman contacts through southern ports. The largest hoard yet found in Britain was dug up from a depth of 6 m by workmen digging at Broadgate, Exeter, in 1810.[7] All the coins belong to the first three centuries BC. Discredited for many years, the coins are nowadays more readily accepted as pre-Roman imports, partly because of other recent finds in Devon.

In the spring of 1941 or 1942 a coin was ploughed up by a farmer in a field near Ridge Cross, east of Holne; he later dug up another in his garden at Court Farm. One is a silver tetradrachm of Alexander III of Macedonia minted not before 326 BC; the other is a silver tetradrachm of the Roman province of Macedonia struck by the quaestor Aesillas, probably at Thessalonica, in 93–92 BC. Both finds are not far from the important native camp at Hembury.[8] Whether the coins have anything to do directly with the tin trade is pure speculation, but the find spots are close to where 'the old flats [of the river valley] containing stream-tin still exist and have been extensively worked' at some time in the past. Unfortunately no Greek coin has been found in any tin stream in Cornwall or Devon.

Pre-Roman coins, native or imports, are rare in Cornwall and none have turned up in an archaeological context. A Cypriot coin of 80 BC is said to have been found about 3 km north of Truro, while 'many silver coins of Syracuse' were claimed by W. P. Shortt, the 19th century Exeter antiquary, to have been dug up in a mine at Malpas near Truro.[9] Neither find inspires much confidence as no such mines exist near Malpas. More acceptable are two bronze coins, now in Truro museum, issued by one of the kings of Numidia (modern Algeria) in the 2nd century BC. The first was found on Carn Brea *c.* 1830, and the second was dug up in a garden near Mount Hawke, St Agnes, 1981.

The scarcity of coinage in Cornwall is explained by the fact that the natives had little interest in coins, which in this period were a new-fangled invention. As late as the early 3rd century AD, Solinus tells in his *Collectanea rerum memorabilium* that the Dumnonii insisted on barter and even refused coinage. It has been proposed that the Roman coin hoards from Cornwall, mostly dated after AD 250, had been accepted as bullion rather than as coin (A. C. Thomas, pers. comm.). This not only explains the lack of Greek coins but also why the Dumnonii minted none themselves. The only Celtic gold coinage in Cornwall, of south-east British types, is almost exclusively confined to the hoard found on the iron age hillfort of Carn Brea in 1749.

Nevertheless Cornwall, always full of surprises, contains one of the most interesting pre-Roman coin hoards discovered in England. The coins were bought in 1909 by a Mr Dixon of Bristol from Thomas Rossiter, a Penzance dealer. In a letter, now in Truro Museum, dated 20 November 1909, Rossiter explained to Dixon that the 'Roman coins' as he called them, had been obtained more than two years before from a woman whose name he did not remember. 'All I know is that she told me they were found in a field in Predannack near Paul while ploughing.' Predannack is on the Lizard peninsula and there is nowhere of that name in Paul, a parish overlooking Mount's Bay west of Penzance. Assuming the parish is correct, the coins may have come from a farm of similar name, perhaps Tresvennack only a mile and a half west of Penzance.

The 45 silver coins are all copies of silver tetrabols of Massalia struck by the Celtic tribes of the Milan and Ticino river region. The first were struck here shortly before 200 BC and continued in circulation in several varieties for about 200 years. Four classes are identified in the Paul hoard, which is now in Truro Museum. Class *a* is the most worn and presumably the oldest, while class *d* is the least worn. All depict the head of Artemis on the obverse, and a lion on the reverse sometimes with the marks ΛΛΛΣΣΛΛ, a poor copy of ΜΑΣΣΑ, Massalia.[10] The four types are identifiable as follows: *a* lion with only one hind leg; *b* lion with two hind legs, Artemis with a pointed nose; *c* Artemis with large pendant earrings, lion bear-like; *d* Artemis with 'cork-screw' curls in hair.

Five copies of Massaliote coins and a Belgic coin, all of bronze, are said to have been found somewhere in Penzance in 1888. However, their origin is in doubt, though Allen believes they may have come from Wallingford in Berkshire.[11] Being Celtic copies, the Paul hoard suggests that trading links between Cornwall and the Mediterranean world were through the Celtic tribes of France and not directly with Greek traders. This view is enhanced by the discovery of a hoard of Massaliote coins on the island of Jersey.[12]

Links between Cornwall and Brittany are well represented in the archaeological record. One example is the Cornish 'fogou' (from Cornish (*f*)*ogo, gogo*, a cave), an iron age structure of disputed use that closely resembles contemporary Breton souterrains. Much Cornish iron age pottery, notably the 'cordoned ware', relates to La Tène pottery of north-west France. These themes are developed by Thomas,[13] who stresses that the Breton connection is only to be expected since:

> The Armorican tribes lay, as their predecessors had lain, directly athwart the traditional trade routes – either the long sea passage, or the route which may have gone overland from *Corbilo* [at the mouth of the Loire] to *Massalia*. The Veneti [of southern Brittany] with their large fleet, clearly controlled some trade with south-west England – tin, slaves, hides, and corn are all possibilities, even if this still involved barter instead of coinage. It was a trade that justified the maintenance of a fleet of several hundred vessels, leading to effective control of the Western Approaches; and it meant enough to the Armorican leaders, mercurial Gaulish temperament or not, to risk, deliberately, a direct clash with Roman arms.

Coins apart, few Greek objects reached Britain. The four Greek vases supposedly dug up at Halamanning in St Hilary parish, Cornwall, *c.* 1931, are real enough. However, the vessels are not contemporary and are in any case funerary *lekythoi* not normally exported. Their 'discoverer', R. J. Noall of St Ives, was a magpie collector of antiquarian objects that he found and art objects that he purchased; he also had a sense of humour. He made known a statue of Osiris said to have been found in St Just-in-Penwith, so the Halamanning find can be dismissed as spurious. In Devon, however, a pair of two-handled, black-glazed cups (*bossae*), together with a jug (*oinochoe*) were found in an artificial cave at Teignmouth and dated to the 4th century BC.[14]

Other British finds regarded as authentic include a 6th century, Attic, black-figure *kylix* from the Thames near Reading,[15] but these need not be detailed here. A fragmentary vase handle decorated with black paint forming the background to a myrtle spray was found on St Martin's, Isles of Scilly, and is displayed in the islands' museum. It is classical (4th or 5th century BC) but could well be a piece washed in from the wreck of the *Colossus*, sunk on 10 December 1798 with Lord Hamilton's rich collection of ancient Greek pottery on board. The rim of a pinkish-buff vase or small amphora was also found at low tide near Tresco, but is virtually undatable.

Nevertheless there are authentic Greek finds in Britain, and a recent archaeological find confirms a Greek presence in British waters. In 1974 an anchor-stock was dredged up off Porth Felen at the tip of the Llŷn peninsula in north Wales (Fig. 58).[16] Such anchors are common in the Mediterranean where they 'litter the roadsteads and ports as well as off perilous

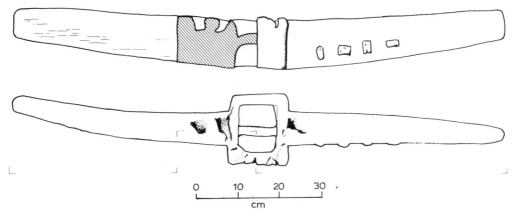

58   Greco-Roman lead anchor-stock from Porth Felen, Aberdaron; (redrawn after Williams in Boon, 1977)

headlands and reefs'. However, in the Atlantic the only other examples are ten from the Bay of Setúbal, just south of Lisbon – emphasizing the lack of Greek finds off the rest of Atlantic Iberia, let alone further north.

The first Greek known to have sailed as far north as Britain was Pytheas, a native of Massalia, whose circumnavigation of the British Isles took place *c.* 300 BC or a little later. Pytheas' principal aim was the furtherance of geographical knowledge, doubtless encouraged by the merchants of his home port alive to the commercial possibilities. To what extent his epic voyage may have opened up British waters to Greek shipping is not known – probably very little. However, there is no reason why a Greek vessel should not have foundered or lost an anchor off the Llŷn peninsula during the late 2nd century or in the 1st century BC, the date of the Porth Felen anchor stock.

The date of Pytheas' voyage is sometimes placed between 310 and 306 BC. Carthage, then in desperate conflict with Syracuse, had little time to pay attention to a single Greek ship running the Carthaginian blockade of the Straits of Gibraltar. Carpenter believes this date too early, preferring the years 242–236 when Carthage was also briefly powerless. For him this is 'the only period during the 3rd century that afforded Massalia, the close ally of Carthage's foe Rome, a feasible opportunity for sending out an expedition whose purpose could not possibly have been approved by Carthage.'[17]

The details of Pytheas' voyage have been discussed by such authors as Carpenter and Hawkes.[18] The question of whether he did or did not reach Iceland, the Thule of Pytheas, is not relevant here. His own account 'Concerning the Ocean' (περι του ὀκεανου) does not survive. All that is known is quoted, or misquoted, or inferred, from the writings of later authors, usually critics. Nothing is known of the tin trade from Pytheas, though the quest for this precious

metal ought to have been one of the mainstays of his expedition.

The Sicilian historian Timaeus (*c.* 352–*c.* 256 BC) is quoted by Pliny in *Natural History* (Bk. IV, Ch. XVI) on the subject of tin – or 'white lead', as Pliny calls it – in a confusing list of islands surrounding Britain.

> Timaeus the Historiographer saith, that farther within at six days' sailing from Britannia, is the island Mictis, in which white lead is produced, and that the Britanni sail thither in wicker vessels sewed round with leather.

The passage is important in describing Celtic craft of a kind well known today in the guise of the Welsh coracle and the sea-going Irish curragh. That apart, the passage is of little value: Mictis is unidentifiable. Speculation is pointless, and Carpenter is probably right in seeing Timaeus' account not as evidence for familiarity with the work of Pytheas, but a complete lack of acquaintance with it. According to Carpenter, Timaeus, who spent 50 years on his native Sicily followed by exile in Athens, would have written his account before the voyage of Pytheas took place. In that event, Timaeus was reporting only the garbled tales of traders, which Pytheas' epic voyage was intended to clarify. The Mictis of Timaeus may have been only another name for the legendary Cassiterides, confusingly equated later with Ictis. De Beer has suggested that Iktis (or Iktin as he prefers to call it as the name does not appear in the nominative case) 'can be recognised in *insulam Mictim* where the initial *M* has been repeated from the final *m* of the preceding *insulam*'.[19]

An account that bears all the marks of authenticity and is generally accepted as derived from the lost work of Pytheas is given by another Sicilian 'historiographer': Diodorus, a native of Agyrium and contemporary of Julius Caesar and Augustus. Anyone who has hurriedly outlined Britain on a blackboard will be aware of its basic triangular shape. Diodorus Siculus did likewise, placing Orka (the Scottish coast opposite the Orkney Islands) at the top, Kantion (Kent) at the bottom right, and Belerium (the Land's End peninsula) at the bottom left. His account is the single most important description by any classical author on the mining and trade of Cornish tin:[20]

> The inhabitants of Britain who dwell about the promontory known as Belerion are especially hospitable to strangers and have adopted a civilized manner of life because of their intercourse with merchants of other peoples. They it is who work the tin, treating the bed which bears it in an ingenious manner. This bed, being like rock, contains earthy seams and in them the workers quarry the ore, which they then melt down and cleanse of its impurities. Then they work the tin into pieces the size of knuckle-bones and convey it to an island which lies off Britain and is called Ictis; for at the time of ebb-tide the space between this island and the mainland becomes dry and they can take the tin in large quantities over to the island on their wagons. (And a peculiar thing happens in the case of the neighbouring islands which lie between Europe and Britain, for at flood-tide the passages between them and the mainland run full and they have the appearance of islands, but at the ebb-tide the sea receded and leaves dry a large space, and at that time they look like peninsulas.) On the island of Ictis the merchants purchase the tin of the natives and carry it from there across the Straits of Galatia or Gaul; and finally, making their way on foot through Gaul for some thirty days, they bring their wares on horseback to the mouth of the river Rhone.

Diodorus travelled widely in Europe and Asia but is not known to have visited Britain. His account reads like an eye-witness description of tin production so is quite possibly derived, perhaps in shortened form, from the work of Pytheas. The translation of the above passage by different authors varies in detail, but the essentials remain the same. The natives dug 'earthy seams' or 'earthy layers', a clear description of sediments above the 'tin-ground' of pebbles in Cornish tin streams. Had the natives been following a lode in hard rock, which would be nearly vertical and not like a 'bed' as Diodorus tells us, some mention would surely have been made of the difficulties of hard-rock mining, such as the use of fire-setting. In any case, in the following section Diodorus talks about Iberian tin mining in which different methods were employed.

> Tin is found in many parts of Iberia, not being discovered on the surface as some have babbled in their histories, but dug and smelted like silver and gold.

Although Diodorus is wrong here, for alluvial tin is common enough in Iberia, the important point is that he knew hard-rock mining was practised in Iberia and that another method – streaming – was used in Cornwall. When the Cornish had obtained their tin ore it was smelted, cast into ingots and carried to an island joined to the mainland at low tide. The implication is that the island was close to the area of tin production.

Diodorus makes a good deal of the way the tide affected Ictis, inserting for the benefit of Mediterranean readers a section on the Atlantic tides, which are unknown within the confines of the Pillars of Heracles.

Concerning the tin ingots, Oldfather translates the passage as 'they work the tin into pieces *the size of* knuckle-bones'. Other authors such as Carpenter and Hencken[21] prefer their ingots *the shape of* knuckle-bones. The Greek word ῥυθμος (rhuthmos) has many meanings concerning regular or recurring motion, from which is derived our 'rhythm'. It also means a measure, proportion, or symmetry, and hence the form or shape made after certain proportions. The verb from it means to form, to fashion, or to mould. 'The shape of knuckle-bones' is certainly a more sensible translation, for no surviving Cornish tin ingot is as small as a knuckle-bone.

The ingots are called ἀστραγαλοι (astragaloi), a plural word commonly meaning dice, which were made of knuckle-bones. The singular word has many technical meanings: one of the neck vertebrae, the ball of the ankle joint, one of the mouldings of an Ionic capital, a physician's measure, and a genus of plants including the milk-vetch. Since Diodorus used the plural, he perhaps implied something the shape of dice which had 'four flat sides, the other two being round'.[22] No surviving Cornish ingot fits that shape, although Borlase in 1758 described from a streamworks in St Stephen-in-Brannel 'now and then some small lumps of melted tin, two inches square and under',[23] but these are undatable.

Is it really necessary to look for an ingot which is either dice-shaped or the weight of a physician's measure? Just as there are the modern terms 'pigs of lead', 'bears' and 'salamanders' of iron, and the French sometimes refer to *saumons* (salmons) of tin, all devoid of their animal attributes, so *astragalos* may simply have been a technical term used to describe one kind of ingot commonly produced in Cornwall. If one is disinclined to assume that an astragalos was bun-shaped or 'pasty' shaped, like so many of the prehistoric Cornish ingots, then one can point to the St Mawes ingot as a possible astragalos. Recent opinion is that the St Mawes ingot (detailed Chapter 26, *see* Fig. 132) is mediaeval as it is far heavier (71·6 kg) than anything older.

Unfortunately, the St Mawes ingot is really undatable. For those anxious to give it a prehistoric date, at least its projecting arms can be likened to the greatly extended epiphyses of a long-bone, and the stamp impressed on one arm can also be likened to a long-bone. A slight resemblance to the much smaller copper 'ox-hide' ingots of the Mediterranean bronze age is likely to be fortuitous, even though Günther-Buchholz considered it a variant of his type 2 copper ingots.[24] Until such time as a St Mawes ingot is found in a datable context, argument will be endless. Until then,

59 'Horse carrying blocks of tin across Gaul' (from James' 'Note on the Block of Tin dredged up in Falmouth Harbour', 1872)

a date contemporary with Pytheas or Diodorus is not much worse than any other.

Concerning the transport of ingots across Gaul, Diodorus relates that the tin was carried 'for some thirty days . . . on horseback' to the mouth of the Rhône and thus to Massalia. James published a delightful picture of a lightly built horse carrying the St Mawes ingot on its flank (Fig. 59).[25] The ingot is well suited to such a mode of transport, but any shaped ingots could have been carried on wooden split-saddles of the type used by Cornish fish-jousters until the beginning of the 20th century. As recently as 1794 Fraser wrote that 'carts are not used either in Devon or Cornwall; everything is carried on horse's backs or on mules', later adding that 'Mules have largely come into great use, particularly for carrying and recarrying the produce and supplies of the mines.'[26] Wooden split-saddles are of considerable antiquity. One was recovered from La Tène at the eastern end of Lake Neuchâtel, Switzerland.[27]

Vigneron[28] worked out a maximum load of about 150 kg for a mule-load in antiquity on level ground and 100 kg in mountainous terrain. Two St Mawes ingots, 143 kg, are at the upper limit set by Vigneron for easy going. Celtic horses were very strong. Assuming their capabilities were similar to the modern Shetland pony which stands only about 1 m high (9–10½ hands) at the shoulder and weighs under 130 kg, they could carry 'a man and his wife 16 miles in a day',[29] or approximately their own weight for 26 km. The distance from the estuary of the Gironde via the Gate of Carcassonne to Massalia and Narbonne is no more than 800 km or around 25 km daily for 30 days. Pack horses were ordinarily used to carry tin and copper from the Cornish mines to the smelting houses in the 18th century and survived in places well into the 19th century. Only as mines became deeper and more extensive did this method prove inadequate.

At Polberro in St Agnes carts were pressed into service only from 1750 because so much ore was being raised.[30] Roads had been too abominable for the use of carts, as Fiennes recorded during her great journey to Cornwall in 1698.[31] Conditions can have been no better in the pre-Roman iron age, which is another very good reason for assuming that Ictis was very close to the area where tin was mined. If it were not so the tin could not have been carried to the island by cart, as Diodorus relates.

The precise location of Ictis has caused considerable – perhaps needless – controversy (*see* Fig. 60). Like Whitley,[32] the present writer believes that:

> The claim of St Michael's Mount to be the island where the merchants bought the tin from the miners of Belerion

(a)

60a St Michael's Mount from the drawing by the Revd Furley (published 1811 and used as the frontispiece to
Hawkins' *Observations on the Tin Trade of the Ancients in Cornwall and on the Ictis of Diodorus Siculus,*
London, 1811)

(b)

60b St Michael's Mount showing the causeway at low tide (engraved by John Pye from the drawing by J. Farington
RA, 1813)

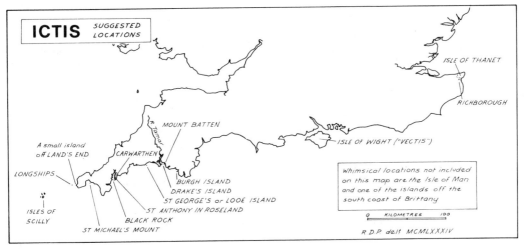

Map 23

is unshaken by any theories which have lately been put forward.

Archaeologists seem bent on disproving that the obvious site for Ictis is St Michael's Mount. The most recent suggestion is that Ictis was Mount Batten, Plymouth,[33] even though there is no evidence that the peninsula was ever cut off from the mainland at every high tide, or even occasionally. Nevertheless, Mount Batten was an excellent site from which to export tin from south-west Dartmoor or the eastern part of Bodmin Moor. Argument is pedantic, for wherever Ictis is placed, it is ludicrous to suppose that every scrap of Cornish tin was trundled there for re-export, since other suitable harbours with safer anchorages occur on the south coast of Cornwall adjoining the very tin streams being worked. Nevertheless, something concerning the debate needs stating here. William of Worcester wrote in his Itinerary of Cornwall in 1478 of an apparition of St Michael at the Mount:

> Apparicio Sancti Michaelis in monte Tumba, antea vocatale Hore-rok in the wodd [sic].

This reference to the Hoar rock, or grey rock in the wood, is a translation of the Cornish *Carrek los yn cos*, which William Camden applied to St Michael's Mount in *Britannia* (1586). This has led to the belief that the Mount remained in a wood until mediaeval times. But this is not the case, for the name was applied to St Michael's Mount in error. Mons Tumba is the French Mont Saint Michel, and Tumbella (now called Tombelaine) is a smaller islet to the north of it. Today the tide recedes more than 16 km, sweeping back in just 90 min. The area was dry land in Roman times, and a paved Roman road of diorite blocks is said to have been found about 3 m below the surface. According to tradition the coastline extended nearly to the Isles Chaussey, the whole area being called the Forêt

de Scissy. A tidal wave in 709 destroyed the greater part of the forest and several villages on its outskirts.[34] The submerged forest in Cornwall's Mount's Bay allowed the uncritical transfer of the name 'grey rock in the wood' to St Michael's Mount.

The coastline of Mount's Bay in the days of Pytheas and Diodorus Siculus was much as it is today, as confirmed by the dated remains of the submerged forest at Lariggan noted in the next chapter (*see below* Fig. 67). There are no geological or historical reasons why St Michael's Mount should not be accepted as Ictis. Many authors have assumed just that, including Barham as far back as 1825.[35] Nevertheless, many have tried to demonstrate that Ictis was elsewhere. Their views have been discussed recently by Maxwell.[36] All the locations need not be mentioned here, though for interest his map of supposed localities is appended, with the latest suggestion of Mount Batten added (*see* Map 23).

Only one alternative to St Michael's Mount has had any lasting support: the Isle of Wight, largely because its Roman name *Vectis* is so similar to Ictis. To equate the two depends upon proof that the Isle of Wight was connected to the Hampshire or Dorset coast as late as the pre-Roman iron age. The distinguished geologist Reid assumed that the connection only broke 'a short time before Caesar's invasion of Britain in the mid-first century BC'.[37] The break is unquestionably geologically recent, though precise dating remains uncertain.

The severing of the chalk ridge which at one time connected the Needles on the Isle of Wight with the chalk cliffs at Studland on the Dorset coast, a distance of 24 km, is comparable to the severance of Dover from Calais in immediate post-glacial times. Throughout its length the Solent exceeds 10 m in the centre of the channel, and where the postulated land-ridge would have linked the island with the mainland, the channel is up to 36·5 m below low tide. That kind of erosion by the once greatly extended River Frome can only have been at a time when the base level of erosion was considerably lower than at present during one of the glacial periods. Its flooding by the rising sea level must have taken place in early post-glacial times, probably during the mesolithic, and cannot have been any later than when sea level in Mount's Bay reached its present height. The whole tone of Diodorus Siculus' description implies the close proximity of Belerion and Ictis. St Michael's Mount is the one locality that fits all the facts.

Cormac's *Glossary*, compiled *c.* AD 900, favours St Michael's Mount as Ictis, for the latter name survived in Irish as *Muir-n-Icht*, 'the Sea of Icht', for the English Channel. Considering the close contact between Ireland and Cornwall in the immediate post-Roman period, proved by the plethora of Irish saints in Cornwall as well as by the archaeological record, it is far more reasonable to assume that *Muir-n-Icht* is named after St Michael's Mount than after the Isle of Wight.

This is not to imply that no Cornish tin ingot was ever exported to the Isle of Wight, for tin was exported throughout Britain and must have been carried up the English Channel to reach northern Europe. There is also a strong probability that Cornish tin sometimes reached southern Europe along a lengthy route via the River Seine. All that is implied here is that in the time of Pytheas and Diodorus, some tin for Massalia was sent straight across the Channel from Ictis to Gaul, perhaps in 'wicker vessels sewed round with leather', as mentioned by Timaeus, but certainly in the ships of the Veneti, which were well capable of the voyage. Strabo (Bk. IV, 4.1) described their ships as having:[38]

> broad bottoms and high poops and prows, on account of the tides. They are built of the wood of the oak, of which there is great abundance. On this account, instead of fitting the planks close together, they leave the interstices between them; these they fill with seaweed to prevent the wood from drying up in dock for want of moisture; for the seaweed is damp by nature, but the oak is dry and arid ... The sails are made of leather to resist the violence of the winds, and managed by chains instead of cables.

The Veneti would then have transferred the tin to horses at Corbilo, somewhere near the mouth of the Loire, or further south on the estuary of the Garonne. Two bronze brooches from the pre-Roman iron age cemetery at Harlyn Bay near Padstow display a connection with this part of France. They were formerly regarded as of Iberian origin, but it now seems likely that they were made in south-west France from where they were exported to both Iberia and south-west England towards the close of the 1st millennium BC.

It is unlikely that this trade route to Massalia survived the lifetime of Diodorus Siculus. From what survives of his account it is known he intended to describe Caesar's invasion of Britain. That section does not exist, but as he implied one expedition had taken place, he may have written his account before the second expedition in 54 BC. In 56 BC, Julius Caesar destroyed the greater part of the Veneti fleet because, as Strabo relates (IV, 194) the Veneti 'were ready to stop Caesar from sailing to Britain, inasmuch as they enjoyed the trade'.

Caesar's action must have curtailed the Cornish tin trade and may even have stopped it for a time. His action may also explain the rapid rise in the importance of Iberian tin in the 1st century BC, and may also be a clue to Caesar's own poor knowledge of the location of British tin mines when he arrived. Caesar knew there was tin in Britain, but described it in his *De Bello Gallico* (V) as *in mediterraneis regionibus*, 'in inland regions' – *mediterraneus* meaning remote from the sea, the opposite of *maritimus*. This disruption of cross-channel trade did not mean an end to Cornish tin production, for there was still the home market and a good deal of northern Europe to satisfy. Although it appears true that the Romans were more interested in Cornish tin from the 3rd century onwards, a later section will give archaeological evidence that tin production in Cornwall was continuous from the beginning of the early bronze age to the present day, a span of some 4 000 years.

NOTES AND REFERENCES

 1   BOARDMAN, 1980, 161
 2   OLIVA, 1981, 96
 3   MEGAW, 1966; WALLIS, 1977
 4   LAING, 1969, 119
 5   BOUSQUET, 1968
 6   BOUSQUET, 1961; GIOT *et al.*, 1979, 253
 7   FOX (ed.), 1973, 135
 8   FOX, 1950
 9   LAING, 1969, 119
10   ALLEN, 1961
11   ALLEN, 1961*a*
12   CARY, 1924, 178
13   THOMAS, 1966
14   FOX (ed.), 1973, 135
15   WELLS, 1977
16   BOON, 1977
17   CARPENTER, 1966
18   HAWKES, 1975
19   DE BEER, 1960, reprinted in De Beer, 1962
20   OLDFATHER, 1939
21   HENCKEN, 1932, 191
22   LIDDEL & SCOTT, 1864
23   BORLASE, 1758, 163
24   GÜNTHER-BUCHHOLZ, 1959, quoted in Piggott, 1977
25   JAMES, 1863 and 1872
26   FRASER, 1794, 36, 46

27   CLARK (ed.), 1965, 312

28   VIGNERON, 1968, quoted in Piggott, 1977

29   SANDERS, 1937, 271

30   LEWIS, 1908, 29; *see also* JENKIN, 1933, 27–31 for the scarcity of wheeled vehicles in mid-18th century Cornwall

31   MORRIS, 1949, 255–6 for the abominable state of Cornish roads, including a pot-hole near Looe in which Fiennes' horse 'had gotten quite down his head and all'

32   WHITLEY, 1915

33   CUNLIFFE, 1983

34   MASSÉ, 1902

35   BARHAM, 1825

36   MAXWELL, 1972

37   REID, 1906; REID, 1913, 75

38   CLARK (ed.), 1965, 292; for ships of the Veneti *see* Muckelroy *et al.*, 1978

24 Tin in Cornwall

[Tin] is in a manner the peculiar and most valuable property of this county, creating at home employment and subsistence to the poor, affluence to the lords of the soil, a considerable annual revenue to our Prince the Duke of Cornwall, and demanded with a great eagerness by all the foreign markets of the known world.

William Borlase, *The Natural History of Cornwall*, 1758

One of the commoner Cornish penny tokens of 1811 depicts on its obverse side a pilchard flanked by ingots of smelted tin and copper. This is a neat portrayal of the well known Cornish toast to 'Fish, Tin and Copper', the three most important commodities that have moulded the county's history. The edible takes pride of place, while copper comes third having only a short history with the bulk of production in the century from about 1750.

Tin has been of greater or lesser importance throughout recorded history, and there is nothing to indicate that prehistory was any different. Cornwall was, and still is, Europe's largest tin producer. The periods when this exalted position was usurped have been few before the development of the Malayan and Australian tin industries in the late 19th century. Tin production in the Erzgebirge outstripped Cornwall in the early 15th century when it reached up to 1 000 t a year, while during the Roman period the Mediterranean world at least obtained more tin from Iberia in the first centuries BC and AD. Cornish production continued throughout the Roman period, and apart from British needs may have remained the main supplier of countries in northern Europe.

In recent centuries, certainly from the 18th century when technical advances allowed ever deepening workings, the bulk of Cornish production has been from underground lode mining. It is as well to remember that as a supplier of lode tin Cornwall is probably second – albeit a poor second – only to Bolivia. Up to the 17th century most of Cornwall's output was from tin streams, and it is only within the last half-century or so that streaming has ceased to be a regular economic proposition. However, even now it continues sporadically with tentative looks at possible development.

There is a vast literature concerning mines and mining in Cornwall. Historical records of the stannaries go back to 1156 and were the subject of a detailed work by Lewis, *The Stannaries*, 'a study of the medieval tin miners of Cornwall and Devon', first published in 1908. Richard Carew's *Survey of Cornwall*, printed in 1602 many years after it was begun, contains invaluable information on methods of tin streaming, which can have changed little in his day from those employed several millennia before. Some of his account is discussed below.

Borlase's *Natural History of Cornwall* (1758) is much less informative about birds and beasts than about the county's mineral wealth, with 11 of the 26 chapters devoted to the subject. During Cornwall's most productive years as a copper producer Pryce produced his influential *Mineralogia Cornubiensis* (1778). Rashleigh amassed an unrivalled cabinet of minerals, publishing in 1797 and 1802 two volumes with plates; these mostly depicted Cornish specimens, describing for the first time such rare minerals as bournonite, clinoclase, connelite, chalcophyllite, bayldonite, and scorodite.

A continuous stream of worthy reports appeared more or less annually from the foundation of the Royal Geological Society of Cornwall at Penzance in 1814. Some of the first work of the Ordnance Geological Survey was undertaken in the south-west by De la Beche, who was the first Director of that institution in 1835. His *Report on the Geology of Cornwall, Devon, and West Somerset* (1839) remains an important source book. Among the papers produced by Henwood, his 'Detrital tin-ores of Cornwall' (*J. Roy. Inst. Cornwall*, 1874) is of particular interest to archaeologists. Hunt's *British Mining* (1887) deserves more notice than it ordinarily receives because of his chapter dealing with prehistoric mining remains.

Study of the geology and mineralogy in the late 19th century was dominated by Collins whose *A Handbook to the Mineralogy of Cornwall and Devon* appeared in 1871 with a second edition and supplement in 1892, and a further list of minerals in *J. Roy. Inst. Cornwall* for 1911. The geological and historical scene was put in perspective in his masterly 'Observations on the West of England Mining Region' which formed volume 14 of *Trans. Roy. Geol. Soc. Cornwall* (1912).

In more recent years there is the important work of

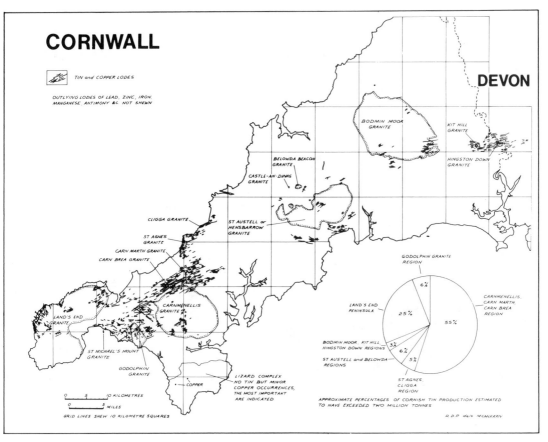

Map 24

Hosking of Camborne School of Mines. Among his many significant publications are 'Fissure systems and mineralization in Cornwall' (*Trans. Roy. Geol. Soc. Cornwall*, 1949), 'Oxidation phenomena in Cornish lodes' (*ibid*, 1950), and 'Primary ore deposits in Cornwall' (*ibid*, 1951). He was also a contributor to *Present Views of Some Aspects of the Geology of Cornwall and Devon*, which he edited with Shrimpton in 1964 for the Royal Geological Society of Cornwall. To a large extent the pioneering work of that distinguished Cornish society has now been taken over by the Ussher Society, founded in 1961 and named in honour of the Devonshire geologist Ussher who, with Reid, Barrow, Dewey and others, was responsible between 1902 and 1913 for the Memoirs of the Geological Society, currently being revised. The geological and mining features of each mine were meticulously traced by Dines in his two-volume *The Metalliferous Mining Region of South West England* (HMSO for the Institute of Geological Sciences, 1956; reprinted with amendments in 1969).

At the time of writing (1984) a paper by Camm and Hosking on 'The Stanniferous Placers of Corn-wall, Southwest England' awaits publication by the Geological Society of Malaysia. It is easily the most important contribution to this aspect of mining com-piled in the 20th century. With such a history of re-search Cornwall is justifiably regarded as the type area for the study of hydrothermal tin deposits.

Mineralization is closely linked to the intrusion of the granites now exposed by the normal processes of weathering in the moorlands from Dartmoor in the east to the Isles of Scilly in the west. These are joined at depth by buried granite ridges, identified in some deep mines and more extensively by gravity surveys and such surface indications as the distribution and width of the metamorphic aureole surrounding all the granites. They form a body of buried granite (a batholith) 200 km long emplaced between *c.* 270 and 290 million years ago into intensely folded 'killas' (mostly clay-slates) of Devonian and Carboniferous age. All the granites were formerly covered by killas to a depth of 1–2 km, though the exact depth is disputed.

The modern tendency is to regard the granites as high emplacements, taking the tin-mineralized Andes as a model; there is evidence that the fine-

61   Narrow working following a near-vertical outcropping tin lode in the granite at Nanjulian Cliff, St Just-in-Penwith, 4 km north of Land's End (photo Herbert Hughes c. 1910)

grained granite 'elvan' dykes in some cases reached the surface.[1] The slowly cooling upper crust of the granite together with the surrounding killas were faulted and fissured. This allowed the passage of late-stage, mineral-rich fluids to fill the fissures, thus producing the swarms of mineral veins with a predominantly ENE–WSW trend in the western half of the region and *c.* E–W in the eastern half. As each mineral crystallizes or solidifies within certain limits of temperature and pressure, those which 'freeze' at the highest temperature, notably tin and wolfram, crystallized first in the veins. The result is that tin and wolfram are found within and close to the granites.

This feature allowed Dines to propose his theory of 'emanative centres', with tin and its associates at the centre surrounded by zones characterized successively by copper, then lead and zinc. These crystallize at progressively lower temperatures enabling them to migrate further from the granite. This is, of course, a gross over-simplification as mineralization took place over a long period with several important resurgences of activity up to *c.* 230 million years ago. Pitchblende from Redruth has been dated to *c.* 130 m.y., while the same uranium mineral from South Terras and Wheal

Owles may be as recent as 50 m.y. Already mineralized lodes often reopened to receive a new suite of minerals to suit the new environment. For this reason the same lode can contain tin and copper at the same level, while at Budnick Consols, Perranporth, for example, lead and zinc were deposited in a re-opened tin lode.

Mineralization is virtually absent from eastern Dartmoor and the Isles of Scilly. The richest areas are linked to the granites and buried ridges of the Camborne–Redruth area with smaller, but none the less profitable, fields in the Land's End peninsula, around St Agnes, in and around the St Austell or Hensbarrow granite, and along the southern side of Bodmin Moor. From here it follows the buried granite that outcrops at Kit Hill and Hingston Down before petering out on the western flank of Dartmoor.

A brief account of the weathering that has affected copper lodes is given in the Introduction. Cassiterite, the only tin ore of importance, is a stable oxide unaffected in the lode by secondary chemical weathering. Thus in some lodes above the water-table it may be the only mineral of economic importance even though it may be accompanied by ubiquitous iron oxides. The iron sulphide pyrite, and iron-rich sulphides such as arsenopyrite and chalcopyrite oxidize to produce an iron-rich 'capping' to the lode known to the Cornish miner as gossan or 'iron hat'. For this reason the discovery in prehistory of an outcropping lode of tin need not have led to the immediate discovery of workable copper, though it could have in some instances. Whether copper or some other mineral such as the lead sulphide galena were worked in prehistory in Cornwall cannot be proved, for any traces of lode mining, open-cast or underground, would have been long since obliterated in the intensive mining of later centuries, especially from the 17th century onwards. Bartholomaeus Anglicus (*fl.* AD 1230–70) implied that shaft-mining had already passed its infancy in England, so it may have been known in Cornwall at this time.

Throughout the first 3 000 years of Cornish tin working it is hard to believe that no copper lode was ever discovered or exploited, even if such exploitation was limited to lode outcrops in the cliffs (Fig. 61). Norden writing *c.* 1584 at a time when copper mining was in its infancy in the county, remarked how the cliffs between Pendeen in St Just and St Ives 'doe glitter as if there were much Copper in them'.[2] Leland, describing the south coast between Dodman and Tywardreath 40 years earlier, wrote that the area is 'replenished with Tynne Werkes with Vaynes yn the Se Clyves of Coper'.[3] Such blatant signs of mineral wealth cannot have been completely missed in prehistory.

There sits in the manager's office at South Crofty Mine a lamp of black-burnished coarse fabric rumoured to have been found in a cliff or 'old mens' working' at Levant or Geevor in St Just, but the story

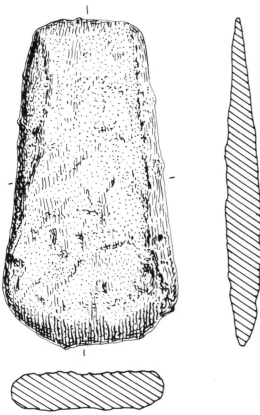

62 Flat axe of copper from Penolva, Paul, 1973; length 8·9 cm (courtesy Penlee House Museum, Penzance)

cannot be verified. Donald Bailey at the British Museum (*in litt.*) has identified the lamp from photographs as Hellenistic of the 1st or 2nd centuries BC, made in Italy or further east, but concludes that it is 'unlikely to have been an ancient import'. In spite of claims to the contrary, the lamp has no value as an indicator of early lode mining in Cornwall.

Early bronze age flat axes of pure copper are rare in the south-west, as might be expected in a tin-rich province. There is one from Bridestow, Devon, in Plymouth Museum; one from 'Cornwall' in the Royal Ontario Museum, Canada;[4] and one found in a field

Penolva, Paul	'Cornwall'
0·005% Pb	0·05% Pb
0·151% Ag	0·005% Ag
0·010% Ni	0·025% Ni
0·003% Zn	— Zn
0·009% Fe	0·019% Fe
0·196% Sb	0·04% Sb
0·001% Bi	0·05% Bi
0·11% As	0·05% As
0·000 4% Mn	— Mn

called 'The Stitches' in Penolva, Paul, now in the Penlee Museum, Penzance (Fig. 62). The latter has been kindly analysed by Dr Hughes of the British Museum Research Laboratory; and its trace elements may be compared with those of the axe in the Royal Ontario Museum. They are not inconsistent with a Cornish origin for the copper, though they do not prove it.

Firmer evidence comes in the form of a late bronze age socketed axe and two pieces of a plano-convex, 'bun-shaped', copper ingot found in 1935 when a garden was laid out in the valley leading to Gillan Creek, St Anthony-in-Meneage. The copper content of the ingot was about 98%. The conclusion of Tylecote, based on the 17 trace elements present, including tin, was that the Gillan ingot had been smelted from a sulphide ore. Other traces included manganese, nickel, antimony, arsenic, bismuth, cadmium, zinc, and cobalt, all of which are typical associates of Cornish lodes.[5] A Cornish provenance cannot be proved, but at least there is no need to claim an imported copper. The minute quantity of tin (0·008%) proves that the ingot was not smelted from a smith's hoard of discarded metal artifacts.

In the summer of 1963 middle bronze age pottery and metalwork were salvaged at Tredarvah in Penzance.[6] The metalwork included a typical south-west England palstave and several pins, including one with a double-spiral head and another with a side-loop. Of more interest is the lump of iron oxide found with them. It is a fine specimen of dark brown to black botryoidal goëthite, typical of material formerly called banded limonite or brown haematite. In Cornwall the finest specimens came from Restormel near Lostwithiel, but good specimens were also obtained in the St Just mines, notably at ɔ˙ ˙ ˙tallack. The Tredarvah specimen would grace ˙ ˙ ˙ ˙ ˙ ˙ ˙ ˙ ˙ineralogist's cabinet, having sharp edges where it was broken from the parent lode. It has all the appearance of having been collected in a fresh condition underground, strongly suggesting the working of a tin–copper lode in the sea cliff in the St Just area only 11 km from Tredarvah.

During the excavations of the Romano-British settlement at Porthmeor, Zennor, several pieces of haematite *c.* 5 cm long were found in hut circle 1. One piece had been scraped, perhaps to form powder for a haematite slip to decorate pottery, traces of which were found on some sherds.[7] A local origin in the cliffs of West Penwith is likely, for massive red haematite and poor 'kidney' ore are common in St Just with notable specimens from Levant and Botallack.

In 1871 an early bronze age dagger and a Cornish urn were recovered when a barrow was removed from near the cliff edge at Angrouse, Mullion. Among the finds was a 'fractured globular specimen of mundick or iron-pyrites'[8] (now in the British Museum), probably used as a strike-a-light. The specimen, 4 cm in

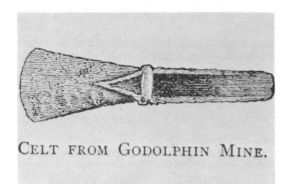

CELT FROM GODOLPHIN MINE.

**63   Middle bronze age palstave now lost, but a side
view in W. C. Borlase's MS *Ancient Cornwall*
shows that it was a high-flanged palstave typical of
those found in the south-west; length *c.* 17·5 cm**

diameter, is not of local origin but is marcasite from the
Cretaceous chalk of south-east England. It is
important here only in showing that the native
population was familiar with iron sulphide, which is
common enough in Cornish lodes.

The most telling discovery was reported by Borlase.[9] He illustrated a middle bronze age palstave
(Fig. 63), which had been found '*with many others* [my
italics], in a coffin, at Godolphin mine, probably between 1740 and 1750'. Borlase assumed it had been
found in a grave, but it is much more likely that all the
axes were found in an old mine working, for the term
'coffin' or 'goffin' was ordinarily used by miners to
describe an open-cast working which followed the
course of a lode. Godolphin Mine, near Breage, is
ancient, reputedly working in 1678, and plans dated
1824 showed a line of 'hypothetical lode outcrops'.[10]
Henwood described the vein[11] as containing cassiterite
as well as copper in the form of chalcopyrite, chalcocite, and native copper, up to 2·4 m thick, and thus an
ideal outcrop for open-cast working in the bronze age.
As proof of the shallow nature of these lodes, Symons[12]
refers to the discovery of a rich copper lode at
Godolphin by tin streamers cutting a leat.

Similar surface workings followed the parallel lodes
of Wheal Vor, also in Breage. James[13] thought they
were worked in Roman times, though without any supporting evidence beyond the fact that contemporary
remains including milestones occur in the area. At
Godolphin 1 600 copper coins of Gallienus, Victorinus, and Tetricus were found in an urn in April 1779,
while a similar smaller hoard was found in Castle
Pencaire in the mid-19th century.

Native copper with small amounts of malachite,
chrysocolla and other secondary coppers occurs in the
extensive serpentinite intrusion of the Lizard peninsula. Mr Ashurst Majendie of Mullion was 'informed
that at low water in spring tides, narrow veins of native

copper may be observed'.[14] Small mines were worked
at Wheal Fenwick near Mullion Cove and at Wheal
Unity to the south-east, while in the centre of the
peninsula lies South Wheal Treasure. Wheal Unity
originated about the 1720s when 'as one Peter Hall was
riding in Predannack Common in the vicinity of
Trenance, his horse kicked up a piece of malleable
copper, in consequence of which a mine was opened
and a quantity of native copper and a small portion of
ore was discovered, both very near the surface'.[15] It is a
reasonable assumption that the Lizard's limited
copper resources were known in prehistory.

Hall's discovery sounds apocryphal, but there are
similar accounts from elsewhere in the county. The
most superficial discovery was at Perran Wheal
Virgin, Perranzabuloe, about 1845 when a lead lode
0·6 m wide was found immediately beneath the turf
covering.[16] Lead was similarly discovered south of
Shepherds in Newlyn East. An area of swampy croft
had been purchased in 1798 by Sir Christopher Hawkins and in subsequent deep ploughing 'a quantity of
very rich lead ore was thrown up', permitting mining
operations for a considerable period.[17] There is evidence from Gwithian, overlooking St Ives Bay, of
bronze age ploughing. This would not have been as
deep as in Hawkins' day, but it could reveal outcropping lodes or even loose pebbles of 'shoad' tin,
reminiscent of the Spanish peasants (already mentioned) who collected pebbles of cassiterite from the
surface of newly ploughed fields.

Tin streaming has revealed lodes, notably on Goss
Moor where:[18]

> There is little doubt that in the course of these operations
> the backs of many lodes were revealed, both in the moor
> itself and in the upper reaches of the Mawgan stream
> which rises on the flank of Castle-an-Dinas. Two such
> lodes were specifically referred to in Carnsewe's notes of *c.*
> 1580, 'one in the north side of Castell Dennyse and rich by
> Bouchard's report [Burchard Kranich]', and one other
> greate loade before John Merryfeldes doure.

At the old Fatwork and Virtue Mine at Indian Queens
on the edge of Goss Moor, the main lode was nearly
27 m wide with rich cassiterite in innumerable short
veins. These were often arranged in horizontal layers
or 'floors', which must have attracted 'the attention of
streamers in the earliest times'.[19] Similarly at Wheal
Maudlin in Lanlivery there were outcropping tin lodes
'immediately under which it had been extensively
stoped by the ancient miners for tin'.[20]

In the face of such evidence it would be remarkable if
all outcropping lodes remained undetected throughout
prehistory. Nevertheless, however common such exposures were or intensive the activity to find them, tin
streaming itself was the principal method of winning
ore. Even on Goss Moor it may have proved more
profitable to leave exposed lodes alone and to concentrate on pockets of alluvial or eluvial tin.

64   Gold working in California; the water-course has been diverted allowing digging behind the protection of the wooden barrier (engraving from Simonin's *Mines and Miners*, 1868)

TIN STREAMING

Tin streams are so widespread in Cornwall that it is pointless to list them all. Any stream coming off the granite and crossing lodes will pick up cassiterite, while shallow basins such as the Goss and Tregoss Moors were rich in much of their extent with coarse cassiterite. Most prehistoric finds in tin streams have come from localities west of a line from the Camel estuary to Fowey, so the finds themselves are a sufficient indication of the areas most attractive to tinners from the bronze age to the mediaeval period. But before these finds are itemized, some account is necessary of the nature of the tin stream deposits and the method of working them.

Archaeologists in general have a false impression of the nature of tin streaming, as the following quotes from recently published works reveal:

> [Tin streams] from which black cassiterite pebbles could be collected with no more equipment than a knowledgeable eye and a pair of willing hands.

> [Tin] supplies before the Roman period were obtained from panning streams rather than mining.

> In early periods the most accessible deposits would have been alluvial gravels along the banks of streams and rivers.

> Tin was found in Cornwall where it was both mined and recovered from stream beds.

Such statements conjure up a picture of an old man with a frying-pan sifting the sands and gravels in a shallow stream much as the 'forty-niners' are popularly believed to have done in California. In fact it is not true of California, where many new arrivals were surprised to discover that gold had to be dug from the ground and was not lying on the surface. Daniel Woods, in his account of gold working at Salmon Falls on the South Fork of the American River written on 4 July 1849, described the reality thus (*see* Fig. 64):[21]

> Prospecting – A spot is first selected, in the choice of which science has little and chance everything to do. The stones and loose upper soil, as also the subsoil, almost down to the primitive rock, are removed. Upon or near this rock most of the gold is found, and it is the object in every mining operation to reach this rock, however great the labor, even though it lies forty, eighty or one hundred feet beneath the surface.... [When found] some of the dirt is then put into a pan taken to the water and washed out with great care.

In Cornwall nowadays, panning would yield a small amount of tin if undertaken, for example, in the lower reaches of the Red River at Gwithian, for here the recent sediments are charged with the effluent of mines in the Camborne area. In spite of a succession of streamers along the valley in the recent past, each one treating the waste of his neighbour higher up the

65   Tin streamer at Porthtowan beach on the border of Illogan and St Agnes parishes about 1902. Such simple methods of treating waste material washed down from working mines inland were regularly seen on beaches in the St Agnes area until the 1930s. While a good illustration of the primitive equipment needed to recover fine cassiterite from tin-bearing sands, the picture gives a false impression of how tinners extracted pebbles of cassiterite from the 'tin ground', which often lay at a considerable depth below the surface, even below sea-level (photo F. and E. Bragg)

valley, fine 'slimes' containing tin escaped capture to end up in St Ives Bay (Fig. 65). Indeed, there are current plans to dredge tin from the Bay. The recovery of 19th century tin dressing floors was said to be as low as 50% in some cases,[22] while as recently as 1913 Thomas[23] confessed that:

> Although matters are improving slightly owing to better mechanical appliances, mine adventurers are still faced by the fact that only from 60 to 65 per cent of the tin ore can be extracted, the remainder going to the tin dressers [ie, streamers] or to the sea.

Such a stream would be charged with other met-alliferous debris from the lode-mines. Collins noted in 1912 how the Red River was 'extremely impure from the presence of pyrites, mispickel, wolfram, chalcopyrite, and other more or less readily de-composed minerals'.[24] Such an assemblage readily distinguishes these recent deposits from what may be termed the natural or 'virgin' stream tin encountered in prehistory. There are rare instances of other minerals (apart from gold) in 'virgin' deposits, such as the 'pieces of cubic galena' at 'Swan Pool, Ladock',[25] and wolfram at Kenton Marsh on Bodmin Moor,[26] but they would have had little impact in prehistory. In Collins' day, the exploitation of 'virgin' stream tin was almost entirely a thing of the past:[27]

> A few straggling tin-streamers may be met with occasionally working over again the old heaps of debris, but so much is the industry a thing of the past that *even the name has been adopted by the people who work over the sands and slimes washed down from the various mines* [my italics].

Only one instance has come to my notice of a prehistoric object turning up in a stream works not treating 'virgin' stream tin. This is a spindle-whorl of mica (presumably from a greisen vein), now in Truro Museum, given to Revd Rundle of Godolphin in the late 19th century. He described it as coming from 'the stream works below Godolphin Bridge'.[28] This was a small concern leased in 1880 by John Stephens for working 'mine sands and slimes'.[29] The spindle-whorl could have come from anywhere in the district and there are several instances of trial workings cutting archaeological sites, such as Castle Pencaire. So its recovery at Godolphin Bridge was pure chance and cannot be taken as evidence of prehistoric tin streaming in that vicinity.

Prehistoric tin streaming in Cornwall was not a labour-saving occupation. It involved plenty of deep digging (Fig. 66), as implied in the description (quoted earlier) by Diodorus Siculus. The natives:

> work the tin, *treating the bed which bears it* [my italics] in an ingenious manner. The bed, being like rock, contains earthy seams, and in them the workers quarry the ore, which they melt down and cleanse of its impurities.

The passage is a little ambiguous but is good enough to be comparable to the accounts of streaming so well documented over the past few centuries, especially in the earlier 19th century when important deposits of 'virgin' stream tin were still being exploited.

The name 'tin streaming' is unfortunate as it bolsters the idea of panning in a stream bed. In reality, 'streaming' refers to the washing of the ore after it has been dug up. This sense survives in Cornish dialect:

66 Digging out the 'tin-ground' on Red Moor, Lanlivery, sometime between 1900 and 1914. The overburden consisted of 'head' (note the angular fragments in the excavated face) and some alluvium which was removed by hand. Only the wheelbarrows distinguish it from a prehistoric working. The area was worked from west to east and the depth of the tin varied from 2 to 8 m. The extracted material was raised by gravel pumps into sluice boxes where the ore was periodically dressed. The low water-table made recycling necessary, the water being pumped out of settling 'paddocks'. Somewhere in this tin deposit was found an iron age La Tène 1 brooch, formerly in Evans' collection and now in the Ashmolean Museum, Oxford (photo and technical data from Consolidated Goldfields Ltd, 1980)

whereas a person up-country will wash or rinse articles such as clothing, a Cornish person will 'stream things out' under the tap.

THE NATURE OF TIN STREAMS

Streams flowing off the moors at the present time obviously carry particles of cassiterite and other heavy minerals, but the transported sediments are so diluted of tin as to be devoid of anything recoverable. In valley profiles the 'tin ground', as the streamers called it, rested on top of the solid rock of the valley floor, as may be seen perfectly in the photograph of the specimen from Pentewan (*see below*, Fig. 80). Above the tin ground were variable thicknesses of sediments, both fluviatile and, near the sea, marine. Sometimes, as on Goss Moor, a poorer band of tin lay higher in the succession resting on what the tinners called a 'false floor', but this was not always present. In any case, all the profitable tin lay on the valley floor. Above the tin ground and occasionally on the moors below it as well were found the remains of wood and other vegetation, as this description of 1847 indicates:[30]

The stream-works in the valley of the Fowey, on the Bodmin moors ... show that twice has the surface been clothed with vegetation. The first time, on a granitic soil, grew large timber trees: a flood laid them down ... with their heads directed down the valley ... and spread a layer of granite pebbles and tin over them: another soil was formed supporting a vegetation of bushes and ferns, the resort of deer, and upon this a finer gravel, the result of slower and longer diluvial action, accumulated: and lastly, on this, a third bed of peat has arisen, crowned with no leafy honours, and whose tallest plant is the low but elegant heath.

The formation of the 'tin ground' occurred principally during the last period of glaciation, though there are indications of older 'palaeo-cassiterite placers'. At Lower Creany tin stream on Redmoor, Lanlivery, at *c.* 137m above sea level, Henwood reported that the tin ground contained 'flints of considerable size'.[31] His observations have been questioned, but if correct they can only represent the vestigial remains of the Upper Cretaceous chalk envisaged as having formerly covered the south-west peninsula, and thus indicating erosion back to the early Tertiary.

According to Camm and Hosking,[32] orange and red cassiterite grains at Red Moor and Breney Common can be matched in the *in situ* lodes only at Gaverigan some 15 km to the west. Unless some other lode source has been completely destroyed, only longshore drift during a period of Pliocene submergence could account for this phenomenon. Such longshore drift would also account for the very rounded cassiterite pebbles on Goss Moor, though others here are not rounded, hence non-marine and presumably Pleistocene in origin.

Recently Charles Smith, geologist at Geever Mine (*in litt*, 1982) has studied palaeo-cassiterite alluvials in the St Just area of Land's End where they occur on the major marine platform between *c.* 114 and 137 m above the present sea level. The streams in the region are so short that fluvial action is insufficient to account for the sub-rounded pebbles of quartz, quartz–tourmaline and so forth which abound. The former coastline, at *c.* 137 to 144 m included the embayment now known as Bog Inn, 1·6 km upstream of Tregaseal. The narrow channel to the sea was eroded by tidal ebb and flow and Atlantic storms along the weak line of kaolinized granite. In this environment the lighter material would have been scoured out allowing the accumulation of heavy minerals, including cassiterite, on and immediately above the bedrock. The tin would have come from the Balleswidden and other lodes outcropping around the Bog Inn depression. Similar occurences are known on the west coast of Thailand where concentrations of alluvial cassiterite can be related to both modern and ancient shore-lines.

For the most part such early alluvials would have been eroded away during the early Pleistocene, though a little cassiterite survives in the Pliocene gravels at Polcrebo in Crowan, flanking St Agnes Beacon, and on Crousa Common in the Lizard peninsula. It is a reasonable assumption that the tin ground, certainly in the lower valleys, was a product of the Devensian or last glaciation. In this period the ice sheets at their maximum extent *c.* 16000 BC reached the Bristol Channel and clipped the Isles of Scilly, leaving the south-west peninsula to suffer the rigours of a periglacial climate. Under these arctic conditions deep permafrost predominated until a relaxation of the intense cold allowed summer thaws to pour cascading melt-waters charged with the debris of erosion into the valleys to a sea level estimated to have been several hundred metres below its present level. Only the heaviest material, including tinstone, could resist the force of the water enough to allow it to come to rest on the valley floors at places still several kilometres inland today. A good deal of very fine tin must have been swept out to sea or lies out of reach beneath Carrick Roads, the Camel estuary and other drowned river valleys or 'rias'.

The speed of weathering and transportation was seen in the Pentewan tin ground where large boulders sometimes contained not only cassiterite but iron pyrite and other sulphide ores. Under oxidizing conditions these would not survive attack by chemical and mechanical weathering. Only during the climatic amelioration following the retreat of the ice would finer particles of tin, and sand and gravel devoid of heavy minerals, settle in the valleys escaping transportation to below sea-level. Support for this idea comes from south-east England where mid-to-late Devensian periglacial phenomena are assigned: 'large quantities of rock waste were scoured off the bare landscape into the rivers that had eroded deep channels in response to low sea-levels'.[33]

Oak and alder grew on top of the tin ground in the valleys, perhaps during the Allerød interglacial around 9000 BC, though this is not certain. If correct, the upper incomplete tin bed of little or no economic value may have formed during the succeeding cold 'snap', which lasted for less than a millennium. From 8000 BC the climate warmed rapidly with a concomitant rise in sea level (the Flandrian transgression) which caused the drowning of river valleys and the deposition of thick sediments, both marine and fluviatile. The major period of deposition is considered to have been between 7000 and 4000 BC.[34]

In the south-west the rise in sea-level has been studied in detail at Start Bay, Devon, where Clarke proposed a shoreward migration of beach barriers and shore lagoons between 7000 and 5000 BC. The rise in sea level was about 1·5 m a century which, on a shallow sloping shore of about 1:500, meant an annual advance of 7·6 m. Clarke believed this advance occurred as a series of 'catastrophic inundations separated by quasi-stable periods'.[35] A similar series of events must have occurred in Mount's Bay in west Cornwall. Dating the final rise of sea level is difficult and not yet resolved, but a mesolithic kitchen midden at Westward Ho!, north Devon, lay at about high tide level when in use about 5500 BC.[36] According to Taylor the rise in sea level is characterized from then on by a succession of minor oscillations,[37] though this is now disputed.[38]

At all events, present sea level was more or less attained during the neolithic between 4000 and 2000 BC. The submerged forest off Penzance at Lariggan (Fig. 67), in common with others around the Cornish coast, grew close to sea level. The amount of alder indicates a wet environment, while oaks also grow down to the tide line at the present time, their lower branches being 'cropped' at every high tide. The final eustatic rise swamped the Lariggan forest around 2000 BC. The calibrated ^{14}C date of oak collected here in the exposure of 1883 is 2315–1655 BC (3656 ± 150 bp).[39]

The point of this discussion is to show that the

67   Submerged forest in Mount's Bay exposed in 1883, viewed from near Penzance railway station looking towards St Michael's Mount (photo Gibson & Sons)

sediments that accumulated in the lower parts of Cornish valleys would have reached approximately their present depth by the beginning of the bronze age and the earliest activities of tin steamers. There has been further deposition by rivers during the neolithic, for according to Taylor,[40] this occurred wherever extensive local forest clearance promoted greater erosion and run-off. The bronze age tinners who concentrated their attention on the lower valleys were thus confronted by a tin ground buried to a considerable depth. An added layer of detritus was deposited rapidly in recent centuries as a result of mining and china-clay extraction. These phenomena are particularly noticeable in the valleys flowing into St Austell Bay and the Carnon Valley at Restronguet.

In the Carnon Valley silting was aggravated by the construction of the County Adit begun in 1748. Nearly 65 km long, including several branches to many important mines in the Gwennap area, it emptied into the valley at Ferny-splat (or Furnace Plat) below Bissoe.[41] Henwood noted that debris from the adit had raised the valley floor at Higher Carnon by *c.* 0·3 m a decade, for the granite slabs of a narrow road that was *c.* 1 m above the ordinary level of the stream in 1815–20 were buried by *c.* 0·6 m of detritus in 1867.[42]

South of the St Austell granite recent silting has been due primarily to china clay production. Whereas the waste from tin streaming was about 1 t of gravel for every 1·3–1·8 kg of black tin, china clay produced 8 t of waste for every tonne of clay produced. Symons noted in 1877 that the Par valley was 'lately an estuary'.[43] Within a century of his writing mooring posts at Pont's Mill, 4 km inland today, were used by vessels of 70 or 80 t. The sediments at Pont's Mill are about 7 m deep, but can exceed 21 m above the tin ground at Par. Everard found that three of the five kinds of beach material in the Par–Pentewan area were largely of river-transported mine waste. Removing 3·6 m of recently derived upper sediments would mean that streamworks at Par and Pentewan were even more prone to inundations by the sea in the bronze age than they were in the winter of 1801 when the sea destroyed the streamworks at 'Poth', as Rashleigh called the workings at Par.

The number of bronze age finds from streamworks near the coast may be partly a consequence of hastily abandoned workings, when self-preservation overcame the tinners' desire to recover their tools. Evidence for this comes from Pentewan and is discussed in Chapter 25. Moorland workings could also be dangerous. The

West Briton of 26 September 1817 reported that operations around Fat Work, Be Lovely, Gilly, and Wheal Grace on Goss Moor had been abandoned 40 years before through 'want of machinery' and because a 'great accession of water drowned a miner'. The workings were only 9 m deep.

The rich tin ground does not extend far up the valleys from the sea. Colenso[44] reported the lack of tin ground in the Pentewan valley above St Austell bridge, while Henwood[45] noted that the search for stream-tin in the upper Carnon valley had met with little success. Here the tin was downstream 'where the Stithians vale and glen called Smelting House vale [the valley through Perranarworthal] opened into the Carnon vale', the pebbles of cassiterite becoming increasingly large towards the sea. For this reason it paid prehistoric tinners to work as close as possible to the sea. At Restronguet in the Carnon Valley an attempt was made in the 1870s to open up a seaward extension below low-tide level. Though scientifically a success it was a financial failure because of technical difficulties, and tin production ceased in 1874.[46]

In 1984 Truro Museum was presented with a silver denarius of Julia Domna (AD 198–211) believed to have been recovered sometime before 1910 in tin streamworks at Reskadinnick on the Red River below Camborne. There is also a possibility that a Roman lamp and a small bone figurine, also in Truro Museum, came from similar workings downstream of Reskadinnick in the latter half of the 19th century.[47] Reskadinnick lies only a kilometre from the Roman 'villa' at Magor. This might lead to the conclusion that these items token a Romano-British interest in Red River tin alluvials. Unfortunately, there is no Red River tinstone in recoverable quantities, although traces of tin were found at a depth of *c.* 10·6 m near the river mouth in a boring 100 m above Gwithian Bridge (Hosking, pers. comm.). The streamworks that lined much of the Red River valley well into the 20th century were only treating the 'tailings' which had escaped the dressing floors at lode mines in the Tuckingmill area. Between 1882 and 1892, over 1 000 t of dressed tin ore were recovered annually in all but three years.

The Roman finds have no certain provenance. The only certainty is that the Red River, and the valley between Redruth and Portreath, contain no alluvial tin, a fact pointed out in 1830 by Carne[48] who wrote that the parishes of Lelant, Gwinear, Camborne, Illogan, St Agnes, and Perranzabuloe have 'no productive streams'. In this part of the county the watershed is near the north coast, so the streams are short and relatively steep. Hence, any stream tin formed during the last glacial period would have been swept out to sea. The Red River also has few tributaries of consequence and is only 13 km long. This should be compared with the Carnon Valley where the important tin stream is fed by a river system totalling 72 km, 40 of

which cross the rich metalliferous districts of Chacewater, Scorrier, St Day, Gwennap, Baldhu, and Bissoe.[49] In this respect the north-flowing streams would resemble those of the Carnon and Pentewan valleys if the south coast had cut them off at Bissoe Bridge and St Austell Bridge, the inland limit of the tin-bearing gravels. The tin alluvials of St Just-in-Penwith are in short valleys, but their origin is different as discussed above, while the tin-bearing valleys of St Columb Minor and St Mawgan-in-Pydar are less steep with sources in the rich Goss Moor area.

Tin-bearing placers on the moors are more correctly termed 'eluvial' and 'colluvial' than 'alluvial' since they have been transported no distance at all, or only short distances from the parent lodes. Surviving evidence suggests that in prehistory the most important moorland tin streams were on the Goss and Tregoss Moors, a Tertiary marine platform 121–52 m above sea-level.

The depth of weathering during the Pleistocene was considerable. Henwood[50] noted that the tin ground at the Merry Meeting streamworks in Roche parish was up to 9·1 m thick under 6 m of overburden, while the natural bedrock or 'shelf' consisted of friable granite up to 6 m or more thick, sometimes impregnated with ore. The same was true of Polskease, Roche, where 3·6 m of tin ground 'enters the friable granite and is seen for some distance beneath it'. The Goss and Tregoss deposits lay in a marshy tract (the name Goss is local dialect for the phragmites reed) of *c.* 26 km² in a depression between the outcropping granites of Belowda Beacon and Castle-an-Dinas to the north, and the northern outcrop of the St Austell or Hensbarrow granite to the south.

Collins took an active interest in the commercial exploitation of the deposits here at the beginning of the present century. He found indications of former tin stream works scattered over an area 'of about two square miles', though no indication of any systematic work. In 1983 Gerrard[51] excavated a 14th and 15th century tin mill at Stuffle, St Neot, on Bodmin Moor on the site of the new Colliford reservoir. He also found that the valley bottom had been worked haphazardly, though the direction was upstream with the waste being carried downstream and dropped in previously worked areas.

The same may be said of the valley alluvials, and only enhances the view that prehistoric workings must have been piecemeal and completely unmethodical by recent standards of working. On Goss Moor this may have been for technical reasons. Henwood wrote in 1873 that 'some forty or fifty years ago the Tregoss Moors exhibited an almost countless succession of low, stony hillocks, and deep, weedy pools, the abandoned scenes of earlier operations. Amongst them, however, many small *tin-stream-works* were still industriously wrought by speculative workmen, either on ancient

68   Stream-tin workings in Kenton Marsh on the north-west side of Buttern Hill, Altarnun, Bodmin Moor, about 1907. Immediately to the left of the men standing in front of the corrugated iron sheet is the 'tye' or waterfall, beneath which the deposit was turned over and the fine material washed away. The tin and wolfram were left partly under the fall and partly in the wooden trough seen behind the upturned wheelbarrow. The depth of the tin-bearing gravel is not stated; the most profitable deposit was deepest, but probably did not exceed about 6 or 7 m. The ground had been turned over in the past, though large areas remained untouched (photo *Q.J.G.S.*, 1908, Fig. 2)

detritus [i.e. on hitherto unworked deposits], or on matter imperfectly gleaned by their predecessors. The works were drained either by *open-cuttings* – by hand-pumps – or by little *lifts* worked by water-wheels, which seldom exceeded, and were often less than six feet in diameter.'[52]

Drainage would have posed problems to the prehistoric tinners, yet at the same time in other parts of Tregoss Moor lack of water would have been a problem. 'Towards the southern margin' of the moor, continued Henwood, 'small quantities of *stream-tin-ore*' are gleaned by a few tinners for 'as long as – but no longer than – the rains of winter and spring supply them with water'. It was the difficulty of maintaining a sufficient and constant supply of water, plus the need for cheap removal and eventual replacement of overburden, which brought all such operations to a close by 1880.[53] The scarcity of water in the summer months may come as a surprise, but these recent accounts all emphasize the difficulties of streaming, which may have placed some moorland deposits beyond the reach of prehistoric tinners. The problem of water is highlighted by a small streamworks south-east of Castle-an-Dinas called 'Wet and Dry'.

The problem of streaming stopped by water-logging

on Bodmin Moor was mentioned by Barrow[54], but in general it should be emphasized that Cornwall's mild climate permitted streaming throughout the year in some workings. This state of affairs would have been impossible in less equable regions such as the Erzgebirge. However, the fortitude of tinners should not be underestimated so long as water for ore dressing continued to flow.

The stream deposits around Wheal Glasson east of Indian Queens are typical of Goss Moor. The overburden was described by Collins as consisting variously of clay, sand, gravel, and 'stent' or the waste of former workings still containing traces of tin with occasional rich pebbles, as well as disrupted bands of peat. The tin ground varied in thickness and rested on a very irregular surface of bedrock. Much of the best tin lay in pits or pockets within the bedrock, perhaps 9 m deep, and it was the tinners' custom to search for these. This must have applied in prehistory as well, as proved by the depth of bronze age finds in streamworks in Roche parish. Barrow described how previous tinners at Kenton Marsh, Bodmin Moor (Fig. 68), had deliberately followed the deeper tin-bearing channels winding across the marsh. They heaped up and left the overburden on one side forming drier ground where small trees grew, so it was possible from a commanding position to trace the courses the tinners had taken.

At Wheal Glasson, Collins reckoned on recovering *c.* 0·2 kg of 'black tin' (cassiterite) 1 m from the overburden which was *c.* 1·8 m thick; the tin ground averaged 5·5 m thick and yielded 1·3–1·6 kg m^{-1}, including the richer 'prills' at the bottom. (Prill referred to any piece of good, rich, solid cassiterite, as found or after ore had been treated or 'dressed'.) Collins described three types of tinstone: 1 Fine grains, clean, and of high quality; 2 Small pebbles and 'prills' of equal or even higher quality; 3 Fragments of ore and veinstone requiring stamping (i.e., crushing). Henwood gave similar accounts of three streamworks operating in 1873 on the north side of Goss Moor: Golden Stream, Wet and Dry, and another unnamed. Here the cassiterite was 'usually in a state of gravel or sand, but sometimes as minute unfractured crystals'.

The foregoing account should not lead to the conclusion that Cornwall was totally devoid of any tin lying on the surface. Small amounts occurred as 'shodes', tin ore 'sprinkled on the surface at various altitudes'.[55] In Henwood's day it was still found 'to some trifling extent', but over a century earlier Borlase noted some shodes 'in great numbers, making one continued course' more than 0·2 km from the lodes.[56] Shode stones were found either on or not far below the surface and sometimes appeared water-worn in spite of the short distance of transport. Since the trend of Cornish lodes is roughly ENE–WSW, deposits of shode were more numerous on the northern and southern flanks of hills, i.e. parallel to the strike of the lodes.

There is little indication of the value of shodes as tin deposits in their own right. They may have been of limited importance as a source of ore in prehistory. This is impossible to prove, but since bronze age tinners exploited the deepest alluvial deposits (as will be made clear in the next chapter), and because so much shode survived into historic times, it can have been of little importance in antiquity. For the mediaeval and later tinners, shodes were principally an indication of the proximity of the parent lodes which were their main objective, as quaintly expressed by Norden about 1584:[57]

> Little stones lye both in and nere the Brookes, and upon the mountaynes wher the metall lyeth; theis stones they call the Shoade, being parcel of the veyne of owre, which being dismembred from the bodye of the Loade, are meanes to direct to the place of profite, as the smoake directeth where the fire lurketh.

An important aspect concerning the distribution of alluvial tin deposits is discussed by Camm and Hosking (Fig. 69).[58] It is clear that some major alluvials cannot be related to any existing lode outcrops. The authors argue that most alluvial tin must have been derived, not from the major tin–copper lodes of the type familiar in the Camborne–Redruth area, but from networks of close-packed small veins ('stockworks') such as that seen at Mulberry near Lanivet. Thus, the existence of tin-bearing alluvials on parts of Dartmoor in Devon, on the north-east side of Bodmin Moor and elsewhere in Cornwall, must have derived from stockworks in the granite roof zone. In those areas they were largely removed by erosion during the Tertiary, particularly between the end of the Mesozoic and the end of the Oligocene when the region was subjected to a sub-tropical climate.

GOLD IN CORNISH TIN STREAMS

Apart from tin, the only other metalliferous mineral recoverable from streamworks of use in antiquity was gold. In an earlier section was mentioned the probability that the importance of Irish gold in antiquity has been exaggerated. Similarly, the importance of Cornish gold has been underestimated or ignored by archaeologists. While gold is outside the central theme of the book, it is likely that its presence acted as a stimulus to streamers to concentrate their attention on deposits well endowed with the precious metal. Almost all Cornish tin streams carried some gold, but not enough for workings to be conducted with gold as the prime quest. Gold was just an added bonus. Carew[59] disparagingly commented in 1602 that:

> Tinners do also find little hopps of gold amongst their ore, which they keep in quills, and sell to the goldsmiths oftentimes with little better gain than Glaucus' exchange.

Glaucus exchanged his golden armour for iron. George Henwood[60] in 1855 had much to say about the gold found in the Carnon Valley tin stream:

> I have seen a few [pieces of gold] as large as half a pea, but they were rare instances; far more frequently half the size of the head of a pin, and even smaller than that ... They did not van for gold [i.e. separate it on a vanning shovel], or the produce would have been much greater. I heard of one nugget weighing 1¼ oz. attached to a piece of quartz matrix, but did not see it; this specimen shows the gold to have come from a vein in the neighbourhood for had it rolled a great distance, the vein would have been detached ...
>
> For the preservation of these prills [of gold] the tinners usually carry a quill about them with one end cut off, to which they fitted a plug of wood, and into this small receptacle they carefully dropped the precious metal, the finding of which was considered a perquisite of the lucky finder. It never was discovered in such quantities as to be worth the notice of the agents, but where gold existed the miners worked for lower wages in consideration of the privilege.

A nugget weighing nearly 56 g and now in Truro Museum was found in the Carnon tin stream in January 1808 (Fig. 70). Few discoveries were anywhere near as impressive as that, but a contemporary account of its discovery added that 'gold has been

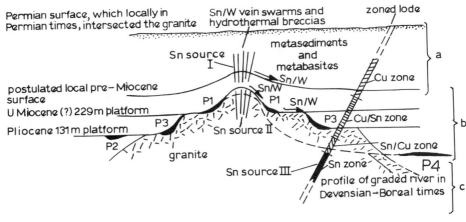

Permian surface, which locally in Permian times, intersected the granite

Sn/W vein swarms and hydrothermal breccias

zoned lode

Sn source I

metasediments and metabasites

Sn/W

Cu zone

a

postulated local pre–Miocene surface

U Miocene (?) 229m platform

Pliocene 131m platform

P1　P1　Sn/W

Sn/W

P3

P3　Cu/Sn zone

b

P2

granite

Sn source II

Sn/Cu zone

Sn source III

Sn zone

P4

profile of graded river in Devensian–Boreal times

c

KEY

PLACER TYPE

HIGH
LEVEL

P1　Probably, in part, pre-Pliocene. In part, at least, Pliocene and Pleistocene (Head). Derived largely from primary vein swarms

P2　Pliocene beach deposits. Not related to present drainage systems. Locally fragments of Oligocene fluviatile deposits (at St Agnes). Probably derived, in part, from early components of P1 placers and, in part, directly from primary deposits (probably largely from vein swarms)

P3　Post-Pliocene deposits. Related to present drainage systems. May be derived from early Pliocene littoral and colluvial placers, from higher level placers, and from primary (essentially vein-swarm) deposits

P4　Late Devensian-Early Boreal. Derived from reworking of early placers and Head in the catchment area and from intersected lodes

LOW
LEVEL

a　Permian to *c.* end of Mesozoic SnO_2 partly or wholly lost to adjacent seas. Dominant source of SnO_2, lode swarms and, possibly, hydrothermal breccias

b　From *c.* Upper Miocene until Late Pleistocene. SnO_2, largely from vein swarms and breccias accumulate in streams and in littoral accumulations. SnO_2 dispersed in marine sediments on platforms, and in head, on platforms and valleys during much of Pleistocene

c　Late Devensian – Early Boreal. Cassiterite accumulated in rejuvenated streams. Source of Cassiterite – Earlier high-level placers. Head and intersected lodes

69　Conceptual diagram indicating relative times of destruction of major types of primary tin deposits in Cornwall, and locations and order of development of tin placers (Hosking, 1983; after Camm and Hosking, 1984)

frequently found in stream works ... in larger quantities than is generally known'.[61] Also displayed in Truro Museum is alluvial gold from the valley below Ladock church, some of it mixed with fine cassiterite, as well as a necklace made in 1802 from Ladock gold (Fig. 70) It was common enough for tinners to find enough gold to make wedding rings. Revd Iago knew an old inhabitant of Bodmin who wore a ring of Boscarne gold;[62] this streamworks was frequently exploited in antiquity. Mr Spoure of Trebartha in North Hill (died 1696) made a signet ring out of 21 g gold from one of his tin streams.[63]

In the 1830s, De la Beche saw gold taken to Truro in quills,[64] and although the amounts were small some tinners made a good haul. Borlase noted in 1758 that tinners in St Stephen-in-Brannel (probably working in the Fal valley) found gold so plentiful that they made a good profit from it.[65] Pryce[66] quoted from the MS 'The

Bailiff of Blackmoor' (i.e. between Bodmin and St Austell *tempore* Elizabeth I) that two blocks of tin being taken to Bordeaux were valued by two Florentine merchants 'to be worth all the rest of the tin there, by reason of the gold contained in them'. About 1750 some blocks of tin were refused by a merchant because of their yellow colour caused by an extraordinary excess of gold in them.[67] Gold similarly occurred in the Dartmoor tin streams where the tinners also carried birds' quills and 'not uncommonly' collected flakes of gold.[68]

Accepting the geological principal that the present is the key to the past, gold would have been found as frequently in streamworks in the 18th and 19th centuries BC as AD. The amount can only be guessed at. An analysis of a mill-float from the Carnon Valley recovered in 1958 contained[69] 1·19% tin and 0·00005% gold. As this material was contaminated by

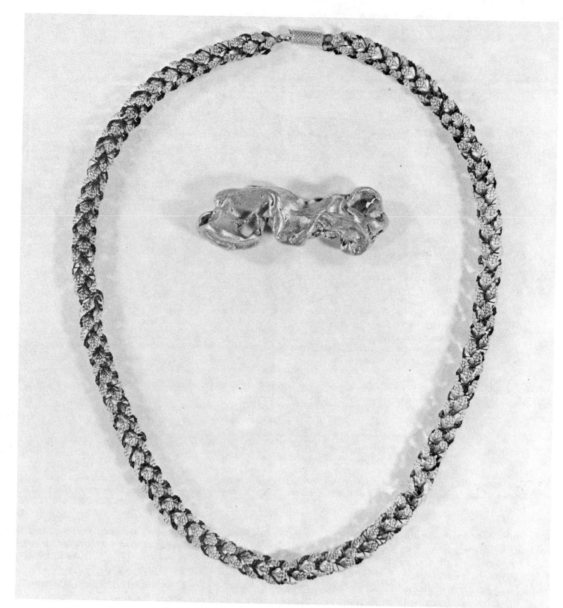

70   Gold from Cornish tin streams displayed in Truro Museum: nugget from the Carnon Valley weighing 59·6 g, length 56 mm; necklace made in 1802 from Ladock gold

the effluent of modern mining it is not accurate in determining the amount of gold recovered in 'virgin' tin ground, but it is better than nothing.

Collins[70] estimated the extent of Cornish stream deposits in valleys and on the moors as totalling some 880 km². Allowing only 0·45 m for the average thickness of the tin ground, he calculated 349 862 265 m⁻³ of ore, or roughly 1 000 000 000 t. Assuming only 0.25% to be cassiterite, he estimated 2 500 000 t of tin

of which 57% had been recovered, the remainder being unworkable according to the technology of the day. If it is assumed that a mere 0·000 05% of that total was gold, then the tin streams contained 1·25 t of gold. How much was capable of extraction in antiquity can be left to the imagination. A good deal would have been too fine to recover, but once flakes or small nuggets had been found, it is reasonable to assume that the utmost pains would have been taken to recover as much as possible. Forbes[71] gives the present economic limit for extraction as 0·000 1% compared with 0·188% recoverable in the 16th century according to Agricola.

Carew's description in 1602 of the use of turf to trap particles of tin is reminiscent of the Near Eastern practice of washing auriferous sands over sheepskins. Agricola believed this method gave birth to the legend of the Golden Fleece. The method is known worldwide; Pryce mentioned that the natives in Brazil extracted fine gold dust:[72]

> ... by laying an ox-hide on the ground, with the grain of the hairs against the water, which passes gently over it. On this they stir and mix the sand and gold-dust; by which means, the small particles sink, and are intercepted in the hair of the hide; while the sand washes off.

The use of turf in Cornwall could well date back to prehistory and have been initiated as a method of obtaining gold as well as the finer particles of cassiterite. A tin stream exploited in Romano-British times that certainly contained recoverable gold is Treloy in St Columb Minor. John Nicholls, proprietor of the workings about the 1830s, told Henwood that the stream-tin 'was frequently mixed with grains of gold; mostly about the size of wheat, but sometimes as large as peas'.[73] The shallowness of the deposits commended themselves to prehistoric tinners. The section seen by Henwood consisted of:

Alternating sand and gravel	2·4–3 m
Vegetable remains	0·7 m
Tin ground	0·1–0·6 m

While the amount of gold in tin streams is small, there is evidence for believing that Cornwall was a supplier of gold in antiquity. In Chapter 18 analyses are quoted of British and Irish gold ores. All are devoid of tin in their trace elements apart from two Highland specimens with barely detectable amounts. Curiously, gold from Cornwall had not been analysed, but through the kindness of Dr Stanley in the Department of Mineralogy at the British Museum (Natural History), Dr Williams has analysed by emission spectroscopy three flakes of randomly selected gold from Cornish tin streams preserved in the County Museum at Truro.

Most British and Irish bronze age goldwork contains traces of tin, some comparable to the amounts in the specimens analysed above; others contain rather more, but the implication is that the gold came from a tin-bearing region. As might be expected, Cornish prehistoric goldwork also contains traces of tin, as the analyses of three early bronze age gold lunulae show:

Harlyn bay (left in photo)	(Au 4539) 0·005%	(50 ppm)
Harlyn Bay (right in photo)	(Au 4540) 0·095%	(950 ppm)
Hennet Marsh, St Juliot	(Au 4541) 0·067%	(670 ppm)

(analyses per Dr Axel Hartmann, pers. comm.)

Continental goldwork also contains traces of tin. Of special interest here is the gold lunula from Kerivoa, Brittany, almost identical in weight and workmanship to the right-hand one in the photograph of the Harlyn lunulae (Fig. 71), which contains 47 ppm of tin (per Dr Taylor pers. comm.). This piece could well have been made from Cornish or Breton gold.

An oval, boat-shaped bowl (18·2 cm long), probably of late bronze age date, found in a peat working near Caergwrl Castle, Clwyd in 1823 has recently been re-examined.[74] Made of Kimmeridgian oil-shale (not wood as formerly believed), it is decorated with motifs representing circular shields below a gunwale, triangular oars, zigzag waves, and the ribs of the boat beneath. The decoration consists of gold leaf applied directly to form the shields, but elsewhere the gold was first wrapped around tin metal before being inserted into the shapes cut into the shale to receive it. The tin, which gave solidity to the fragile gold, has altered to a mixture of the tin oxides cassiterite (SnO_2) and romarchite (SnO). The gold contained among other impurities 0·13% (1 300 ppm) of tin. It is not known whether all the tin was originally part of the gold or has resulted from alloying. Nevertheless, it is possible that both the gold and the tin it was wrapped around are both products of Cornish tin streaming.

The formerly held view that Cornish prehistoric goldwork was made of Wicklow gold traded for Cornish tin is no longer tenable. Ireland was unquestionably an important metallurgical centre in the bronze age, but just as the high quality metalwork of southern Scandinavia depended on imported minerals, so too the Irish industry relied far more on imported raw materials than has been assumed until now.

Analyses of gold from tin streams

BM Lab No.		Tin in ppm	Other traces qualitative only						
			Fe	Cu	Ag	Ni	Hg	Pb	Bi
8266	Ladock	25–50	m	m	m	tr.	tr.	tr.	—
8267	Crow Hill, St Stephen-in-Brannel	25–50	m	m	m	tr.	tr.	—	tr.
8268	Carnon Valley	25–50	m	m	m	—	tr.	tr.	—

m = major trace; tr. = trace

71   Two early bronze age gold lunulae and bronze flat axe found at Harlyn Bay near Padstow in 1863 and now in Truro Museum. The axe (length 11·4 cm) contains 9·7% tin; the gold collars (left to right) contain 0·005 and 0·095% tin (photo Nat. Mus. Antiquities of Scotland)

THE ART OF TIN STREAMING

> Streamers – a mining avocation peculiar to itself, and requiring a hardihood of frame not often found in underground miners, as well as a perfect knowledge of tin in all its varieties.
>
> Henwood, *Mining Journal*, April 1860

Streaming in the mediaeval period and earlier was unsystematic and inefficient, so a great deal of ore was left to be exploited in later centuries. Carew relates that some Elizabethan tinners, instead of looking for new lodes or stream deposits:[75]

> do take in hand such old stream or lode works as by the former adventurers have been given over, and oftentimes they find good store of tin, both in the rubble cast up before, as also in veins which the first workmen followed not.

He also says that streamers looking for new deposits did the same as tinners searching for a lode:[76]

> They discover these works by certain tin-stones lying on the face of the ground, which they termed shoad.

As mentioned above, many of the important tin streams are well away from any outcropping tin lodes, hence shoad, so Carew oversimplified his account. It is not impossible that shoad played a part in the discovery of alluvials in the Pentewan valley not far from the Polgooth lodes. However, this could not have happened at such localities as Treloy in St Columb Minor, or even the Carnon valley where the most productive alluvials were several kilometres downstream of outcropping lodes. Remembering how tin was found on the beach at Piriac in Brittany in 1817, it might be assumed that stream tin eroded from the deposits would have been detected on Cornish beaches. But this can hardly have been the case when the tin ground at estuary mouths was buried beneath 12–24 m of overburden. The lodes outcropping in Mount's Bay when 'fretted off' yielded a small amount of cassiterite that storms threw up on the beaches near Penzance,[77] while Klaproth noted that at Perranporth cassiterite:[78]

> lies some yards under the sea-sand and therefore can be collected on at the time of low-water.

However such occurrences were unusual, of minor importance, and even if appreciated in antiquity would have been no use in detecting the major valley tin streams in the county.

How the first tinners became aware of the buried tin ground of alluvial deposits can only be guessed at. Trial and error and a build-up of local knowledge are obvious factors, but perhaps the use of the divining rod (*virgula divinatoria*) in antiquity should not be discounted. Both Agricola and Pryce had a good deal to say about its use. Pryce (1778) obtained much information from William Cookworthy, better known for his discovery of china-clay in Cornwall. The present writer has seen a Redruth man in the presence of geophysicists from Exeter University locate a mineral lode. To allay all doubts of sceptics, each of us present in turn held the divining rod while he held our wrists. All of us were amazed by the force with which the rod twisted in our hands, while the geophysical equipment confirmed not only the presence of the lode but the depth which he had predicted. Dowsing for minerals 'has always been at least sympathetically received in Camborne',[79] and one native of the town published a book on the subject in 1913.[80] Simonin considered the use of the divining rod by *les sourciers* in France. While regarding an operator he saw in action as just a clever imposter, he was compelled to admit that the renowned Madame Rey of Dauphiné did discover mines by this means.[81]

Carew says little about the method of tin streaming. Of later accounts, the one by Pryce in 1778 is the best.[82] The tinners first sank a 'hatch' (shaft) of up to 13 m onto the tin ground, took a shovel-full of the ore, washed off all the waste and:

> from the tin which is left behind on the shovel, he judges whether that ground is worth the working or not.

This is the technique known as 'vanning', for which a shovel with a large broad blade and short handle is employed. It is especially useful in skilled hands for sifting very fine particles from the waste sand and clay. It requires practice, but a good operator quickly concentrates the tin near the shovel's handle with progressively lighter waste on the blade beyond. The short-handled oak shovel, made from a single piece of wood, found in the Carnon streamworks *c.* 1795 is unlike the typical digging shovel. It may well have been for vanning, though its date is unknown. It is not known whether pre-mediaeval tinners understood vanning, but any attempt to recover fine tin and particles of gold, or to estimate the value of a piece of ground, should have led to the principle of vanning at an early period.

The normal shovel for digging, still used in Carew's day, appears very primitive to modern eyes, but as the handle is easily detached new wooden blades could be quickly replaced. The earliest were entirely of wood, but by Elizabethan times Carew reports that 'the utter part [is] of iron, the middle of timber, into which the staff is slopewise fastened'. These resemble the implements described by Champaud from the Gallo-

Roman workings at Abbaretz in Brittany. Most of the wooden shovels in Truro Museum are probably mediaeval, but two from Boscarne, Bodmin, of oval shape, appear more ancient (*see below*, Fig. 121). One has now been dated by ^{14}C to about the 9th century AD; the implications are discussed in Chapter 27.

Until the very last days of tin streaming in the 20th century, the only method of digging to the tin ground was with a shovel and a wheelbarrow. In the 18th and 19th centuries it was usual to work on a broad front across the width of a valley, proceeding up or down the valley, keeping the river at bay with embankments, and backfilling as they went. In the Carnon valley streaming proceeded in this way as near to the sea as Mellingey Creek, and complicated embankments were needed to keep the workings free of water.[83] No indication of any prehistoric embankment was found, suggesting that in antiquity tinners dug a 'hatch' and no more. This is discussed further below.

Having reached the tin ground and proved its value, the tinner proceeded by going:

> to the lowest and deepest part of the valley, and digs an open trench ... which he calls a level, taking utmost care to lose no levels in bringing it home to the Stream.

In this obscure sentence, Pryce meant that the level – in fact inclined to allow the flow of water – was dug from the tin working to a point lower down the valley where waste water could discharge into the river. There is no evidence to suggest that this was done in antiquity. In any case it was not possible in those parts of the lower valleys where the tin ground was not just deep, but below sea level. This was true in the Happy Union tin stream in the Pentewan valley when described by Lipscomb in 1799.[84] Here water was collected from several small streams and allowed to trickle over the side of the tin workings which were 15 m deep. The water spilled into small basins or troughs into which the tin ground was shovelled and washed. The refuse was thrown out, presumably over the bottom of the workings, while:

> the pure tin or ore are put into baskets or boxes slung upon ropes, and conveyed out of the mine by the assistance of an engine, which alternately raises the superfluous water and these boxes.

This, on a grand scale, must be the principle used in prehistory, though probably without the refinement of allowing water to trickle into the workings in the first place. Apart from using hand-pumps and water-wheels, Pryce noted how the streamers disposed of excess water by 'teeming' it out with a scoop. It is not impossible that the late bronze age cauldrons from Broadwater in Luxulyan, each of 181 l capacity, were used to keep the workings dry. Diodorus Siculus relates that the natives treated the tin ground 'in an ingenious manner', though whether their skill in drainage bettered the use of a bucket will never be known. In the lower valleys, even excluding the problem of excessive tides, superfluous water could have presented such a problem in the wetter months that tin working became in places a seasonal activity. The Archimedean screw described by Diodorus Siculus for draining mines in Spain, seems far too exotic for use in Cornwall, even if the natives of Belerium were civilized 'because of their intercourse with merchants and other people'.

There is no practical reason why in Roman times a water wheel could not have been used in Cornwall, just as it was used for drainage at Dolaucothi gold mine near Pumpsaint in Wales. However, at present there is no evidence for anything so sophisticated in the county. Water wheels were used in Cornwall and Devon at least as early as the 14th century to work machinery on the dressing floors, but they may not have been used for drainage until later. Lewis recorded that at Bere Ferrers, Devon in 1480 the ordinary method of drainage was by windlass and bucket, several men being paid as 'water winders'. A small water wheel was used, as well as 'suction pumps', which required the undivided attention of three semi-skilled mechanics.[85]

The rag-and-chain pump, described and illustrated by Agricola, is not referred to before the reign of James II in Cornwall. The horse-whim, in which an animal pulled around a vertically pivoted drum onto which the rope was wound, was also a mediaeval innovation according to Lewis. It seems, therefore, that the only drainage methods likely in prehistoric Cornwall were possibly to dig a level where practical, and more probably to raise water in wooden buckets, leather bags, or bronze cauldrons. It is also easy enough to envisage children spending all day hoisting water in small containers.

In antiquity the tinners, in most cases, would probably have removed all tin ground from the workings to be treated at some short, convenient, distance away, just as Swete described in 1780 for a small working employing about a dozen men somewhere on Goss Moor:[86]

> Their business was in one part to dig up the rubble and carry it to be placed underneath little water falls [obviously man-made] of about ten to twelve feet high, in another to separate the stones which were purified from the sand and clay by the force of the stream and to lay them in distinct heaps.

The most important indication of how bronze age tinners proceeded is provided by Stocker in 1852.[82] The workings at Wheal Virgin tin stream in the Pentewan valley were at that time about 2·4 km upstream from Pentewan harbour in the vicinity of Levalsa Meor. The streamers found an old shaft that extended from about 3 m below the surface to the base of the tin ground 4–6 m below. The shaft was, therefore, no more than about 4·5 m deep. It was square in section, but Stocker gave no indication of its size, nor did he sketch the find. It was constructed of an oak framework, the timbers joined by mortices and tenons (Stocker called them 'tenets') with the spaces between filled with hurdles of interlaced oak twigs. Similar hurdling was supposedly employed to consolidate loose ground in the upper parts of shafts in the Rycholt–St Geertruid flint mines in Holland, dated to between 3750 and 3650 BC (Felder, pers. comm. 1983). If clay-lined, the hurdling at Wheal Virgin would have prevented water seepage. Colenso wrote in 1829 that the silt above the tin ground at Pentewan:[88]

> is of so fine a texture, that it will scarcely admit air or water to pass through it ... It is therefore often applied by the streamers to resist water, and if formed into a basin, will hold it like earthenware.

The top of Stocker's shaft was presumably the bronze age level of the valley floor, the sediment above being the historically recent build-up of alluvial sediments already referred to. The shaft was also filled with sand. It is reasonable to assume that it was filled in antiquity after being swamped by the river bursting its bank, or by an exceptional rise in the tide. In this way the workings at Par were inundated by the sea in December 1801, and the workings in the Carnon valley were temporarily halted by a similar misfortune in the summer of 1799.

Violent gales swept much of Europe in early December and were reported in the *Royal Cornwall Gazette* (19 Dec. 1801):

> On the ninth instant we had the wind from S.E. to S.W. it blew very strong, and the sea rose two feet perpendicular higher than has been known for many years .... Porth stream tin work [*see* Fig. 72], which had an embankment at Par was washed away, and the tin work quite destroyed. The captain of the work being on the embankment was washed into the sea, and narrowly escaped drowning.
>
> The adventurers viewed the embankment and work on the 7th instant, and thought it impregnable for years to come; they sold it for £1 300, to be paid for on the 15th. How uncertain (says our correspondent) are all terrestrial possessions.

How much more vulnerable must have been the insubstantial diggings of bronze age tinners. That the shaft at Pentewan was abandoned in a hurry is supported by the discovery at the bottom on the tin ground of a bronze 'chisel' (now lost), and a middle bronze age socketed spearhead preserved at Penlee House Museum, Penzance. How tin working would have proceeded in the shaft had it not been swamped is not clear. Maybe the tin ground was removed only from the area covered by the shaft, but there is the possibility that several metres of tin ground could have been undercut from each side of the shaft.

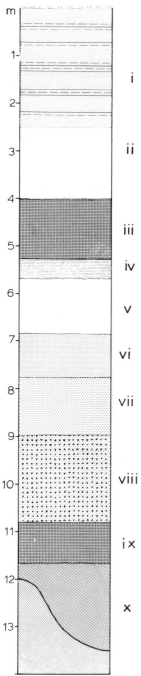

In Nigeria, certainly as recently as 1950, a method used to obtain quick, cheap tin from deep alluvial deposits, and locally known as 'lotoing', was to sink pits 1 m in diameter to the bedrock up to 12 m below. When the base was reached, a series of tunnels was driven to link up the pits in a grid pattern. The disadvantage was that it left as supports sections of tin-bearing ground representing over 30% of the ore body. Some lotos collapsed overnight owing to the weight on the pillars.[89] Nothing so complicated as 'lotoing' was likely in Cornwall. Even Stocker's simple shaft is the only one recorded, a survival due to a happy – or unhappy – accident; for under normal circumstances, after removal of the tin ground, the timbering would probably have been removed for re-use.

Apart from wooden shovels, Carew wrote that the first miners used pick-axes made of 'holm' (holly), box, and 'hartshorn', which tinners in his day commonly found in their workings. By Elizabethan times, picks were made of iron. How recently they were made of wood is anybody's guess, but in 1873 Worth noted that 'there are men still living in Cornwall who in their younger days used to cut their plough shares from the hedges'.[90] One oak pick from St Just, now in Truro Museum, is probably mediaeval, though it merits dating by [14]C. In spite of Carew's allusion to other woods, all wooden tools in Truro Museum are of oak.

Antler picks, Carew's 'hartshorns', are of great antiquity as mining tools both in this country and abroad. Since the animals shed them annually, they were plentiful and could be used with a minimum of preparation. An antler pick cannot have been used by a tinner earlier than the bronze age, but an early bronze age date is more probable than anything later. The two surviving examples in private collections in Cornwall should be dated by ^{14}C, as they could be among the earliest datable objects from any tin streams anywhere

---

i   surface of grass and other vegetables; *c.* 50 mm below is found micaceous sand, mixed with a little earth, of a light colour, in strata of various shades and thickness, according to the quantity of matter driven down by successive floods; total depth 2·5 m

ii   bed of light dove-coloured earth, with less mica than level i; 1·6 m

iii   peat, burns well; 1·2 m

iv   clay, ash-coloured, that adheres strongly to tongue,

    mixed with vegetable substances; 0·3 m

v   fine earth, with small blue spots and decayed vegetables; 1·1 m

vi   very fine sand; 1 m

vii   coarser sand; 1·2 m

viii   larger sand; 2 m

ix   very black fen, with small decayed vegetables, giving a disagreeable smell when on fire, and containing some bituminous matter; 0·86 m

x   tin ground, with which the ore is found in various thicknesses, 0·3–1·8 m, upon very light-coloured fragments of slate and yellow clay, on which the tin ore lodges in various sizes, from minute particles to stones weighing a few kilos. Various sorts of tin ore are found in this work, brought down by floods from different lodes in the high ground at considerable distances

72   Section of the stream work at 'Poth', in the Parish of St Blazey, about 0·4 km from high water mark (redrawn from Rashleigh, 1802)

(Figs. 73–4). Deer remains occurred as sub-fossils in various horizons in the overburden of tin streams, and it is not impossible that one of these was used by a tinner. Certainly in recent times tinners are known to have used sub-fossil wood, both as firewood and, in one instance, for making cartwheels, though it usually crumbled away.

Both surviving antler picks were found in the Carnon Valley tin stream, as will be described in the section dealing with the finds from these deposits. Both have been kindly examined for the author by Dr Turk. The larger pick is from an 8-pointer red deer (*Cervus elephas*). The bay and tray tines, as well as the tops, have been removed leaving the brow tine constituting the pick. The tip is damaged, possibly subsequent to its discovery. It is stouter and would have been shorter than most present-day antlers. The smaller specimen, also from red deer, is a 6-pointer with all tines removed and a hole bored through the bone of the antler close to where the brow and bay tines had been attached. Inserted in the hole is the tine of a single-pointer. The antler is lightweight, partly owing to leaching of calcium phosphate, and it may have come from an individual living in woodland. The darker colour compared with the other pick is largely owing to coats of varnish, but if some staining were the result of immersion in the tin stream, it is strange that the other pick was not similarly affected.

The larger pick with the brow tine in place clearly could have been used with a normal swinging action, but the other pick could not have been used in this way without dislodging the tine, which fitted very loosely. Picks of this type are rare, but examples from flint mines survive in the mining museum at Bochum in West Germany. To use them the tine must have been hammered with stones or wooden mallets. In a flint mine, the efficiency of this method can be appreciated in levering open-joint planes in the chalk. However, at first sight its use in a tin stream is less understandable, as brute force and a good swing should suffice.

At Aibunar in the Balkans, where antler picks were used in the 4th millennium BC,[91] they were liked because of their lightness, elasticity, and robustness. The great disadvantage was their lack of weight, so that to use them to best advantage the butt of the antler was hit with a heavy hammer stone. It must be concluded that this action was used with one, if not both, of the Carnon antler picks.

The only detailed description of the physical character of the tin ground in any stream deposit is given by Colenso for the Happy Union streamworks at Pentewan.[92] The tin ground, generally 1–2 m deep, but occasionally as much as 3 m thick, was divided by the streamers into what they called 'loose ground' and 'tough ground':

the loose ground consists of sand, stones, and pebbles; the tough ground has an additional inter-mixture of yellow clay, which cements the whole so firmly together as to render it difficult to be separated and washed.

The specimen in the Royal Geological Society of Cornwall photographed for this book (Fig. 80, *below*), shows 'tough ground' resting on the bedrock. Similar specimens of conglomerate, largely consisting of tin-stones from the Carnon Valley are in Truro Museum, and these must also have been hard. The deposits would have been far too consolidated to be scooped up with wooden shovels. Most probably the ground was loosened with the picks before shovelling, a process subsequently undertaken using bronze axes.

On neolithic sites such as Cissbury in Sussex, shovels made of ox scapulae were found. None is known to have been used by tin streamers, even though ox bones were recovered as sub-fossils at Pentewan and Par. The Overton Down experiment showed ox scapulae to be rather poor tools, so their use may have been impractical in alluvial tin deposits, especially in 'tough ground'.

PREPARING THE ORE FOR SMELTING

In 1816 Polwhele produced an excellent account of tin streaming in St Austell Moor, the Pentewan valley immediately below St Austell.[93] The tinstone lay at a depth of about 5.5 m and varied:

from the bigness of a goose egg and larger down to the size of the finest sand ... This stream tin is of the purest kind, and a great part of it, without any other management than being washed on the spot, brings thirteen parts for twenty [65 per cent] at the melting house.

The advantage of stream tin was its purity, and such deposits as Polwhele described would have presented few problems to tinners in antiquity. In Carew's day the treatment of lode tin was quite simple:

For being once brought above ground in the stone, it is first broken in pieces with hammers, and then carried either in wains or on horses' backs to a stamping-mill, where three, and in some places six great logs of timber, bound at the ends with iron, and lifted up and down by a wheel driven with the water, do beat it smaller. If the stones be over moist they are dried by the fire in an iron cradle or grate.

From the stamping mill it passeth to the crazing mill which between two grinding stones, turned also with a water wheel, bruiseth the same to a fine sand. Howbeit, of late years they mostly use wet stampers, and so have no need of the crazing mills for their best stuff, but only for the crust of their tails.

Stamps were probably a late mediaeval invention introduced to Britain from Germany where they were also known as *stamfer* or *stamper*.[94] At first they were used dry, but wet stamping was introduced at Joachimsthal by Paul Grommestetter in 1519. A crazing mill similar to that described by Carew formerly existed in a ruined building 6 × 4 m at Retallack farm, Constantine, on the Carnmenellis granite. There were

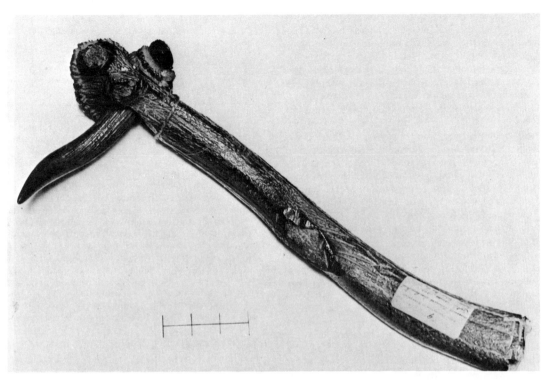

**73, 74   Antler picks from Carnon Valley tin stream; (73) length of handle 48 cm, tine 14 cm (74) length
53 cm (private collections)**

several millstones of 1–1·2 m in diameter, all 'more or
less grooved on the face in a circular direction', and
worked by a water wheel.[95] The crazing mill did not
differ in principle from the rotary quern for grinding
grain, so the earliest examples would probably have
been indistinguishable, assuming that crazing mills
were introduced as early as the iron age or Romano-

British period. It would have been realized at an early
period that smelting was improved if large pebbles
were first crushed, especially if, as has already been
noted in the Pentewan deposits, the largest pebbles
sometimes contained sulphide ores as well as large
amounts of unwanted quartz and killas.

Widespread in the mining districts are large slabs of

both granite and 'greenstone' (dolerite and related types) pitted with hollows *c.* 15 cm in diameter. These are interpreted as the anvils on which large lumps of tinstone were pulverized by hand (Fig. 75). Several of these were found at Retallack farm, indicating that hand-crushing continued until quite recent times. Also at Retallack was a rough slab of granite 1·2 m long and 35 cm wide and deep. On each of its sides were four hollows 18–23 cm across and 5–13 cm deep, of very regular shape, suggesting crushing with a set of machine-driven stamps. Normally such 'cup-marked' stones are undatable. Many must be mediaeval, like the fine granite example set on end by the churchyard wall at St Neot on the edge of Bodmin Moor. There is 0·9 m standing above ground; it is 0·6 m wide × 0·4 m thick with three depressions about 15 cm across on both surfaces, one side having a couple of smaller secondary depressions.

A grinding mill and pounding stones assumed to have been used for tin ore were found in an iron age or Romano-British structure 'accidentally destroyed by workmen' at Vyneck, Boscawen-Un, in St Buryan in 1867.[96] The finds comprised the fragments of a granite basin, a quern and fragments of three others, several mullers or pounding-stones, and various domestic items including pottery, which Hencken described as iron age and inferior Roman.[97]

The separation of the fine particles of tin from the 'gangue' or waste was, in the historical period of mining, commonly done in a 'tye' or 'strake', which Pryce described as 'frames made of boards, fixed on or in the ground, where they wash and dress the small Ore in a little stream of water; hence termed Straked Ore'. The tye was *c.* 2·7 m long, 1·2 m wide and 1·2 m high, and inclined so as to allow the water to run off. As the ore was shovelled in the workers 'turn it over and over again under a cascade of water that washes through it, and separates the waste from the Tin, till it becomes one half Tin'. The best tin, by its heavier weight, collected at the top of the tye nearest the inflow of water, 'and by degrees becomes more full of waste, as it descends from that place to the end or tail of the tye, where it is not worth the saving'.

No prehistoric tye is known from Cornwall, but as has already been seen from the Gallo-Roman remains at Abbaretz in Brittany, the large number of pieces of broken planks and other fragments were interpreted by Champaud as the remains of sluices constructed to direct the flow of water over the dressing-floors. Given the close links between Cornwall and Brittany, it is a reasonable assumption that the Tye had a similar antiquity in Cornwall going back to the pre-Roman iron age if not before.

A similar more primitive method is described by Carew:[98]

> The stream, after it hath forsaken the mill, is made to fall by certain degrees [in steps], one somewhat distant from another, upon each of which at every descent lieth a green turf, three or four foot square and one foot thick. On this the tinner layeth a certain portion of the sandy tin, and with his shovel softly tosseth the same to and fro, that through this stirring the water which runneth over it may wash away the light earth from the tin, which of a heavier substance lieth fast on the turf. Having so cleansed one portion, he setteth the same aside and beginneth with another, until his labour take end with his task. The best of those turfs (for all sorts serve not) are fetched about two miles to the eastward of St Michael's Mount, where at a low water they cast aside the sand and dig them up; they are full of roots of trees, and on some of them nuts have been found, which confirmeth my former assertion of the sea's intrusion. After it is thus washed, they put the remnant into a wooden dish, broad, flat, and round, being about two foot over and having two handles fastened at the sides, by which they softly shog the same to and fro in the water between their legs and they sit over it, until whatsoever of the earthy substance that was yet left be flitted away. Some of later time, with a sleighter invention and lighter labour, do cause certain boys to stir it up and down with their feet, which worketh the same effect. The residue, after this often cleansing, they call black tin, which is proportionably divided to every of the adventurers when the lord's part hath been first deducted upon the whole.

Carew's excellent account describes a method which could hardly be more simple, but it is not likely to be revealed in the archaeological record, at least not in Cornwall. The handled tin bowl from Treloy in St Columb Minor is sufficiently similar to Carew's wooden bowl for 'shogging' to assume that this method was practised during the Roman period. There is little reason to doubt that 'turf' was employed then and at an earlier period. Its use has already been compared to the ancient use of animal skins for trapping fine particles of gold. The tye is also described by Pliny for gold washing in Egypt. His account is taken from the lost work of Agatharchides, a Greek geographer of the 2nd century BC:[99]

> At length the masters of the work take the stone thus ground to powder, and carry it away in order to perfect it. They spread the mineral so ground upon a broad board, somewhat sloping, and pouring water upon it, rub it and cleanse it; and so all the earthy and drossy part being separated from the rest by water, it runs off the board, and the gold by reason of its weight remains behind; afterwards they draw off any earthy and drossy matter with slender sponges gently applied to the powdered dust, till it be clean, pure gold.

A brief account by Diodorus Siculus (Bk. V, 27, 2) on the dressing of gold ore in Gaul about the turn of the 1st century AD, confirms that all the elements detailed by Carew were already in use, though unfortunately he does not give a picture of how the work was conducted. 'They grind, or crush the lumps which hold the [gold] dust, and after washing out with water the earthy elements in it they give the gold-dust over to be melted in the furnaces.'[100]

75 Mediaeval tin-crushing stones; *bottom right* part of a crazing mill found at Vorvas farm, Lelant (photo Herbert Hughes, 30 July 1907)

NOTES AND REFERENCES

1 GOODE, 1973

2 GRAHAM, 1966, 13

3 LELAND, *c.* 1540, 'Itinerary' in Chope (1967 ed.), 41

4 PRYOR, 1980; the Cornish axe is No. 918.33.108 in the Sturge collection

5 TYLECOTE, 1967

6 PEARCE & PADLEY, 1977

7 HURST, 1937

8 *Proc. Soc. Antiquaries*, 1873, 429–30

9 BORLASE, 1872, 41

10 DINES, 1956, **1**, 191; CAREW, 1602, recorded that Godolphin Mine yielded at least £1 000 annually to Queen Elizabeth

11 HENWOOD, 1843, Table 13

12 SYMONS, 1884, 166

13 JAMES, 1945

14 MAJENDIE, 1818, 33

15 HITCHENS & DREW, 1824, **2**, 503–504

16 SYMONS, 1884, 166

17 JENKIN, 1963, (7), 24–5

18 JENKIN, 1964, (9), 25

19 JENKIN, 1964, (8), 58

20 JENKIN, 1964, (9), 48; DAVIDSON, 1926, 97, lists outcropping cliff lodes in St Agnes at Trevaunance and Trevellas, at Pendour Cove in Zennor, and the Perran Great Iron Lode at Perranporth Bay. Tin stockworks which he considered could have been visible outcrops in prehistory are at Mulberry near Lanivet, Wheal Fortune in Sithney, and Magdalen Mine at Ponsanooth

21 SALWAY, 1978, 54–5

22 MORRISON, 1980, 36

23 *Royal Cornwall Gazette*, 10 April 1913, 4, col. 1; report of Address to Mechanical & Metallurgical Association by William Thomas

24 COLLINS, 1912, 381

25 KLAPROTH, 1787, 13

26 BARROW, 1908; HOSKING *et al.*, 1962, note that 'whereas tungsten (as wolframite) tends to concentrate in progressively finer fractions as the distance from the source increases, tin (as

cassiterite) does not. Unlike wolframite, cassiterite is neither readily abraded nor disintegrated, and so both coarse and fine fragments can be found far from the source.'

27   COLLINS, 1887, 22–3
28   RUNDLE, 1892
29   HUNT, *Mineral Statistics 1880*
30   PATTISON, 1847
31   HENWOOD, 1873, 215
32   CAMM & HOSKING, 1984
33   WYMER, 'The Palaeolithic', in Simmons & Tooley (ed.), 1981, 78
34   MERRIFIELD, 1982
35   CLARKE, 1970
36   CHURCHILL, 1965
37   TAYLOR, 1980
38   HEYWORTH & KIDSON, 1982; they discuss the difficulties of correlating evidence of sea-level changes from disparate areas and conclude that 'More intensive studies are needed to determine sea-levels at specific dates'
39   BARKER & MACKEY, 1959, item BM–29
40   TAYLOR, 1980
41   JAMES, 1949, 176–8; COLLINS, 1912, 210–11
42   HENWOOD, 1870, xvii–xviii, 'Presidential Address', *JRIC*; Whitley, 1881
43   SYMONS, 1877
44   COLENSO, 1829
45   HENWOOD, 1828
46   TAYLOR, 1873; BARTON, 1970, 159–71
47   THOMAS, 1972
48   CARNE, 1830; STEPHENS, 1899, notes that in the Red River valley 'Poor tin gravel was met with constantly, but no rich pockets or patches of payable gravel.' He also makes the interesting observation that 'modern tin streaming' – the recovery of tin washed down the valley from the dressing floor of mines – did not begin in Cornwall until about 1860.
49   HILL, MACALISTER & FLETT, 1906, 97; REID & SCRIVENOR, 1906, 83, on the St Agnes area state that 'the valleys there are short and steep, and have not yielded much [alluvial] tin'
50   HENWOOD, 1828
51   GERRARD, 1983
52   HENWOOD, 1873, 216–17
53   COLLINS, 1909
54   BARROW, 1908
55   HENWOOD, 1873, 244–7
56   BORLASE, 1758, 161
57   GRAHAM, 1966, 13
58   CAMM & HOSKING, 1984
59   HALLIDAY, 1953, 88
60   HENWOOD, 1855, 8–9
61   *The Annual Register*, New Series, 25 January 1808, **8**, 12

62   IAGO, 1890–1, esp. 222 footnote
63   HENWOOD, 1873, 231 footnote quoting from the MS *The Book of Spoure*, 1690
64   DE LA BECHE, 1839, 613
65   BORLASE, 1758, 214; Ch. XIX 'Of Gold found in Cornwall' (213–17)
66   PRYCE, 1778
67   STEVENS, 1928
68   BRAY, 1838 (1879 ed., **2**, 375)
69   HOSKING & OBIAL, 1966
70   COLLINS, 1912
71   FORBES, 1964, Ch. 5, 155
72   PRYCE, 1778, 246
73   HENWOOD, 1873, 219 footnote
74   GREEN *et al.*, 1980
75   HALLIDAY, 1953, 90–1
76   HALLIDAY, 1953, 89
77   BORLASE, 1758, 164
78   KLAPROTH, 1787, 13
79   THOMAS, 1950
80   FIDDICK, 1913; the Camborne poet John Harris described the use of a dowsing rod cut from a white thorn for finding a lode (*The Mountain Prophet, The Mine, and other Poems*, London, 1860, 49)
81   SIMONIN, 1868
82   PRYCE, 1778, 131–7
83   JENKIN, 1967, (13), 43, notes how the alluvial workings were destroyed by the tide breaking through the barrier in the summer of 1799
84   LIPSCOMB, 1799, 251–5
85   LEWIS, 1908, 195
86   SWETE (1780), 1971, 196
87   STOCKER, 1852
88   COLENSO, 1829
89   SOUTHWOOD, 1946; FAWNS, 1905, 52, notes similar methods used by the Chinese in Malaya when the overburden was too thick for open-cast working
90   WORTH, 1873
91   ČERNYCH, 1978
92   COLENSO, 1829
93   POLWHELE, 1803, **2**, 10
94   DONALD, 1950, 312, notes that Burchard Kranich (*c.* 1515–78) must have been using stamps in Cornwall; HOOVER & HOOVER, 1950, 282–3
95   BRYANT, 1882; when he visited the site *c.* 1855 he found the remains of an old buddle 'made with rough stones', the sand between them yielding 'a good van of tin'
96   *JRIC*, 1868, 71, 'Chronological Memoranda' for 1867
97   HENCKEN, 1932, 159, 294
98   HALLIDAY, 1953, 94
99   HOOVER & HOOVER, 1950, 280 footnote
100   HEALY, 1978, 144

25 Prehistoric finds from Cornish tin streams

There are also taken up in such [tin] works certain little tools' heads of brass which some term thunder-axes, but they make small show of any profitable use.

Richard Carew, *The Survey of Cornwall*, 1602

Cornwall has been fortunate in the attention it has received from chroniclers of topographical and antiquarian detail ever since William Bottoner, better known as William of Worcester, made the hazardous journey to the western tip of England in 1478. Less happy is the scant attention paid to the discoveries in tin streams before the second half of the 18th century. However, John Leland reported about 1540 that streamers at Marazion Marsh had unearthed what can only have been a hoard of middle or late bronze age metalwork. Carew, whose observations on tin streaming in Elizabethan days are extensively quoted in the present work, was in a position to report a good deal more about the relics left by the 'old men'. However, apart from the concise quotation that introduces this section he mentions only a 'brass coin of Domitian's' without even saying in which tin work it had been found.

Many discoveries which would now be regarded as important must have been cast aside or destroyed by numerous tinners from the mediaeval period onwards. The discovery of ancient ingots of tin was just a bonus to many tinners who sweated long hours for little return. The Roman tin bowl found in the streamworks at Treloy, St Columb Minor (*see below*, Fig. 107*a*), in 1830 was only saved from the smelter by the intervention of Dr Boase, a partner in the smelting company, who happened to be present when it arrived. An ingot from Trethowel Wood, St Austell, was sold to Carvedras Smelting Works in Truro for £3.6s.5d; fortunately part of it was saved by the Daubuz family, owners of the Works, and presented to Truro Museum in 1922. In 1898, of the five large mediaeval tin ingots recovered from Fowey harbour, only one escaped the smelter's flames to survive in Truro Museum. The Revd Le Grice called the smelter at Angarrack a 'Goth of a refiner' for smelting down an ingot found in Gwinear about 1820.

In spite of such destruction a significant amount survives either as physical objects or as drawings, and Cornwall is fortunate in possessing a greater wealth of material from tin streams than from any other tin field known to have been worked in antiquity. In dealing with this material, rather than listing all the items chronologically, it is convenient to treat each tin streaming area in turn with all the objects known to have been recovered from the workings.

A question sometimes posed by archaeologists about ancient finds in Cornish tin streams is 'What is the depth at which the find was made?' It is clear from the question that it is supposed that a section through a tin stream would give an archaeological stratification similar to that seen in the silted up ditch of an iron age hillfort, with the oldest objects at the bottom and the youngest at the top. But this is not the case. Tin streams do exhibit *geological* stratification of post-glacial layers of fluviatile and marine sediments with horizons rich in sub-fossil plant remains, deer horns, whale bones, etc., as were seen especially well in the Pentewan valley (*see* Fig. 83, *below*). The human artifacts recovered from tin streams bear no relation to the geological horizons, which are increasingly old with depth. Excluding such things as the bottoms of wooden stakes driven into the surface (which could be several metres below the present surface), all human artifacts recovered by tinners, mediaeval shovels or bronze age axes, were from the same horizon – the 'tin ground', which varied in depth with location.

Whenever the depth of the artifact was recorded it was invariably 'on the tin ground' or can be deduced as such, if the depth is given in feet, from independent geological reports that state the approximate depth of the tin ground. In no case is there a record of a prehistoric artifact being found at a shallower level in a tinless horizon. The depth at which some finds were made was not recorded, doubtless because the tinners thought this unnecessary as it was obvious to them that it was on the tin ground.

Archaeologists have also enquired whether or not an object found in a tin stream could have 'filtered down'

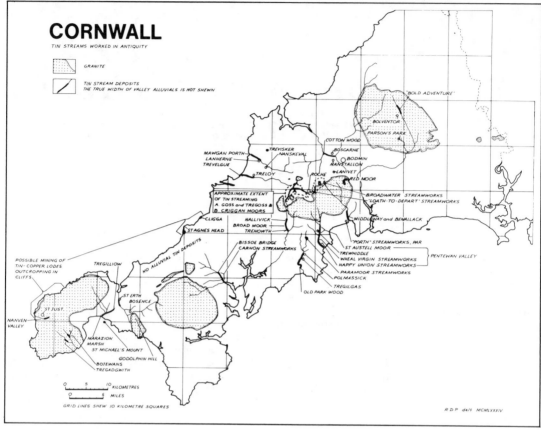

Map 25

from the surface. The answer is no. The fact that the geological sub-fossils have maintained a constant level and not dropped to the bedrock is proof enough that man-made objects would not fall through the clays and sands overlying the tin ground. Nor is it at all likely that a prehistoric man would simply dig a pit to bury something, like the Trewhiddle hoard, on the tin ground so far below the surface. For that reason it is safe to conclude that the Trewhiddle hoard was buried in a working tin stream for safety. All objects, therefore, whether bronze age or mediaeval, would turn up on the tin ground. Sometimes they did so in fairly close proximity, like the 'dark age' tinner's shovel and Roman remains in the Boscarne tin works.

The discovery of prehistoric artifacts in Cornish tin streams is a thing of the past. The chances of additional objects coming to light are remote. Apart from the use of a Malayan-type dredge on Goss Moor in the 1920s, steam shovels and similar machinery at St Erth (*see below* Fig. 126) and on Red Moor, Lanlivery, in the 1930s, tin streaming had always been a pick-and-shovel affair with rarely anything more labour-saving

than a wheelbarrow. Tinners in such close proximity to the overburden and the 'tin ground' had a reasonable chance of finding the discarded artifacts of the 'old men'.

Some parts of Cornwall still attract mining companies interested in potentially workable stream deposits, as on Bodmin Moor and near Wendron. Even if permission were granted for future work in the face of stern opposition from conservationists (who frequently seem to forget that many sites such as Red Moor and Marazion Marsh are entirely, or in part, the product of previous tinners), it is doubtful whether anything of great interest to archaeologists would be recovered. The use of modern machinery to cut high labour costs makes it unlikely that prehistoric artifacts would be detected *in situ*. Nor could mining companies be expected to stop work while an archaeologist inspected the deposit before its total destruction.

A good deal has survived from the operations in the late 18th and 19th centuries, though insufficient attention was paid to them at the time of their discovery. Stocker gave a tantalizing account in 1852

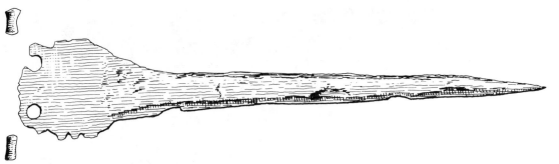

76   Middle bronze age rapier found, with a palstave, in 1796 in Benallack tin streamworks (original lost; redrawn from Borlase, 1871)

of what must have been a bronze age tinners' wooden shaft at Pentewan. However, no measurements or even a rough sketch of the structure were made at the time. This lost opportunity is unlikely to be repeated.

ST BLAZEY AND PAR

Borlase illustrated in his *Naenia Cornubiae*[1] a 'dagger' (Fig. 76) that had been found 'in close proximity to a celt, similar to the one from Godolphin' at Benallack *c.* 12 m deep. The Godolphin find is discussed in Chapter 24 (*see above* Fig. 63); it is a typical middle bronze age palstave. More details of the discovery were not published by Borlase, though the find circumstances had been published by Lysons.[2] From this and Borlase's MS *Ancient Cornwall* (1871) it is known that the

> brass Weapon [was] found about 40 feet under the surface at Benallack above Par, in the parish of St Blazie, in streaming 1796. With this was found a Celt, exactly resembling that found at Godolphin Mine.

The dagger fits into the class 1 rapiers of Rowlands,[3] which appeared late in the middle bronze age. They have a mean length of 30–35 cm, against 40–45 cm of earlier types. The Benallack rapier is thus long at 40·7 cm.

Benallack is not a name surviving on Ordnance Survey maps, so the provenance of the rapier and palstave has been in doubt. Benallack is, however, the name of a tin streamworks in the area known as Middleway where tinners are depicted working on a map of 1794 (Fig. 77). The articles of agreement were drawn up on 20 April 1791:[4]

> Owners of land and tin bounds in St Blazey and Tywardreath, with Chas. Rashleigh of St Austell, gent., on behalf of co-adventurers in streamwork intended to be worked in St Blazey Moor and Par green to be called Benallack. Chas. Rashleigh to cut leat over fields of Thomas Carlyon and others, and to work ground in proper manner, paying dues to lords of the soil: Chas. Rashleigh to make good the bridge over St Blazey river 'for horses, carts, and carriages to pass and repass on ... and finish the same with good timber or stone properly

built and railed on each side for the safety and accommodation of Travellers.'

The land was thus owned by several families and there had been disputes here earlier in the century. In 1733 Francis Scobell, in conflict with Philip Rashleigh, 'renewed his acts of hostility and drowned some of the [tin] works' at Par, as he had done the previous year.[5] This local dispute is important for two reasons. Firstly it shows the ease with which tin workings close to the sea could be flooded, though natural disasters are sufficient to account for bronze age losses. Secondly it explains the imprecise location given for another archaeological find of some importance. Britton and Brayley in their description of Philip Rashleigh's collection at Menabilly, Fowey, referred to 'several British instruments found at Benallack'.[6] One must have been the rapier, another could have been the now lost palstave, while a third can only have been the so-called Fowey pin (Fig. 78), formerly in a private collection and now in Bristol City Museum.

In January 1796 Philip Rashleigh read before the Society of Antiquaries in London a short paper describing an object that he believed to be a Druid's hook for gathering mistletoe. The accompanying illustration (Fig. 78) depicts a bent bronze age pin with a piece of amber set in the end of the head. As a mineralogist, Rashleigh was familiar with amber, having specimens in his cabinet of minerals, and would have realized its implications for ancient trade. It was clearly something he desired for his own collection, and as it might well have been dug up on someone else's land this could account for his uncharacteristically vague details of its location. Similar pins are known from elsewhere in southern England and are close in style to pins from western Bohemia and Slovakia.[7] According to Rashleigh the pin was found

> at the bottom of a mine near the River Fowey, ten fathoms under the surface of the earth, where a new work was begun for searching after tin ore.

There are no tin lodes or tin alluvials on the Fowey river below Lostwithiel, so a find spot in a valley

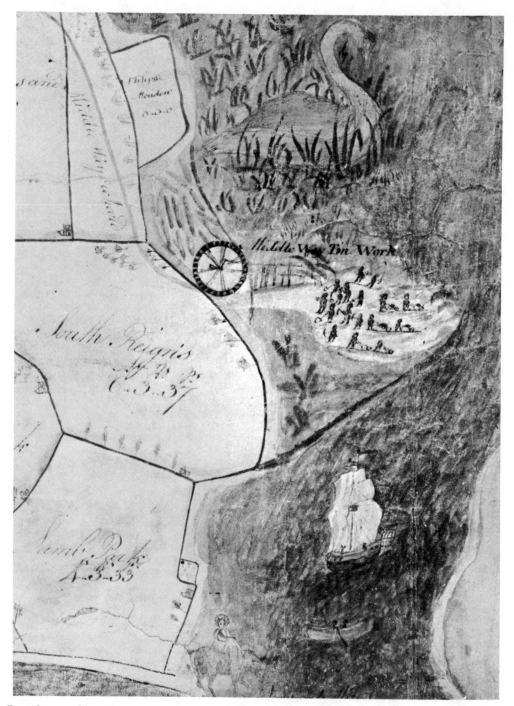

77   Part of a map of Roselyon in 1794 showing the Middle Way Tin Stream. The site is now occupied by rough ground and the railway line by the Par Loop (photo Greenham/Woolf collection)

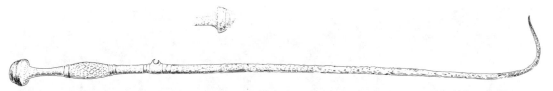

78 **'Fowey' pin, middle bronze age (*c.* 1200 BC) from tin streamworks at Middle Way, St Blazey; length 33 cm** (from *Archaeologia*, 1796, *12*, Plate LI, Fig. 8)

leading to St Austell Bay is a geological necessity. It is out of the question that miners were sinking a shaft on a lode and happened to break through into a bronze age level. The miners had to be tinners who sank a 'hatch' onto the tin ground, which they knew had to be there at some depth, and discovered the pin in the totally silted up excavation of bronze age tinners. Over 18 m is unusually deep for Cornish tin streams. At Sandrycock and Porth, the valley between St Austell and Par, the bottom of the tin ground lay *c.* 13·5 m below the surface. However, there has been unusually heavy silting in recent centuries in the Luxulyan valley between Pont's Mill (above St Blazey) and Par. In 1877 Richard Symons gave the depth of the tin ground as 7·3 m at Pont's Mill and a massive 22 m in the lower reaches of the valley.[8] Considerable silting took place in the century before Symons' survey; even so, a depth of 18 m could easily have existed at Middleway in the 1790s, and no other tin stream fits the circumstances.

The Roselyon Map, part of which is reproduced in Fig. 77, was drawn in the very year that the Fowey pin was discovered. It shows a ship sailing alongside the Middleway workings. In the bronze age when the surface of the land was lower, the spot must have been at the seaward limit of exploitation, when the river would have been navigable for about 2 km above St Blazey. The depth of tin ground at that period can only be guessed at, but is unlikely to have been more than about 12 m. Other items of prehistoric or mediaeval date were also recovered according to Pryce in 1778,[9] whose description illustrates the variable depth of the tin ground in this stretch of the valley:

> In St Blazey Moor, at the depth of twenty feet, they have what they call Stream (Tin Ore) about five feet in thickness in the bottom, great part of which had been anciently wrought before Iron tools were known, several wooden pick-axes of oak, holm, and box, having been lately found therein. Over this they have a complete stratum of black mud, fit for burning; on this a stratum of gravel, very poor in Tin; on this another stratum of mud; and uppermost gravel again.

Before turning to the bronze age and later finds from the Pentewan valley, it is relevant to note here the evidence of early bronze age tin working discovered in a ring-banked enclosure excavated in 1972 on Caerloggas Downs. This is an area of high ground in the middle of the St Austell granite close to the aptly named Penwithick Stents ('stents' means tin-bearing

ground). The finds included two cassiterite pebbles in the ditch silt, seven fragments of glassy tin slag (*see* Fig. 79), and a small, fragile fragment of a dagger of the Camerton-Snowshill type with pointillé decoration datable to *c.* 1500–1400 BC. There was also an amber fragment, one of the very few recovered in Cornwall apart from that on the 'Fowey' pin.

The association of these finds at Caerloggas is casual; nonetheless, the tin slag is important as the first to be recognized from such an early context in Britain, or indeed in Europe. The slag consisted of silica glass containing numerous spherical blobs of white metallic tin, some visible to the naked eye.[10] The amber is equally important as it may well have reached Cornwall as part of a trade in Cornish tin. This suggestion should be seen in the light of the trade suggested by the Exloo necklace discussed in an earlier part of this book.

THE PENTEWAN VALLEY

> The most considerable stream of tin in Cornwall is that of St Austell Moor, which is a narrow valley about a furlong wide (in some places somewhat wider) running near three miles from the town of St Austell southwards to the sea.
> Revd William Borlase, *Natural History of Cornwall*, 1758

The part of the valley worked in Borlase's day was close to St Austell where the tin ground lay at no great depth, the overburden being *c.* 3·3 m thick and the base of the tin-ground on average *c.* 5·5 m. As Borlase relates, the stream tin was 'of the purest kind' (*see* Fig. 80) varying in size from the finest sand to pebbles 0·3 m in diameter. After washing on the spot, it yielded 5·9 kg of metal out of 9 kg of tinstones.

On Plate XX of his *Natural History of Cornwall*, Borlase illustrates the front and back views of a tin ingot (Fig. 81), one of two 'lately' found in one of the streamworks of the Moor *c.* 2·4 m below the surface, hence not on the tin ground but some way above it. Each weighed *c.* 12·7 kg and was unlike any ingot known to Borlase who believed, because they bore no stamps, that they belonged to the time of King John 'when the Jews had engrossed the tin manufacture'. Borlase's engraving shows a very regularly shaped ingot, though he described them as 'much corroded by the sharp waters in which they have layn, a kind of rust or scurf-like incrustation inclosing the tin'.

The ingots do not survive; even if they did no closer

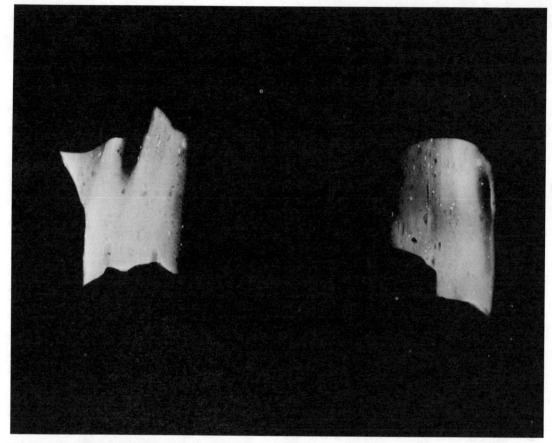

79   X-radiograph of a slag fragment showing white spots due to metallic tin in a glassy matrix from Caerloggas Downs (photo AM Lab., Dept of the Environment. Crown copyright, reproduced with permission of the Controller of HMSO)

dating than Borlase's might be possible. Their semicircular handles made them easy to carry on horseback. They are unlike any known mediaeval ingots, and their superficial resemblance to those recovered from the wrecks at Port Vendre, France, and Capo Bellavista, Sardinia, makes a Romano-British date possible.

William Borlase's descendant William Copeland Borlase[11] illustrated two views of an object of polished jet (Fig. 82) found on 13 November 1790 in Pentewan tin streamworks 14·6 m below the surface 'on the natural soil and on a level with hazels, oaks &c'. The object has long since been lost. The streamworks were either Wheal Virgin or the Happy Union begun in 1780. The vegetation indicates that the find was made at the top of the tin ground on which Colenso[12] and Winn[13] described a layer of vegetable matter 15–30 cm thick.

The importance of the find lies in its great antiquity.

It is a 'jet slider', an object of uncertain use but perhaps a belt fastening. In Britain 17 others have been found scattered widely as far north as southern Scotland.[14] Most fall into a late neolithic context, though the early bronze age has been suggested for some. Of necessity the Pentewan example belongs to the bronze age, presumably to the earliest period and not much later than *c.* 2000 BC. The depth of 14·6 m indicates a find spot well down the valley. The depth of the tin ground increased downstream, and when Colenso described the Happy Union streamworks in 1829 (*see* Figs. 83–4), 1–3 m of tin ground lay under 16 m of overburden, a good 9 m below low water of spring tides. The jet slider was probably found within 2 km of Pentewan.

The works begun in 1780 progressed in two directions; the Happy Union worked downstream and the Wheal Virgin tin stream moved upstream, the two being *c.* 1·6 km apart in 1829. Wheal Virgin has yielded most, perhaps all the ancient artifacts, doubt-

80   Rare specimen of stream tin from the Pentewan Valley in 1829, showing the base of the 'tin ground' resting on
a Devonian slate bed. The specimen is 'tough ground'; cassiterite pebbles cemented by a yellow clay make it
difficult to mine, 22·5 × 20 cm (photo reproduced by permission of the Royal Geological Society of Cornwall)

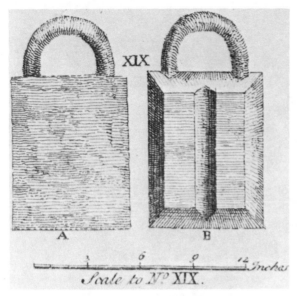

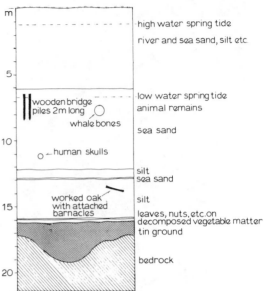

81   Front and back view of tin ingot found in St Austell Moor (from Plate XX of Borlase, 1758)

83   Happy Union Tin Streamworks, Pentewan Valley (redrawn from an annotated section in Colenso, 1829)

less because the shallower depth of the tin ground was more attractive to 'the old men'. When the Wheal Virgin finds were made in the early 1850s, the workings were *c.* 3 km south of St Austell and 2 km north of Pentewan in the vicinity of Levalsa Meor.

Borlase[15] sketched a 'celt', nowadays described as a socketed chisel of the middle bronze age (*c.* 1100 BC, Fig. 85). The tinners found it in April 1851 'at a depth of 28 feet'. The decoration of three converging ribs suggests that the implement may be the forerunner of the late bronze age socketed axes common in south-west England and south Wales and popularly known as South Wales axes.

On 15 October the following year, Stocker[16] described two bronze artifacts found at a depth of 4·5–7·6 m at the bottom of the wooden shaft already dis-

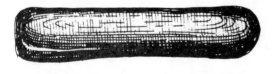

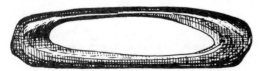

82   Top and side views of a 'jet slider' from Pentewan streamworks, probably dating to *c.* 2000 BC; length 6·7 cm (redrawn from Borlase, 1871)

cussed in the section dealing with the methods of tin streaming in Cornwall. The objects are a spearhead and a 'chisel'. The chisel was 'still in the hands of some one of the adventurers or the captain' of Wheal Virgin. It has long since been lost. Stocker described it as 'about eight inches long' and 'about an inch in width'. Assuming these measurements are about right, the object may have been part of a rapier blade, examples of which have been found in tin streamworks at St Blazey, Paramoor in St Ewe, and St Mawgan-in-Pydar. The spearhead survives in the Penlee House Museum at Penzance (Fig. 86). The point is broken off, as is the greater part of the socket below the blade. When complete the socket probably bore two side loops; in any case, the shape and section of the blade is typical of middle bronze age spearheads.

On 15 May 1851 a splendid wooden tankard (Fig. 87, now displayed in Truro Museum) was found in Wheal Virgin works. The depth of the find is not recorded, but the excellent state of preservation tokens burial at a good depth in anaerobic conditions. It is therefore unlikely to have been found anywhere except on the tin ground at a depth comparable to the 8·5 m at which the socketed chisel had been found the previous month. The tankard is built of eight ash staves of unequal width held together by three bronze strips equidistantly spaced with the top band, for obvious reasons, overlapping the rim and extending for 2 cm down the inside. A bronze handle is rivetted to one of the wooden staves and decorated with simple semi-circular discs at the centre of the handle and mouldings at the ends. The wooden base is missing, but grooving

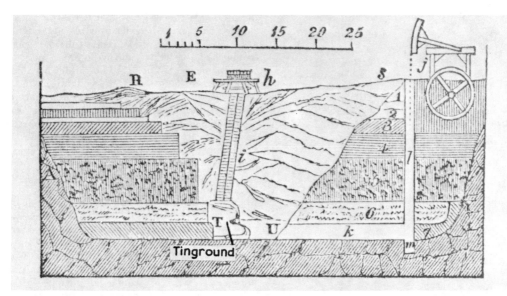

84   Cross-section of the Happy Union streamworks from Leifchild, 1854. Operations ceased in 1837 after working the ground immediately behind Winnick sand-dunes adjoining the port of Pentewan. The illustration highlights the problems of drainage, hardly less formidable in Wheal Virgin further up the valley and from which probably all the prehistoric artifacts were removed

on the inside of the staves just above the base suggests it was probably like the single piece of turned wood in the late 1st century AD tankard from Trawsfynydd in Gwynedd.

Apart from Pentewan, other tankards reasonably complete are those from Trawsfynydd, Kew, and a late Roman example from Shapwick in Somerset. The Pentewan tankard is difficult to date accurately, though a date sometime in the 1st century AD is generally accepted.[17] Wooden tankards were popular for centuries for, as Corcoran explains, wood makes the best containers for storing beer, using modern beer barrels as an analogy. It is unlikely that the Pentewan tankard was used for baling out water from the workings; rather its ample size emphasizes that tin streaming was thirsty work. Until well into this century underground tin miners kept their water cool in small wooden casks. The Pentewan tankard certainly merits dating by the ^{14}C method when techniques require minimal destruction.

The most impressive discovery in a tin stream is the hoard of Saxon silver now in the British Museum (Figs. 88–9). Wilson called it 'one of the most important finds of the whole Anglo-Saxon period'.[18] It is certainly an unexpected find in Celtic Cornwall. The discovery was first made known to the Society of Antiquaries by Philip Rashleigh of Menabilly on 8 May 1788.[19] The objects had been found by tinners in a streamwork near St Austell on land belonging to John Rashleigh of Penquite. Rogers[20] gave the date and location as 8 November 1774 'in a tenement, parcel of the manor of Trewhiddle, in the valley below St Au-

stell'. Trewhiddle house lies only 457 m west of the St Austell river so the find was made in that section already referred to as St Austell Moor.

The metalwork is of outstanding artistic merit and contained a gold filigree pendant lost with three other objects before 1866. Among the three objects was a small gold ingot of tapering hexagonal shape. It is tempting to see this as a by-product of a local tin stream. The first analysis of Cornish gold was published by Forbes in 1869 of gold from St Austell Moor.[21] The hoard contained 114 coins only two of which survive, though Rogers described them all. One of the survivors is a previously unrecorded silver penny of Ethelwulf struck at Canterbury by the moneyer WFA. The other, a coin of Coelwulf, shows that the hoard cannot have been deposited before 874.

The animal decoration, particularly well seen on the two strips interpreted as drinking-horn mounts, has given rise to the name 'Trewhiddle Style' for this art form in which the silver relief is set against a black composition known as niello. The most important pieces include a silver chalice similar to one from Hexham, Northumberland, and the earliest in Britain (the 4th century 'chalice' from Water Newton, Cambs. in 1976 is a two-handled cup used as a chalice). Unique is the scourge of plaited silver wires, an ecclesiastical object otherwise known only from literature. Further artistic details have been fully explored by Wilson and Blunt.[22]

The chalice, which contained the rest of the hoard, had been placed 'in a heap of loose stones, the refuse of

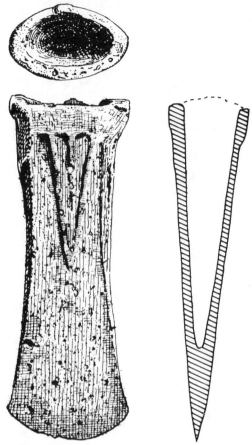

85 Middle bronze age socketed chisel from Wheal Virgin streamworks; length 11·5 cm (courtesy County Museum, Truro)

an old tin working, and covered with a common slate' at a depth of 5 m below the surface, the stones clearly intended as a marker to facilitate later recovery. It was about 875, as Hencken[23] relates, 'the first great series of Danish attacks upon England was nearing its climax'. However much the Trewhiddle hoard was prized in the late 9th century, no one would have dug a hole 5 m deep to hide it from marauding Danes, and

the inevitable conclusion is that it was hidden in a working tin stream where disturbed ground would not attract unwanted attention. This is the only closely datable find from a tin stream in the 'Saxon' period – if such a name can be applied to Cornwall where Saxon as opposed to Celtic objects are very rare. What the hoard was doing in Cornwall in the first place is a matter of unresolved conjecture, though it is worth remembering that the only other hoard of Saxon silver coins, half a century younger than Trewhiddle, was found only 11 km to the east near Fowey in 1953.[24]

A mediaeval interest in the Pentewan tin is proved by the discovery of a silver penny of Edward III in 'a streamwork at St Austell' and presented to Truro Museum in 1842–3 by a Mr Borrow.

Through the kindness of Prof. Rapp, two specimens of oak from the Pentewan streamworks were radio-carbon-dated shortly before this book went to press. The first, a 'piece of oak from Pentewan stream-work', now in several very black carbonized fragments, was presented to Truro Museum about the mid-19th century by Mr Couch, but the find circumstances are not recorded. The specimen, UCR 1829 is dated 3120 ± 100 bp, 1665–1115 BC, the middle of the bronze age.

The second specimen is far more intriguing. On 5 July 1839 Winn[25] described wood recently obtained from the Wheal Virgin tin streamworks then at Leval-sa Meor and progressing upstream at a rate of about 1·8 m a month (*see* Fig. 90). Much of the vegetable matter was sub-fossil, but specimens of nuts and wood that Dr Winn presented to Truro Museum have not survived. The barren sediments 9 m thick overlay a tin ground *c.* 1·5 m thick. The sediments differed from those at Happy Union in that they lacked the marine deposits encountered further down the valley. A 'large' trunk of a semi-carbonized oak was 'imbedded in the upper portion of the tin ground in a horizontal position and on its upper surface about 11 feet from its extremity there was a large angular notch which must have been formed by a hatchet or some other sharp instrument'. A portion containing one side of the cut notch was presented to Truro Museum where it is still preserved.

Sub-fossil wood resting on top of the tin ground ought to have a date of *c.* 10 000 bp, or even earlier.

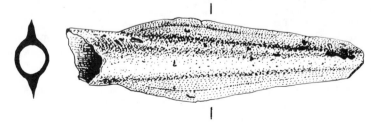

86 Middle bronze age spearhead from Wheal Virgin streamworks; length 10 cm (courtesy Penlee House Museum, Penzance)

87   The Pentewan tankard, 1st century AD; height and dia. *c.* 15·5 cm (courtesy County Museum, Truro)

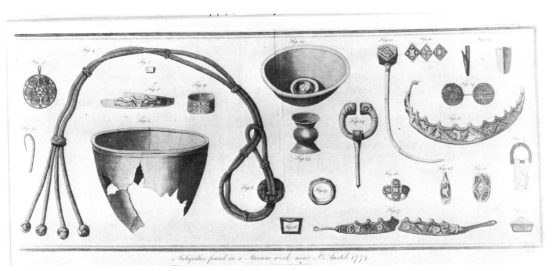

88   Trewhiddle hoard of late Saxon metalwork (*Archaeologia*, 1789, *9*, Plate 8; photo Society of Antiquaries, London)

89   Reconstructed silver chalice from the Trewhiddle hoard; height 12 cm (*Archaeologia*, 1794, *11*, Plate 7; photo Society of Antiquaries, London)

Wood found within the tin ground cannot have arrived there at the same time as the cassiterite, and it can hardly be doubted that Winn's oak trunk had been put there by man. It has been dated at the University of California, Riverside (UCR 1828), to 4140 ± 100 bp, giving a time-range of 3015–2415 BC. This is uncomfortably early, even if the precise calendar date lies at the end of the range. It suggests that tin streaming in Cornwall began much earlier than hitherto suspected.

It is just possible that the oak trunk is a neolithic sub-fossil dug out of the upper level of the alluvial overburden. Sub-fossil wood from tin streamworks and from coastal submerged forests has been used in recent times for the manufacture of trinkets and pieces of furniture. Thurston[26] notes how pieces of Mount's Bay submerged forest were used to peg out a tennis court, while a Penzance lady had a music stand made by her father from one of three oaks exposed during a gale in November 1873. All this is speculative and highlights the need for further radiocarbon-dating of wood surviving from tin streams.

Before leaving the Pentewan valley, the important animal sub-fossils recovered from the sediments overlying the tin ground should be noted. The finds listed by Colenso in 1829 and Winn in 1839 include deer antlers, the jaws of a pig, the humerus and scapula of an ox and, more importantly, an imperfect radius, two phalanges and two large bones of a whale noted by Colenso and still preserved at the Royal Geological Society of Cornwall's museum at Penzance. The whale bones were described by Flower[27] in 1872 as *Eschrichtius robustus*. They have not been examined recently by an expert, but Turk (*in litt.*, 1983) is confident that they are the same as *Eschrichtius glaucus* (syn. *Rhachianectus glaucus*), the California gray whale, the sole survivor of a family that formerly contained many species. Winn, however, assumed the animal was a young *Balaena mysticetus* or bowhead whale of circumpolar distribution.

The bones came from the Happy Union streamworkings in a bed of sea sand at a depth of *c.* 7·6 m, and their discovery in a former estuary mouth makes the

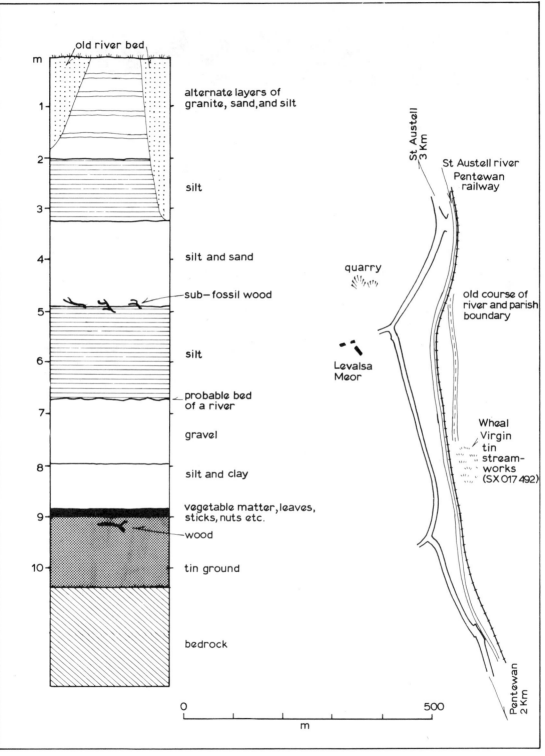

90 **Wheal Virgin Tin streamworks, Pentewan Valley, 1839 (section redrawn from J. M. Winn, 1839; location from the Tithe Map of St Ewe Parish, 1839)**

91   Middle bronze age rapier blade from Paramoor streamworks; length 20·9 cm (photo Ashmolean Museum, Oxford)

California gray whale more likely. It is the only baleen whale that enters shallow water to calve, and it will head for the shallows if chased by a killer whale (*Grampus rectipinna* in the Pacific). Walker[28] described them as having 'been observed rolling in the surf in water barely deep enough to float them'. Although confined now to the Pacific coast of north America, sub-fossil remains have been found in the north-east Atlantic in the Netherlands and Sweden. The Pentewan remains merit further examination because other species have been stranded in estuaries, notably the rorqual or sei whale (*Balaenoptera borealis*) and the blue whale (*Sibbaldus musculus*) in the Firth of Forth in a mesolithic context.[29]

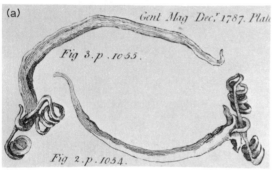

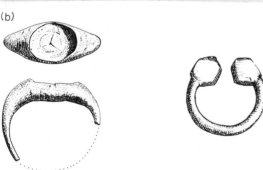

92   a Two views of fibula from Tregilgas (from *The Gentleman's Magazine*, 6 Dec. 1787); b Finger ring, dia. 27 mm; c Penannular brooch, dia. 20 mm from Polmassick streamworks (courtesy County Museum, Truro)

THE PORTHLUNEY VALLEY

The small stream rising in the tin-rich ground 2 km west of the renowned Polgooth mines flows to the sea at Porthluney near Caerhays Castle. The valley has been worked for stream tin which was gold-bearing. In 1758 Borlase related how James Gaved, a tin streamer at 'Luny, in the parish of St Ewe', found gold in 'a blue sandy flat' as well as gold in fragments of quartz veinstone.[30] Klaproth was much impressed with the gold at 'Pensagillis tin stream' on the border of Creed and St Ewe parishes, the same deposit found less than a kilometre away, and which was 'remarkable' for the native gold:

> now and then . . . met with . . . though very rare in pieces of the value of two or three pounds sterling.

Klaproth had a grain in his own collection 'the size of a flattened pea',[31] but in September 1756 a nugget of *c.* 20 g was found in a streamworks in Creed, almost certainly at Pensagillis.[32]

Only a year before the publication of Borlase's *Natural History of Cornwall* (1758), a tinner at 'Parmoor' (now Paramoor) dug out of his streamwork a dagger blade 20·9 cm long (Fig. 91). Because of its neat workmanship Borlase considered it to be Roman; it is now known to be a middle bronze age rapier, though shorter than the weapons usual in this class. The two rivet-holes and rounded hilt place it early in the sequence somewhere between 1300 and 1000 BC.[33] In 1760 the same streamwork also yielded up a wooden shovel 'cramped and edged with iron', about 2 m below the surface[34] and clearly of the type described by Carew in 1602 as then in current use.

'Cornubiensis' of Liskeard[35] submitted a drawing of a fibula 'lately dug up by some stream tinners in Tregilgas Moor, St Ewe', *c.* 2 m below the surface (Fig. 92*a*). It had been offered to a goldsmith who clipped off 'part of the tongue at one end'. The brooch does not survive. Gold would be an unusual metal for a fibula, but whether gold or 'bright bronze' as Haverfield supposed, both Haverfield and Hencken accepted it as belonging to the 1st century AD, as it closely resembles the simple bow brooches, many of which predate the Claudian invasion of Britain. A close parallel was found beneath the Roman bank at Cirencester[36] and comparable to La Tène II and III types from Camulodunum.

In 1787 two further objects were dug up by tinners streaming in the valley close to Polmassick in St Ewe parish. They were presented to Truro Museum by Rogers in 1919. The first is part of a finger ring (Fig. 92*b*) of a type common in the 2nd century AD. (Butcher, pers. comm. 1982). Analysis by Justine Bayley (*in litt.*, 1982) showed it to be brass with only 1–2% tin and a trace of lead. The casting was poor with blow-holes and cracked metal around the setting. The second, a penannular brooch (Fig. 92*c*), is unusual if Romano-British, though one from Lydney, Gloucestershire, has similar terminals,[37] and it could be of post-Roman date. The composition of both objects is found in Roman jewellery. However, the workmanship is so poor, according to Bayley, that they could be either bad Roman workmanship, perhaps of local manufacture, or more recent in date. The latter is more likely for Dickenson confirms that the penannular brooch belongs to Fowler's type G group G1 8 (plain hoops and plain terminals) dated to the 6th century AD.[38] It compares well with a silver example from Worlebury, Avon. As both objects were found in approximately the same place they could be contemporary, and it is safer to consider them as both post-Roman.

It is quite likely that the valley was streamed for tin down to the vicinity of Caerhays Castle. This has not been proved, but it would be unusual if alluvial deposits were not found here. In November 1869 two labourers digging a ditch on the edge of 'Beechtree Wood' (Old Park Wood) skirting the valley, cut through *c.* 1 m of soil to a deposit of sea sand. Set in the sand between three stones was a jug closed with a wooden bung (Fig. 124*c*). The finders battered off the neck and handle to reveal no less than 2 500 third brass coins, some of them with a silver wash. Some were quickly dispersed but 2 289 were described.[39] The hoard covers the period AD 253–82 with most of the coins belonging to the reigns of Victorinus (512 coins), Tetricus I (878 coins), and Tetricus II (396 coins). Most are in good condition with only 22 illegible. Examples are distributed in several museums with a large number in Truro. More important from our point of view is that the jug is made of almost pure tin (96%) and closely resembles the one found at Bosence camp in St Erth. The high tin content may suggest a local manufacture, though it could easily be an import of the Somerset pewter industry, which will be discussed in a later section.

Although the hoard was not found in a tin streamworks, it may legitimately be regarded as proof of Roman interest in Cornish tin, though the jug was presumably buried by a local person and not an up-country trader. There could well have been post-Roman interest here, not only because of the Polmassick jewellery, but because the present writer was fortunate enough to find on a molehill on the east side of Porthluney Cove a fragment of imported Bii amphora.

The burial of the jug in sea sand has suggested to some that the valley was tidal as far as Old Park Wood in Roman times.[40] The hoard was a few metres above high-tide level. It is far more likely that the valley floor was dry in Roman times. Porthluney Cove is backed by a considerable sand bar, comparatively insignificant only because of the narrowness of the Cove. There can be little doubt that its formation is contemporary with other Cornish bars, such as Loe Bar, Helston, and the submergence of the forest in Mount's Bay *c.* 2000 BC.

THE FAL RIVER VALLEY

The middle reaches of the Fal valley in the parish of St Stephen-in-Brannel have been profitably streamed for tin in the past and were well known for the small amounts of gold recovered from the workings. Grains of gold from 'Crow Hill', the streamworks north of Trenowth, recovered in the early 19th century, are displayed in the County Museum in Truro. Borlase wrote in 1758[41] that in the streamworks of the parish:

> they find now and then some small lumps of melted tin, two inches square and under ... this kind cuts with difficulty, and more harsh and gritty than the common melted tin ... These fragments I look upon as fragments ... scattered from the Jewish melting houses.

None survives, so there is no way of dating them. They could well be mediaeval, but those determined to identify ingots with the 'astragalus' of Diodorus Siculus may care to liken these small pieces to dice (the meaning of astragalus in the plural).

The antiquity of streaming in the valley was proved in 1793 when a metal bowl weighing 1 kg was dug out of the streamworks at Hallivick (Fig. 93), immediately west of St Stephen's churchtown.[42] The bowl is in good condition, suggesting that it is made of pewter rather than of pure tin as originally believed. Pure tin bowls from the Carnon Valley and from Parson's Park in St Neot suffer badly from corrosion. The Hallivick bowl, with its distinctive octagonal collar, has several parallels. With it in the British Museum are one from Lakenheath and two from Icklingham, Suffolk.

By far the finest example is from the Isle of Ely, Cambs., perhaps from a hoard found at Sutton.[43] It offers important dating evidence because the flange is decorated on its upper surface with Christian motifs – peacocks for immortality, an owl (?) for divine wisdom, nereids for baptismal regeneration, and a chi-rho symbol flanked by A W, alpha and omega. The chi-rho is of the early Constantinian form indicative of a 4th century date, and certainly not much later than AD 400.[44]

Bowls of the Hallivick type are not known on the Continent, but are native products. This interpretation is made all the more likely by the

93   Pewter bowl from Hallivick; height 11·5 cm, dia. at mouth 14 cm (photo British Museum)

discovery in 1928 of another, squatter example with a thin flange, on Meare Heath in Somerset.[45] This, surely, is a product of the local pewter industry using lead from Charterhouse in the Mendips and Cornish tin. The industry expanded after the mid-3rd century to fill a gap created by the decline of imported pottery such as samian ware, which had virtually ceased to be imported after the end of the 2nd century. Most pewter vessels belong to the late 3rd or 4th centuries.[46] The links between Cornwall and Somerset were close at this time, and the Hallivick bowl would be a likely import from Somerset.

A selection of pewter vessels was found during the 1878–9 excavations of the baths at Bath (Fig. 94). They include a flagon, superior in workmanship to those from Caerhays and Bosence in Cornwall, a fine bowl of Hallivick type, a flat dish and candlestick, and a remarkable mask of pure tin from the culvert in 1878. In all there were 13 pieces of pewter tableware from the baths' reservoir and outfall drain. It is thought they may represent 'a single offering deposited together perhaps by one of the many local pewter manufacturers'.[47] One of the main centres of production was at Lansdown, north of Bath, perhaps a large Roman village, where many stone moulds were excavated in the early 20th century. As well as the use of tin at the major pewter factories, some tin was used at villas and must have added to the demand for Cornish tin. Some small-scale smelting carried out at Chedworth, Gloucs. where crucible fragments analyses at the

(a)

a tin mask, height 33 cm

94 Selection of pewter vessels and mask of tin found at Bath, 1878 (photos by permission of the Institute of Archaeology, Oxford, and the Roman Baths Museum, Bath)

(b)

94 b Pewter flagon, height 19·5 cm

British Museum showed deposits of bronze alloy composed mainly of copper, tin, lead, and zinc.[48]

Borlase[49] described 'a collar, or other ornament, of brass, once gilt, and jewelled', dug up in streaming for tin in 1793 at 'Trenoweth, Broad Moor, St Stephens in Branwell' (Fig. 95). Later authors wrongly assigned it to Trenoweth in Lelant, where there is no tin stream. Borlase was not entirely correct either, for the earliest published reference to the collar describes it as 'found in a stream work called Trenoweth in the year 1802'.[50] Borlase probably confused the date with that of the discovery of the Hallivick bowl. Assuming the collar was dug up in 1802, it would be consistent with the steady progress downstream of the tinners, working systematically from Hallivick in 1793 to Trenoweth

(now spelled Trenowth) nearly 2 km away. Such steady progress up or down a valley was typical of the period, as in the Pentewan valley already discussed.

The Trenowth collar has been detailed by Megaw.[51] Analysis proved that the back plate of the collar is tin-bronze, but the front plate is brass with *c.* 30% zinc, a technical innovation that dates it to within a century or so following the Roman invasion of AD 43. The decoration of 'formalised palmettes constructed by branching comma motifs, the centre of each element being picked out with the glass inlays', is strikingly similar to that on a bronze neck-ring from Portland in Dorset, also presumed to belong to the 1st century AD. The Trenowth collar further resembles a much finer piece from Wraxall in Somerset. Megaw considers the conservative Celtic mode of punched decoration and the use of glass insets as typical of the west of England in the 1st century, perhaps even indicating its manufacture within Cornwall, rather than an import from the Somerset area. Ornaments were made locally, for at Castle Gotha, only 10 km east of Trenowth, a stone mould for casting penannular armlets was excavated. The main period of occupation of the site falls within the first two centuries AD.[52]

Finds were also made in a streamworks 'about half a mile east' of St Stephen-in-Brannel, that is at or near Gwindra.[53] They include two pieces of gold, one bearing a striking resemblance to the twisted 'spring' part of the Tregilgas fibula discussed above (*see* Fig. 92). Unfortunately, there is no proof that the finds are prehistoric. They could be mediaeval as they were found in 1780 with a quarter noble of Edward III and a coin of Henry II, *c.* 4·5 m under the surface of the streamworks.

The Fal rises on the Goss Moors, but prehistoric finds from this extensive tract of ground are dealt with in a later section.

THE CARNON VALLEY

> The Carnon tin stream-works, which for the present are abandoned, were here conducted on a large scale, and in a very spirited manner – the water having actually been banked from the works, which were carried on for a distance in the bed of the estuary. The space of ground thus streamed exceeded a mile in length by 300 yards in width. In this the tin stratum, which varied in thickness from a few inches to 12 feet, was found at a depth of from 40 to 50 feet below the surface, under accumulations of marine and river detritus, consisting of mud, sand, and silt. One of these beds contained the trunks of trees, and the horns and bones of deer, and in the tin-ground grains of gold and pieces of wood-tin were occasionally discovered.
>
> Murray, *Handbook for Devon and Cornwall*, 1859

The Carnon Valley (*see* Fig. 96) contained the best known of Cornish tin streams. Small-scale operations have continued here intermittently until recent years, and in 1980 a feasibility study examined the possibility

(c)

94 c Pewter dishes dia. (*left to right*) 13·4, 15·2, 15·3 cm; dish on right contains a coin of the house of Constantine, early 4th century

of recovering tin from the tidal zone of Restroguet Creek.[54] An excellent account of the richness of the Carnon deposits was given by Henwood in 1855.[55]

> The tin was found in boulders or rounded lumps, varying from the size of a man's fist to a grain of sand, the smallest generally the richest, as the quartz or prian with which it had been associated had been washed away, leaving the parcels very pure. I have seen lumps as large as a man's fist nearly pure black tin, a state to which all tin is reduced before it comes to the smelters' market, and a most expensive process it is for some varieties of ore.

Tonkin[56] informs us that 'On the wastrell of this manor [Arworthal] have been large quantities of tin dug up from time to time.' What this means is revealed by Pryce[57] who was told that in about 1708:

> The low lands and sands under Perran Arworthall, which are covered almost every tide with the sea, have, on its going off, employed some hundreds of poor men, women, and children, incapable of earning their bread by any other means.

In this instance the tin was recovered from the surface, as Pryce explains:

> occasioned by the refuse and leavings from the stamping mills &c which are carried by the rivers down to the lower grounds; and after years lying and collecting there, yield some money to the laborious dressers ...

No prehistoric artifacts were, or could have been, recovered from these simple operations. The intensive and systematic workings that revealed evidence of prehistoric tin streaming began eight years after the publication of Pryce's book. A notice in the *Sherborne Mercury* for 10 July 1786, advertising the sale of 1/16th share in the 'Carnan [sic] Stream Work', described it as 'discovered very lately'. No precise position is given for the workings, but the description of them can only fit a location well upstream, probably close to Bissoe:

> The bed of tin is found to be about four feet thick. There have been two shafts sunk about six fathoms distant from each other, and the bed of tin is found equally promising in each. There is one very great advantage that will arise in working the aforesaid stream; that is, whereas in other streams the overburne or different stratas of sea mud and gravel is found to be from 40 to 50 feet, it is here not above 18 or 19 feet only, at most.

As work progressed downstream the tin ground became deeper, so that in 1801, when the streamworks were described as nearly a mile in length, the bed of tin pebbles was 1·2–1·8 m thick below an overburden of 11 m.[58] On 11 May 1796 the workings were visited by Maton and Hatchett,[59] the latter describing the workings in rather more detail:

> The present workmen have found Oaken Shovels and a Pick Ax of Stags Horn which latter is in the possession of Mr Fox at Falmouth ... The present stream works are about ¾ of a mile long by nearly 300 yards wide. 190 Men and Boys are employed in these works. The work was begun towards the North and now goes on towards the

95   Trenowth collar, probably 1st century AD; dia. 15·5 cm (photo British Museum)

South. They advance about 1½ feet in one day in the whole of the work.

Even this far upstream, probably near the present railway viaduct, the workings 'if not banked would be overflow'd by the tide'.[60] This must have presented difficulties in antiquity when the level of the land must have been several metres lower. By 1798 the workings were 'much failed owing to the junction of two creeks', those passing through Perranwell and Perran Wharf at Devoran 1·6 km downstream of the railway viaduct. Indeed, in the summer of 1799 the workings were 'greatly injured by torrential rain showers, whilst an embankment which had earlier been erected to keep back the tidal waters was also damaged'.[61] Even the drier conditions of the bronze age would not have

allowed working further downstream than this.

These accounts allow the find spots of the prehistoric artifacts recovered c. 1790 to be placed somewhere in the neighbourhood of Higher Carnon and the present railway viaduct, the stretch of river shown in Fig. 96.

Collier, Hatchett, and Maton commented on the continual recovery of 'some gold, generally in small particles', Maton adding that 'we actually saw several particles among some tin-ore that had just been washed'. The nugget discovered in January 1808 has already been referred to (Fig. 70). Similar amounts of gold must have been found by prehistoric tinners.

The stag's horn pick owned by Mr Fox of Falmouth (Fig. 97) has already been discussed in the section

96 View down the Carnon valley to Brunel's railway viaduct, photographed *c.* 1908 from the auxiliary tin dressing plant downstream of Bissoe owned by the Falmouth Consolidated Mine Ltd

dealing with methods of tin streaming. It must be early and merits dating by the ^{14}C method. Confirmation of working in the early bronze age rests on 'a celt of yellow metal found AD 1790, 26 feet under the surface near an human skeleton in Carnon stream work' (Fig. 98). It was drawn half-size 'from a copy by Mr Hennah'.[62] The axe is now lost, but the drawing clearly shows a flat axe with a thin, narrow, rounded butt, a marked

97 Tinners' tools of uncertain date (redrawn from Henwood, 1855). Note that the shovel, which could be mediaeval, still has 'string' attached to it for tying to the separate handle. The deer horn pick is likely to be bronze age. Although a poor drawing it is probably part of the one photographed in Fig. 73, *above*

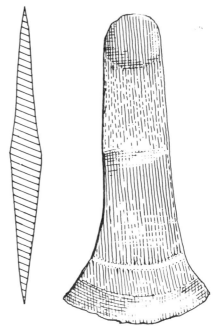

98 Early bronze age flat axe from the Carnon streamworks; length 15·3 cm (redrawn from Borlase, 1871)

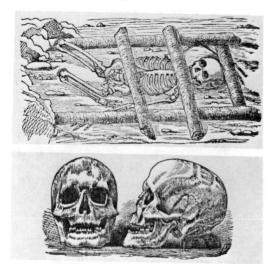

99   *a* **Skeleton and** *b* **its skull, found in 1823 resting on the tin ground in the Carnon streamworks (from Henwood, 1873)**

central thickening and a splayed cutting edge, features placing it with the typologically later group *b* flat axes which need not be earlier than *c.* 2000 BC.[63] The depth of discovery is a sure sign that it was found on the tin ground and not at a higher level. Whitaker referred to this axe, and to *two* antler picks from the 'Carne' tin stream which he saw at Tregothnan house in November 1792.[64] One of the picks was 'tinged strongly at the picking end with the stain of some metallic matter on which it had been employed' – probably oxide of iron. Whitaker continues:

> Not far from them was found a brass instrument, that had clearly, from the shade still remaining upon the covered part, once had a handle clipping it round the middle, and leaving out the two ends for striking. July 19, 1794, was promised by Lady Falmouth a sketch of all three, done by the hand of the Rev. Mr Hennah, Rector of St Austell; but, as he had pronounced the brass instrument to be no celt, and as I proved it to be one, he never sent the sketch.

This was the sketch later copied by Borlase.

The same stretch of valley below Bissoe Bridge was reworked on subsequent occasions. On 29 March 1823 another skeleton was found 'about half-way from Tarron-dean to the Arsenic manufactory', between the confluence of the vale through Perranwell and the arsenic works at Bissoe Bridge, so probably not very far from where the flat axe and skeleton had been dug up 33 years earlier. It is possible that the 1790 skeleton and axe were contemporary. It is tempting to regard the 1823 skeleton as equally ancient, though it is not datable. Relatively recent finds have been made in the Carnon workings, notably the bronze figure from a mediaeval crucifix at a depth of *c.* 9 m under the riverbed *c.* 1812.[65] Although presented to Truro Museum it has long since been lost, a fate which befell

the 1823 skeleton (Fig. 99). First notice of the discovery appeared in the *West Briton* of 4 April 1823:

> On Saturday as the labourers employed at Carnon streamworks were removing a quantity of mud, they discovered a heap of stones, under which were four pieces of oak inclosing a human skeleton, the teeth and larger bones of which were in nearly a perfect state. The tomb was covered with a deposit of mud; 17 feet in depth, and was 22 feet below the present low-water mark, on what is denominated 'tin-ground', namely, stones mixed with gravel amongst which tin is found. The four pieces of oak are each about 8 feet in length, roughly hewn, and about 8 inches in diameter. One of these pieces lay on each side of the body: the other two were laid across these, over the breast; the stones were piled over the whole. The wood is more decayed than the timber found in these stream-works generally is. The body must have been interred many centuries since.

Jago reported to Henwood[66] that the 'pelvis, other bones, and undecayed – but much worn – teeth, showed the remains to be those of a man, not exceeding five feet five inches in height, and, probably, much beyond middle age'. The roof and side of another skull (now in the museum of the Royal College of Surgeons) was found in the Carnon tin stream lying 'immediately on the tin-producing stratum' at a depth of 11 m in 1809. Keith published two sketches of it in 1915 in the mistaken though commonly held belief that it predated the deposition of the overlying strata.[67]

Keith also drew and described another supposed neolithic skull, formerly in the collection of the anthropologist Buck,[68] which had been found at a depth of 9 m in 'an alluvial tin mine at Sennen' below sea level. The nearest tin stream of which there appears to be any record lies north of Sennen churchtown at Porth Nanven – or Pornanvon, as Henwood called it – 1·6 km south-west of St Just, but 'long since exhausted' when Henwood wrote in 1873. The depth of working here is not recorded, but 9 m seems very deep for this part of Cornwall.

A brief account of objects 'lately found' in the Carnon stream works is given by Gilbert in 1817.[69] These included 'several ancient coins, and other articles of remote date', some of which were in the possession of Sibley of St Neot, the particulars of which Gilbert was unable to ascertain. Sibley was agent for the Carnon streamworks. Two coins in his possession were described[70] as having been 'lately' dug up 'a full 15 feet beneath the present surface' adjacent to where the workmen had also dug up pieces of gold, human bones, deer's horns, an oak shovel, and other articles 'evidently very ancient'. The coins, or medals as the report called them, were of brass. The better preserved carried the inscription CAES DOMITAV C GERM COS XIII CLMS, probably a misreading for IMP CAES DOMIT AVG GERM COS XIII CENS PER P P. On the reverse was S C and a figure believed to be Mercury, though more likely Apollo or some other.

100   **Bead-rim bowls of nearly pure tin from *left* the Carnon tin streamworks, and *right* from Parson's Park, St Neot (courtesy County Museum, Truro)**

Domitian was Augustus from 81 to 96. The second coin in poorer condition bore the name COMMODVS (177–92); the obliterated reverse bore a figure identified, probably wrongly, as Mercury.

Other Roman coins from the tin works were presented to Truro Museum by Bullmore in 1843–44, but are now lost and no record of them remains. Of the other finds recorded by Gilbert, the oak shovel may have been the one with the handle and blade in one piece and now in Truro Museum. The 'pair of stag's horns' found in 1811 at a depth of 11 m sounds like a sub-fossil, similar to others from the Pentewan valley, and not ancient tinners' tools.

Another lost hoard of Roman coins from the Carnon tin works was described by Sainthill in 1852.[71] Writing from Cork, he was also impressed by a piece of 'elk's horn, and of its skull', similar to finds made in Ireland, from the Carnon streamworks and which he had in his Cornish home. Notwithstanding his Cornish links, he understood nothing of tin mining, wrongly believing that the coins came from a copper mine 'near Perrenworth [sic], which mine is situated in the centre of a small creek in the Falmouth harbour'. He was confused by the miners' usual practice of giving depths of underground workings in fathoms and the common practice of measuring the depth of streamworks in feet, for it is clear that the coins were found in the Carnon works at a depth of around 30 feet (9 m) and not an impossible '30 fathoms [55 m]'.

His note, dated 23 December 1851, states that the coins came from the 'late W. Williams Esq of the 54th Regt' who presented him with 'about sixty coins, usually termed small Roman brass, though these were struck in copper'. They measured only 8 mm in diameter and weighed 0·26 g with a few double that size. The coins do not appear to have survived. Of those dated by Sainthill, all belonged to Tetricus I (AD 270–3) except for one of his son Tetricus II who was Caesar at the time. The coins may have been radiate minims. Whether the coins, 'about two or three handfulls' in all, had been found in a container is not stated, and

accounts are too vague to link them with the discovery of a tin bowl in the streamworks.

Writing in 1828, Henwood[72] gives the finds in the Carnon valley as 'a shovel made of wood, and a pick of horn'. The tin bowl (Fig. 100), now displayed in Truro Museum, was presented by Henwood sometime between 27 August 1831 and 29 November 1832, which narrows considerably the most likely date of its discovery. No details of the find circumstances survive. It is only a likely supposition that it was found 'on the tin ground', but the bowl and the coins are enough evidence in themselves that tin streaming took place in the Carnon valley in Roman times.

Dating the tin bowl is difficult. Another, its twin in size and shape, was found on Bodmin Moor in the summer of 1932 by a peat-cutter 'beside the stream which divides Colliford Downs from Parson's Park, on the high moor of St Neot parish, about two miles north from the churchtown', at a depth of 0·9–1·5 m where the peat was no more than 1·8 m deep.[73] This part of the moor – 'Fowey Moor' of the mediaeval tinners – has been extensively streamed in the past. Only in 1979 was the first mediaeval tin mill in Cornwall excavated on the St Neot river upstream of East Colliford on the site of the proposed Colliford reservoir. The tin bowl was found only *c.* 0·8 km to the east in marshy ground that now forms the entrance to Parson's Park china clay pit. The area forms a depression in the granite where small amounts of stream tin are likely to have accumulated. Indeed, several small pools shown along the stream on the 1907 edition of the six-inch Ordnance Survey map (before the expansion of the clay works) are typical of those produced by tinners on Goss Moor and elsewhere. When analysed, the St Neot bowl proved to be 99·1% tin. Typically of objects of such purity it has oxidized badly and a good part of it is missing. The Carnon tin bowl is also badly preserved. The opinion of Hawkes in 1934 (who was shown the St Neot bowl by Croft Andrew), which need not be disputed today, is that the bowls are more likely to be Roman than anything else. A late date fits in with other

101   **Middle bronze age pin from a streamworks in St Columb; the shaft is bronze and the bulb-head is a tin–lead alloy over a clay core; in the schematic section the clay core is stippled; it was held to the shaft by a peg, now missing; length 14 cm (courtesy Cambridge Museum of Archaeology and Ethnology; redrawn from a photo and section by Colin Shell)**

datable tin and pewter vessels discussed above such as the Hallivick bowl (Fig. 93) and Caerhays jug (Fig. 124). Not unlike the Carnon and St Neot bowls is the silver bowl of 178 mm diameter, the first item discovered in the Water Newton hoard in February 1975 near Peterborough on part of the Roman town of *Durobrivae*. Coin evidence dates this hoard to *c.* AD 350,[74] a period when the British pewter manufacturers would have been imitating a good range of silver articles.

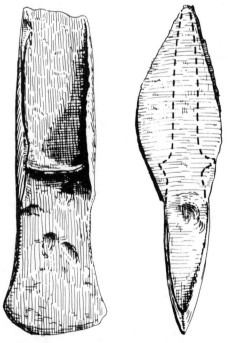

102   Middle bronze age palstave; length 10·8 cm (courtesy County Museum, Truro)

ST COLUMB

Copeland Borlase's private museum contained an interesting find from a tin stream described in his MS catalogue of 1882 as

> 'Item 109. Bronze Pin, perforated with leaden head, ornamented with horizontal and perpendicular lines, found in a stream work near St Columb' (*see* Fig. 101).

At the sale of the museum in October 1887 the pin was purchased by Cambridge Museum, where it was recently restudied by Shell.[75] The find spot could have been in St Columb Major, as either 'Minor' or 'Lower' was usually appended if the smaller, adjoining parish was meant. The rule was not always kept, and the most important tin stream in the area was Treloy in St Columb Minor. It is more likely to have come from here than from some other forgotten working. A number of items from Treloy have been dispersed and lost, and the pin does sound like the 'instrument of bronze, constructed apparently for piercing wood' which was found at Treloy before 1839.[76]

Colin Shell has shown that the head of the pin consists of a thin sheet of tin–lead alloy cast over a clay core. It is much worn with a good deal of the decoration obliterated. The shaft is a normal tin-bronze. The lead was added deliberately to the alloy, though whether or not it was obtained from one of the local lead lodes is impossible to say. Lead occurs widely in this part of Cornwall. Not only does it occur in many minor lodes in and around Newquay, but it may be very superficial as at Perran Wheal Virgin, near Perranporth, where a lead lode 0·6 m wide was found immediately beneath the turf cover in 1845.[77]

A similar bulb-headed pin came from the famous Wall Mead barrow at Camerton, Somerset, opened up in January 1818.[78] The finds also included a whetstone, an 'incense cup', and a dagger that has given its name to the Camerton–Snowshill series of 'Wessex' early bronze age daggers. If the St Columb pin is contemporary, a date *c*. 1800–1600 BC would be appropriate. However, it could be younger, for as Shell points out, similar pins are found widely over central and western Europe from the early bronze age down to 'Late Tumulus and Early Urnfield' times. Borlase had a European example in his collection, one 'of great beauty' from the 'lake dwellings, Switzerland', which had tin inserted in one of the perforations in its head.[79] It is likely, as Borlase thought, that the tin was of Cornish origin.

Another find from near St Columb may well have come from a tin stream (Fig. 102). It is a rather worn axehead of bronze, containing 18% tin, found in 1809 'in a pool between Blue Anchor and St Columb', certainly meaning St Columb Major. It is a low-flanged palstave with a small stop-ridge, early in the development of middle bronze age palstaves. Natural pools are extremely rare in Cornwall, so there is a strong possibility that the axe came from one of the many tinners' excavations in the vicinity. The area was certainly suitable for early exploitation, lying on the edge of the very productive Goss Moor. For example, at Fatwork and Virtue Mine, only 100 m south of Indian Queens and contiguous to Blue Anchor, an arrangement of horizontal tin 'floors' in the upper levels would originally have been exposed at the surface. According to Jenkin,[80] it was a site likely to have been exploited by streamers 'in earliest times'. A possible location between Blue Anchor and St Columb Major is at Halloon where there are two flooded tin pits on the opposite side of the road from Halloon Barton farm. They were working around 1800, both for tin and later for tile clay.[81]

RED MOOR, LANLIVERY

Helman Tor lies 5 km south of Bodmin at the northern end of a granite ridge flanked on the west by Breney Common and on the east by Red Moor (*see above* Fig. 66). The tin ground on these moors, at a shallower depth than in most other parts of the county, was detailed by Henwood in 1873.[82] He gave two sections of workings on Red Moor.

103   **Bronze La Tène brooch from Red Moor; actual size (photo Ashmolean Museum)**

Lower Creany	
Peat	60–90 cm
Granitic clay	30–90 cm
Tin ground with some flints and gold	120–150 cm

Upper Creany (Wheal Prosper)	
Peat	15 cm
Granitic clay	30–90 cm
Tin ground with microscopic gold particles	120–150 cm

This was a profitable area last worked, with the Breney deposits, during World War II. Between 1900 and 1914, Red Moor was mined in the old style using shovels and wheelbarrows. Recent investigation by Consolidated Goldfields showed the depth of the tin ground to vary between the extremes of 2 and 8 m. Pollen analyses of the overlying peat provided ages of *c*. 6000 BC,[83] so prehistoric tinners had the same shallow depth of overburden to contend with as recent streamers.

An iron age fibula (Fig. 103), formerly belonging to Evans,[84] is now in the Ashmolean Museum, Oxford. Accompanying the brooch is the handwritten account:

> Found at Redmore nr St Austell in Cornwall under 6ft of Peat & 20 inches of River Gravel. Beneath the sand lay another deposit of Peat 2½ feet in thickness which had been partially cut as fuel. Mixed with the cut blocks of this second peat deposit were the remains of a smelting hearth and pieces of Tin slag.

The back of the card carries the location and the name W. W. Smyth Esq. (Sherratt, *in litt*. 1982). As Hencken pointed out,[85] 'This is certainly one of the clearest indications of prehistoric tin-working that has yet been encountered.' The brooch is a well known La Tène type dating between *c*. 400 and 200 BC. Evans illustrated a similar one from Cowlam in Yorkshire, while among others are two from a bronze-smelting site at Merthyr Mawr on the Glamorgan coast.[86]

Evans had obtained the brooch from no less an authority on Cornish mining than Warington Smyth with whom he had discussed the antiquity of man as evidenced by finds from tin streams.[87] Smyth, eight times President of the Royal Geological Society of Cornwall between 1871 and his death in 1890, had been from 1852 Mineral Surveyor to the Duchy of Cornwall. He doubtless obtained the brooch in the course of his work. Henwood did not know of the find, though it is likely to have been discovered early. According to Edward Skewes[88] no one was then working on Red Moor, the greatest activity having been 'about 40 years ago' when 100 men were employed.

Red Moor was probably streamed in the bronze age. There is no direct evidence from the tin ground, but at

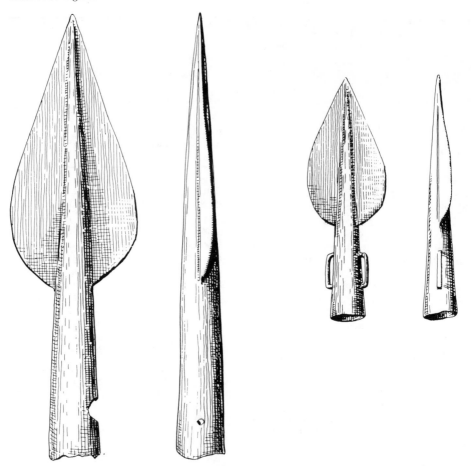

104   Middle bronze age spearheads from streamworks in Roche; *left* unlooped spearhead, 18 cm *right* spearhead with side-loops, length 12 cm (redrawn from Borlase, 1871)

Bodwen above the southern edge of the alluvial deposits there is evidence for a smelting site,[89] notably through the discovery during ploughing in 1971 of two fragments of a 'greenstone' mould for the manufacture of rapiers (Rowlands' class 2) datable to between 1400 and 1200 BC.[90]

Skewes[88] wrote that 'Jew's house tin' had been found at Red Moor. This, of course, is undatable. It probably refers to the discovery by a tinner, John Hare, 'at Lanlivery' in 1855. Hare found the remains of a smelting site with specimens of ore and refined tin,[91] and perhaps an ingot from this discovery was the one sent to Hunt at the Royal School of Mines in London. Hunt presented it to the Museum of Practical Geology. It weighed 21 kg.

Even if the tin used at Bodwen in the bronze age did not come from Red Moor, the site lies only 5 km east of the tin stream at Criggan Moors where important discoveries were made.

GOSS MOOR AND CRIGGAN MOORS

The Goss and Tregoss Moors form the largest alluvial tin deposit in Cornwall. The 19th century tinners were well aware of the activities of the 'old men' as they:[92]

> sometimes find portions of the peat [cover] wanting, and they always pass over the ground, for they know by experience that search for ore would be labour in vain; as the disturbed state of the *overburden* shows that the ground has already been worked.

Nevertheless, streamers still unearthed a number of bronze age and later antiquities, though it is not always possible to pinpoint the find spots.

A characteristic of the moorland alluvials was their varied character, the best quality tin lying beneath the alluvial cover at the bottom of irregular depressions of variable depth in the solid bedrock. Barrow found the same on Bodmin Moor[93] where the richest part of the tin-bed lay in a distinct channel below the general level

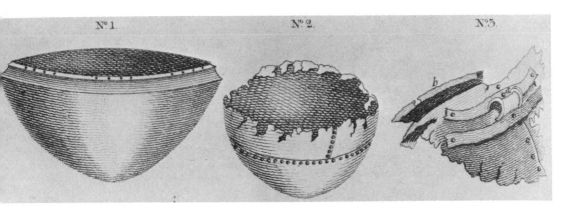

105   Bronze cauldrons of Hallstatt type (late bronze age) from Broadwater streamworks no. 1 is 270 cm circumference at bulge, 70 cm dia. at top, and 50 cm deep; no. 2 was damaged after its discovery, though when found was less perfect than no. 1; no. 3 is a fragment broken off 2, showing the rim *b* and loop *a* which would have held a ring handle (from the *Gentleman's Magazine*, July 1795 Plate III)

of the 'shelf' or bedrock, and it was these channels that the 'old men' steadily followed. They also worked the less rich gravels on either side, but many patches were left, supposedly because they were too poor or because, in many cases, they were waterlogged. Barrow noted that the water table was sometimes many metres above the level of the tin ground, and the haphazard and short-sighted methods of extraction 'tended to obstruct the escape of water from the tye, so that the working-face of the deposit became waterlogged'.

The same must have been true of Goss Moor. The deepest workings here appear to have been on the Fore Moor between Bilberry and Bugle where the tin ground lay under an overburden of 4·5–6 m.[94] These geological circumstances must be partly responsible for the great number of tiny disjointed operations undertaken piecemeal throughout the area in recent centuries. Such a streamwork was Loath-to-Depart in Roche – a rich working if the name can be relied upon as a good judge. In July 1803 tinners found here at a depth of 9 m, clearly deep enough to be on the tin ground, a side-looped middle bronze age spearhead (Fig. 104). This 'brass' artifact does not survive, but is figured in Borlase[95] together with another unlooped spearhead 'found deep in a streamwork' also in Roche parish. The quality of the tin here was excellent. Klaproth commented that in the St Dennis–Roche area the cassiterite was 'larger and less rounded' than in the lower valleys, with many pieces of tin stone 'still showing their crystallized angles'.[96]

'Loath-to-Depart' was a fairly common name for a rich tin stream. This one lay on the south side of the Roche to Bugle road less than a kilometre east of Roche Rock, one of a number of setts leased by the Rashleigh family about 1800. The other spearhead must have come from one of Rashleigh's adjoining setts between Loath-to-Depart and Bugle; Cost-all-Lost, Great St

George, Woon, Crown Filley, Gracca (alias Grâce à Dieu), or Wheal Hope, all of which are in Roche parish (Penderill-Church, *in litt.*, 1982).

Other bronze age finds came from Broadwater streamworks just inside Luxulyan parish at the headwaters of the Par river immediately south-west of Savath farm. The depth of the tin ground here was often more than 9 m,[97] though the section of the deposit published by Henwood[98] was shallower than this.

Granitic sand	1·8–2·1 m
Mud, granitic sand and gravel with vegetable remains	1·2–1·5 m
Vegetable remains and silt	1·2–1·5 m
Tin ground: small pebbles some of which are but little rounded	2·1 m
Granite bed or 'shelf'	

The potential of this rich tin ground had not escaped the attention of late bronze age tinners, for two bronze cauldrons (Fig. 105) were found here on[99] 'March 28, 1792, with their mouths upwards, and full of gravel . . . about 28 feet under the surface of the earth'. Both were in good condition, though the second was 'much battered and mutilated, having fallen into rude hands, and being used as a common utensil', perhaps for baling out water, the use they may have had in the bronze age workings. In spite of giving the date of the discovery as 28 March, a later part of the account states that they were found 'about 9 feet apart, within three weeks of each other' by John Nichols and John Stevens of Luxulyan. It was hoped that more finds would be made 'as they unburthen the ground in search of tin'. Whether the vessels were Roman or Phoenician was beyond the writer's understanding. It is now known that they belong to a series of late bronze age cauldrons with strong Hallstatt affinities datable to within the first half of the 1st millennium BC.

106 Bronze penannular brooch, c. 6th century AD from a streamworks near Lanivet; length of pin 55 mm (courtesy County Museum, Truro)

Vessel number 1 held 180 l and had been 'worked out of one entire piece, in a manner equal, if not superior to any modern skill in workmanship, and weighs 14 lbs'. The base was rather pointed with the rim turned in. A similar rim, more sharply turned in, is seen on the cauldron from the River Thames in London. This also resembles the Broadwater cauldron in being formed from a single sheet of bronze. Its height (34·5 cm) and diameter (66 cm) are also similar to the Broadwater cauldron. The degree of in-bending of the rim varies considerably in this class of vessel, and is more pronounced on the London example than on Continental ones. Such in-bending may be a late feature found among Greek and Etruscan smiths from *c.* 650 BC.[100]

The second Broadwater cauldron is built up of tiers of bronze sheeting riveted together, typical of nearly all the British–Irish series. Well known examples are from the River Cherwell in Oxfordshire,[101] from Battersea in London, and a late type with iron rim and a bronze body from Spettisbury Rings in Dorset. Arguably the finest are two bronze cauldrons found with five socketed axes of the type common in south Wales and south-west England, three socketed sickles (one of iron), and three socketed chisels, at Llyn Fawr near Rhigos in north Glamorganshire. These cauldrons are dated to *c.* 650 BC. A similar example from Sompting, Sussex was dated by Curwen to between 750 and 500 BC.[102] The position of the 'tunnel-shaped' staple, which originally held a bronze ring handle on the second Broadwater cauldron, is unusual. Ordinarily the staple was on top of the rim, but in the Cornish example it was placed well below it.

Hawkes and Smith[103] believed such cauldrons were of Continental origin, perhaps traded here from Massalia *c.* 600 BC, though they are now considered to be purely native products.[104] Nevertheless, it may be no coincidence that the London and Broadwater

number 1 cauldrons have a Continental distribution to the north of and partially embracing the area of origin of the silver coins of the Paul, Penzance hoard already discussed.

In 1835 two tin streamers working 'near the parsonage at Roche, came down upon an enclosure of rude stones, wherein was found a block of tin of peculiar shape', though what it was is not recorded, unfortunately. 'Nearby lay several ancient coins, a number of which were sent to London "to be inspected by the learned".'[105] Whether the coins were Roman or mediaeval is not known, but the tinners were fired with sufficient antiquarian vigour to open up a barrow 180 m away where they unearthed 'several pickaxes and spear-heads of brass'.

One of the few instances of proof of tin streaming in the post-Roman period comes from this part of Cornwall, though the precise area is not known. It is a bronze penannular brooch (Fig. 106) presented by Fox to Truro Museum between 1821 and 1829, recorded in the accessions as coming from a streamwork 'near *Lanivet'*. Museum records also state that it was found 3 m below the surface. Barham mentioned the find in 1860,[106] giving its location as 'ten feet below the surface, on Goss Moor'. Hencken also assumed it was from Goss Moor,[107] but unless Barham had more information, now lost, it may not have come from so far west. Other streamworks were operating closer to Lanivet in the early 19th century. The churchtown is central between tin works at Boscarne to the north, Red Moor to the south-east, and Criggan Moors and Broadwater to the south-west; all of these are equally possible find spots.

This is the only penannular brooch in the museum found in such circumstances, so there can be no confusion. There has been debate about the date of the brooch, which falls within Fowler's ill-defined group 'G'.[108] Dickenson describes the Lanivet type as having 'bull's eye' decoration on the terminals, a feature current between the late 4th and 6th centuries AD.[109]

TRELOY, ST COLUMB MINOR

In the summer of 1939 the late Croft Andrew excavated the large and impressively defended cliff castle of Trevelgue, St Columb Porth, close to the outskirts of the modern town of Newquay. No report was published, though the finds make it clear that it was an important iron age and Romano-British site notable for its remains of smelting, chiefly of iron but also of bronze. Six recognizable bronze objects were recovered, including a lynch pin, and over 20 scraps of metal with 'a substantial lump of wrought bronze'.[110] Crucible fragments were found and a rectangular pit was interpreted as a kiln for the preparation of charcoal. Curious round-bottomed holes 'very neatly cut in the rock and provided with slate lids' were thought to be moulds 'for forging metal cauldrons and

(a)

107 a Romano-British tin bowl and cover; dia. 35·5 cm from Treloy streamworks

bowls'. Objects of pure tin were evidently made at Trevelgue, for a tin spindle-whorl, nearly 25 mm in diameter, was found by chance years later in an exposed face of Trevelgue and presented to Truro Museum in 1954. Tin is very rarely used on its own, but it may not be surprising at Trevelgue since the site lies close to the tin-rich deposit in the stream, which has its source on Goss Moor near Ruthvoes in St Columb Major.

A brief account of the tin deposit at Treloy, within 3 km of Trevelgue, has already been given in Chapter

24, together with a note on the gold which was also recovered. A tin bowl from Treloy (Fig. 107a) and the tin spindle-whorl may well have been made at Trevelgue out of locally won ore, though probably not in one of Andrew's rock-cut holes.

The simple but unique bowl, complete with a flat cover on which are scratched the Latin numerals XX, was found by tinners about 1826 at a depth of c. 3·6 m. It is difficult to date but probably belongs to the 3rd or 4th centuries. The well known series of late Romano-British stone (greisen) bowls, widespread in Cornwall

107 *b* Roman disc-brooch of bronze with enamel inlay from Treloy streamworks; dia. of disc, 30 mm max. (courtesy County Museum, Truro)

and recently discovered in a 4th or 5th century context at Pudding Lane in London,[111] sometimes carry decorative and non-functional handles carved in low relief. Good examples were found in the settlement of Trethurgy near St Austell, occupied between about AD 300 and 600.[112] The bowls are certainly close copies or skeuomorphs of metal prototypes, of which the Treloy bowl serves as a good example with its 'soldered-on' handles.

Not far from the bowl was found, about the same time, an oval disc-brooch of a type common in the 2nd century AD (Fig. 107*b*), though some types are known

to have lingered into the 3rd century.[113] Indeed, the Treloy brooch closely resembles one of those found on Nornour, Scilly, and like the Scillonian example has lost its central setting of stone or enamel, leaving only the blue and white enamel panels in the surround. These are the most spectacular finds from Treloy, but others were made in the 1820s or '30s, though most have long since been lost.

The tin bed at Treloy was 'of small thickness',[114] but as it was covered by only 2·4–3 m of overburden, working it would have been far easier than the deeper but richer south coast deposits at Pentewan and elsewhere. It was a deposit that was likely to have attracted bronze age tinners. Henwood, in his 1873 account of Cornwall's 'detrital tin ores', mentions '*celts* corresponding in shape with some figured by Borlase'[115] (Fig. 108). Borlase's figure 8 is specifically mentioned – some kind of socketed chisel with a widely splayed cutting edge – though as Henwood referred to 'some figured by Borlase', other palstaves or axes may have been found. Henwood may have seen them when a young man at Treloy, for another geologist, De la Beche,[116] also wrote in 1839 of 'Celts' found at Treloy, a term unlikely to have been used for anything other than bronze age axes of some sort or another.

De la Beche also refers to 'Roman coins, ancient rings and brooches'. 'The latter', the rings and brooches, 'are now in the Museum of Economic Geology. One ring is made to fit the wrist . . . and one of the brooches would appear to be Roman.' Unfortunately, the bronzes are no longer at the Geological Museum, for all their metalwork was divided up between other London museums in 1901 and the Treloy items cannot now be traced. However,

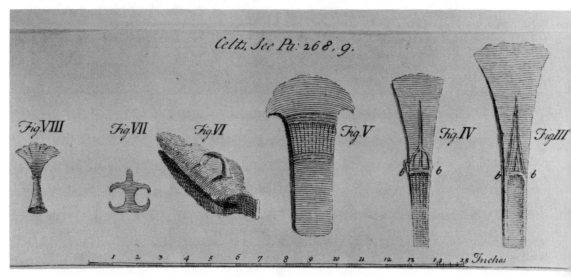

108   Unprovenanced bronzes comparable to finds from Treloy streamworks

ST MAWGAN-IN-PYDAR

The River Menalhyl flows 5 km north of the Treloy tin stream through the wooded Vale of Lanherne from its source in the tin-bearing country of Castle-an-Dinas, St Columb, to the sea at Mawgan Porth. The unsuspecting visitor to St Mawgan would never imagine that this pleasant spot had been streamed for tin. William Leddicote, sometime superintendent of the streamworks, told Henwood that there were:[117]

> between Lanherne and Mawgan Porth several alternations of mud, sand, and gravel, here and there mixed with large stones [which] overlie a mingled mass of branches, leaves, nuts, and other vegetable remains, which rests on a tin bed of poor *tin ground*.

This confirms the local tradition that the valley was streamed as far as Mawgan Porth. The Menalhyl is flanked by a series of long, narrow meadows between Mawgan and Gluvian farm, all of which are likely tin bounds. The deposits may have seemed poor to tinners familiar with the richness of Carnon, Pentewan, and other south-coast valleys, but the shallowness of the Mawgan deposits attracted bronze age tinners. The Revd Francis Vyvyan Jago,[118] Rector of Landulph, wrote a letter to Samuel Lysons dated 20 May 1813, describing how 'About a month since, some miners streaming for tin in a meadow called Long Moor, immediately under and belonging to Lanhearn [sic] House . . . found a number of Celts and a sword in high preservation . . . at a depth of twelve feet under a bed of black mud, and scattered at small distances upon a bed of smooth pebbles much resembling those on the beach a short way off' – in fact, 2·5 km away. The pebbles clearly formed part of the tin ground, and 3·6 m is the expected depth, being comparable to that at Treloy in a valley of similar size and length. None of the finds was illustrated by Jago. Most of the celts were socketed axes 'generally of the common form', *c.* 12·5 cm long, 'with a square socket and a ring at the side', and probably the normal south-west England socketed axe.

Jago obtained one axe himself that was not socketed but had 'a groove on each side' (Fig. 109), a misleading way of describing a palstave that was later given to John Jope Rogers of Penrose. Borlase later sketched it for *Ancient Cornwall*, and in 1919 it was acquired by Truro Museum. The palstave has an extremely good cutting edge, confirming the report that the 'celts were apparently fresh from the mould, and supposed to have been manufactured at the spot'.

Two objects were presented to the Society of Antiquaries by Revd Conybeare on 24 November 1814:[119] a thin bar or rod of bronze 'a foot long and ¼ inch thick and bent' and a fragment 'of a rudely formed saw of bronze', which is still in the Society's collection. Evans,[120] for unstated reasons, doubted that it was a saw although he gave several parallels from French

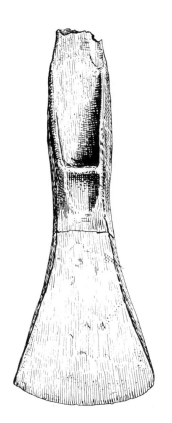

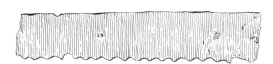

109 Middle bronze age palstave, length 15·4 cm, in Truro Museum; part of bronze saw, length 10·4 cm; both from Lanterne streamworks (courtesy Society of Antiquaries, London)

one item may still exist. De la Beche also mentions 'an instrument of bronze, constructed apparently for piercing wood, or substances of that hardness . . . curious in as much as it would seem to show that iron was then scarce.' The identity of this intriguing object is uncertain, but the description easily fits the bulb-headed pin (*see above* Fig. 101), formerly in the collection of Borlase and now in Cambridge, which came from 'a tin stream in St Columb' and is described elsewhere in this volume.

Mediaeval streaming is confirmed by the discovery of 'Coins of the early English Kings' in the streamworks near Treloy, though 'not with the Celts &c' according to De la Beche's astute informant.

110   Bronze age rapier comparable to one from Lanterne streamworks (redrawn from *Archaeologia*, 16)

bronze age contexts. A saw might be expected in a tin works, remembering Stocker's description of mortice and tenon joints in the wooden-framed shaft at Pentewan already referred to.

'Two or three axes' unfamiliar to Jago, were *c.* 13 cm long 'and of one uniform width, about one inch, with a square socket three parts of the length.' This description admirably fits the Breton type axe. Such an axe helps date the hoard to *c.* 600 BC as well as proving that traditionally middle bronze age axe types, the palstave now in Truro Museum, had a long life late into the bronze age. The same also applies to the sword found in the St Mawgan hoard, for it is a rapier with an ancestry going back to 1300 or 1400 BC (Fig. 110).

Jago likened the 'sword' to Fig. 2 (*Archaeologia*, **16**). The St Mawgan example was 53·5 cm long and in excellent condition being 'yet sharp'. The rapier and some of the axes were said to have been claimed by the nuns at Lanherne on behalf of Lord Arundel, but there is no record of them ever having been at the convent. When Evans[121] noted the rapier it was in Trevelyan's collection. Its present whereabouts are not known.

The lack of iron age finds from the Menalhyl should occasion no surprise, for metalwork of this period is rare anywhere in Cornwall. But interest probably continued in the St Mawgan tin stream, for there is

circumstantial evidence for its exploitation during the Romano-British period. Above the river on the south side of the village lies Carloggas camp, occupied from the early years of the 1st century until about the middle of the 2nd.[122] Evidence for metalworking came from hut A, apparently a smelting workshop, used between *c.* AD 25 and 50–70. A pebble from the hut contained 78·71% tin metal (1·45%Si, 1·4%FeO + Al, and lesser amounts of Cu, Ti, and Ca), while a piece of tin slag yielded 79·71% tin with similar amounts of Si, Fe, Al, Ca, Pb, and Cu, as the pebble. The slag was only partially reduced.

Hut A also yielded 'droplets of bronze' with *c.* 10% tin and traces of added zinc or some other impurity. There were also pieces of crucible, a tiny bronze bar (? ingot), and a folded piece of bronze which proved to be a finely decorated strip, perhaps one end of the bronze 'spine' extending from the central boss of an oblong shield, though something smaller, perhaps a scabbard decoration, has been suggested since. Whether it was made up-country or locally, there is no doubt about the origin of the tin. The finds from hut A do not prove the cassiterite came from the Menalhyl. Even if it did not, Treloy was close by, but one further object suggests a continued interest in the Menalhyl tin.

The *West Briton* of 27 July 1821 reported that 'a short

111   'Carnanton' tin ingot (courtesy County Museum, Truro)

time since', people digging in a field belonging to James Willyams of Carnanton 'cleared the head of a spring for the convenience of drinkers' and found a broken pitcher and pieces of silver. A search revealed several hundred coins from the reigns of Elizabeth, James I, Charles I, and a few of Charles II. This apparently superfluous information is important since the report continues:

> The vessel which contained the coins was hidden within a few yards of the spot where an ancient block of tin was discovered about two years since: and which Mr Willyams then presented to the Museum of the Cornwall Literary & Philosophical Society.

This was the early title of the Royal Institution of Cornwall; the ingot (Fig. 111) was presented before 27 August 1821. The Institution was also given 68 of the silver coins, the published Report for 1829 (pp. 29, 30) stating that they came from Nanskeval, a farm (demolished in 1980) less than 2 km east of Carnanton house. Hogg wrote in 1825 that the ingot had been found 'two feet and a half under the surface, in swampy ground, and contiguous to what is usually called a Jew's House'.[123] Had this really been a Romano-British smelting house, it is a tragedy that no better record of it exists. Memory of the site lives on, for Mr Jack Jones, an elderly retainer at Carnanton, was told by his grandfather that 'a treasure' had been found in the field still called Jew's House Meadow.[124]

The find spot reveals nothing. A small lily pond, doubtless the area cleared out in 1821, has now been filled in. When Henderson visited the spot in about 1916,[125] he saw the stream issuing from a little square building with a semicircular arch, and was told that the stream was 'richly impregnated with tin'. This cannot be true and must be a garbled recollection of tin in the Menalhyl a short distance away. There is no reason why the Carnanton ingot (as it has long been called) should not be composed of smelted Menalhyl cassiterite.

The wedge-shaped ingot weighs 17·9 kg, is 53 cm long, and at the thick end 18 cm wide by 7 cm deep. Its convex underside cooled in a simple open mould, probably of granite judging from its texture. When cold it was stamped in the central area of the flat upper side with at least nine poor, small impressions, not one of which is at all clear. Interpretation is doubly difficult because the tin has blistered, one large blister having 'burst' since Haverfield took the close-up photograph (Fig. 112) in 1900.[126]

One stamp below the centre of the photograph shows a head with a 'crest' interpreted as a helmet, facing right, with a 'small shield or buckler' slightly to the right, though to see it demands a certain eye of faith. Another head is on the stamp near the top centre of the photograph. It looks to the present writer quite unlike the other head, having curving lines which suggest a female, looking right, with a hairstyle not

unlike that of Faustina junior – though it would be unwise to draw any conclusion from that.

Haverfield was in no doubt that the helmeted head could be 'safely ascribed to the fourth century', buttressing his argument with a favourable interpretation of the faint inscription stamped twice below the helmeted head. Haverfield's final view, published posthumously in 1924,[127] was that 'the inscription has three or four letters, of which the first one or two may be D or D D, and that the last two are plainly N N run together as NN. It is perhaps permissible to explain them as abbreviations for *dominorum nostrorum*, "the tin of our lords the emperors".' This formula appears on Continental silver ingots of the 3rd and 4th centuries. When the present writer examined the ingot in 1967 with Warner,[128] it was clear that there was insufficient room for D D, and I E appeared more plausible, though only the vertical strokes are definite. Curiously, when Haverfield first published the ingot[129] he also read the letters as I E NN or I F NN.

The most that can be safely said about the ingot is that it is Roman. Because it is assumed that the Romans took more interest in Cornish tin production from the late 3rd century, it is tempting to give the Carnanton ingot a late date. However, the evidence for Boscarne, presented below, indicates that they were interested in local production in the 1st century, so on present evidence it would be unwise to be dogmatic.

Less than 3 km north of St Mawgan-in-Pydar lies the settlement of Trevisker occupied during the bronze and iron ages, and famous for its excellent sequence of Cornish-style bronze age pottery. It was excavated in 1955 and 1956 before the construction of a school on the site.[130] A fragment of a bronze knife from House A and 'mineralized lumps' from structure B were examined and shown to have similar compositions. The lumps were 'spilt waste metal or dross' and indicate bronze working on the site. Trace elements included lead, arsenic, zinc, silver, nickel, iron, and manganese, all consistent with the use of local ores. There was dramatic proof of the use of local tin by the discovery of pebbles of stream tin *c.* 1 cm in diameter. Over 20 were found in the silt of structure B, one in House C and two from a post-hole in House A. The latter two are the most important as oak charcoal from House A gave a ^{14}C date of 3060 ± 95 bp (*c.* 1500 BC).

The cassiterite pebbles, now in Truro Museum, vary between 5·5 and 7·0 specific gravity, except for one of 5·2 which probably contains little or no tin (Fig. 113). Dr Bromley of Camborne School of Mines (*pers. comm.* 1983) kindly examined two selected pebbles. One was devoid of characteristics that might indicate a provenance, but the other showed 'a bright orange-brown colour and sharply defined oscillatory zoning', a type of cassiterite diagnostic of the area to the north-west of the St Austell granite, between Indian Queens

0   5
|_____|
                        cm

112   Carnanton tin ingot: close up showing impressed marks made when the ingot was cold; just below the centre is the supposed helmeted head (photo *Proc. Soc. Ant. Rep.*, 1900, 2nd series, *18*)

113 **Pebbles of stream tin recovered from archaeological excavations: the two larger ones above are from Carn Euny, Sancreed; the smaller ones are selected from those found at the Trevisker bronze age settlement in St Eval;** *top left* **24 mm weight 12·35 g** *top right* **27 mm weight 23·7 g (courtesy County Museum, Truro)**

and Belowda Beacon. This form is 'rare elsewhere' in Cornwall. The pebbles at Trevisker could have come from the nearby alluvial tin deposit in the Vale of Lanherne at St Mawgan, whose headwaters lie between Castle-an-Dinas and Belowda Beacon, but no specimens of cassiterite from St Mawgan can be found for analysis.

BODMIN MOOR

The parish of Bolventor, created in 1846 largely out of the parish of Altarnun, has the distinction of being

named from a tin streamworks: 'Bold Adventure' lying in the upper reaches of the Fowey between Trezelland farm (formerly Trezelling, Tresellyn, or Tressillon) and Palmersbridge, 1·2 km north-east of Jamaica Inn. The whole area, in the middle of the mediaeval tinners' Foweymoor, has been extensively worked. 'Deep Hatches', west of Jamaica Inn, recorded by Leland *c.* 1540,[131] and 'Drywork', or 'la Driework' in 1272, on the Fowey just below Palmersbridge, are most probably the names of streamworks.[132] William Hals (*c.* 1700) noted how the parish of Altarnun 'hath in it tin loads [sic] and streams'.[133] It is not known what ancient tinners' relics might have been unearthed at that time, for comparatively little was recovered from the 19th century workings.

Confirmation of early workings comes from Colliford, St Neot, where Sandy Gerrard (pers. comm., 1984) excavated streamworks before the completion of the reservoir. No early artifacts were recovered, but a pollen profile provided a ^{14}C date of AD 960 ± 90 for 0·5 m peat, which had accumulated over earlier workings by tinners.

An open mould of 'greenstone' for casting early bronze age flat axes (now in Truro Museum) was dug up with a 'small hollow basin' of Polyphant stone (now lost) 'appearing like a crucible', in or before 1849 in the meadow below Altarnun vicarage.[134] The discovery could indicate the use of locally won cassiterite, and proof of later bronze age working appeared in Malan's report in 1889 that 'some years since a bronze spearhead was found under six feet of gravel in the stream-work below Jamaica Inn: it is now in Tavistock'.[135] Unfortunately the spearhead cannot now be traced, but Malan's location is good enough to equate it with Bold Adventure, which was working in the 1840s and throughout the 1850s (Fig. 114). Pattison described the workings in 1847.[136]

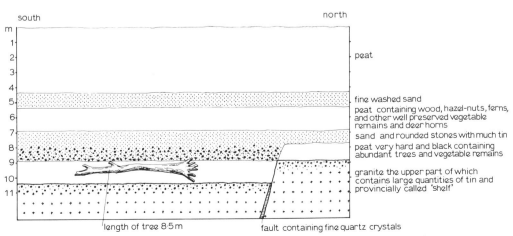

114 **Section of the tin streamworks at Bolventor, Bodmin Moor (redrawn from Pattison, 1847)**

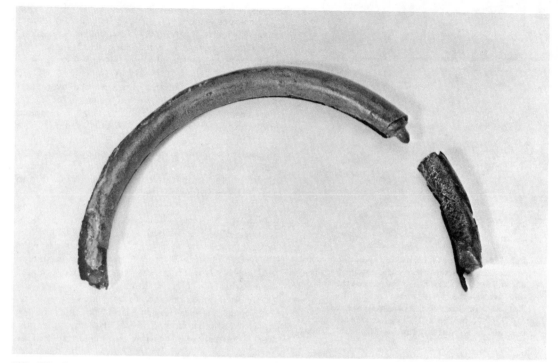

115 **Two fragments of La Tène torc, or collar, *c.* 400 and 200 BC found near Bodmin (courtesy County Museum, Truro)**

	Thickness/m
Peat	4·3
Fine sand	0·9
Peat with wood, hazel nuts and deer antlers	1·5
Tin-bearing gravel and sand	*c.* 2·1
Peat and trees with occasional tin at base	1·5
Granite shelf	

The workings were clearly close to Jamaica Inn, for Pattison described how the wood in the deposits was so abundant that it was commonly used for fuel by the tinners who stacked it in large piles near Bolventor. Malan's spearhead came from the base of the tin-bearing gravel and sand, though probably higher up the valley where the overburden would be thinner. The deposits were rich; Henwood noted remarkable tin crystals here 25 mm in length and over 37 mm base circumference.[137] In April 1860 he described how the previous month a streamer had dug up an iron-tipped wooden shovel of the type described by Carew in 1602, and noted how the finder's father a few years before had dug up one made entirely of wood. The 1860 find was upstream of the workings described by Pattison, for the shovel was at a depth of only 2·75 m 'beneath a peat bog, on the upper layer of 'tin ground', and the

present streamers are working the lower, or next to the base rock.' The find spot was 'near the entrance of the adit of the Tresellyn Consols Mines'.[138]

There were extensive streamworks in the mediaeval period in St Neot parish, and from the Romano-British period there is the tin bowl found at Parson's Park in 1932 (*see above* Fig. 100). This has already been discussed with the similar example from the Carnon tin stream.

An iron age La Tène style torc (Fig. 115) was found sometime before 1864 in a peat-bog 'near Bodmin'.[139] There is no suggestion that it was found in a tin stream. Its interest lies in the fact that the surviving pieces comprise tubes of thin bronze covering a core of tin metal that had expanded to split the bronze. Analysis suggested that the tin had partially converted to a chloride. The two fragments fit together with a bronze mortice and tenon joint, similar to one on a bracelet from Cowlam in Yorkshire.[140] The torc has been dated to between 400 and 200 BC. It is probably of Cornish manufacture, for iron age ornaments were certainly being made in east Cornwall about this time.

John Thurnam exhibited at the Society of Antiquaries in 1857 drawings of a stone mould from Camelford made of 'soft sandstone' for casting bronze harness buckles similar to examples from Polden Hill, Somerset, Stanwick in Yorkshire, and Westhall,

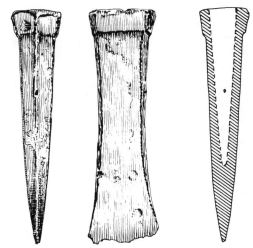

116 Socketed bronze chisel, *c.* 1100–1000 BC; length 12·5 cm (courtesy Bodmin Museum) from Boscarne streamworks

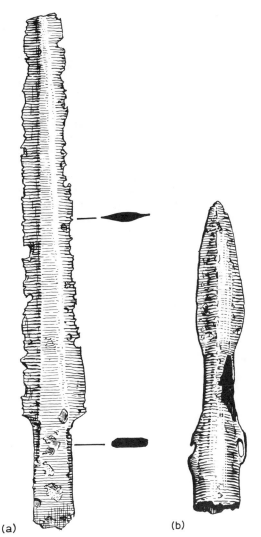

(a)   (b)

117 Tanged knife (length 17·3 cm) and socketed spearhead (length 10·4 cm), both of bronze, probably from Boscarne streamworks (courtesy Bodmin Museum)

Suffolk.[141] A tin core is also known from an example of Irish 'ring money' dated to the transition between the bronze and iron ages. The penannular rings of *c.* 25 mm diameter are of gold or a gold sheath encasing a base metal such as copper. In 1926 some were found as part of a hoard from near Abbeyfeale, Co. Limerick on the Kerry border. One of them was unique: the sheath was gold with traces of lead and copper, but the core was pure tin, 'more pure than the tin electrodes used for comparison purposes'.[142] By casting doubt on the existence of recoverable tin in Ireland, it is likely that the tin was Cornish, and maybe the gold as well.

THE RIVER CAMEL

The tin stream that has yielded most of the antiquities in east Cornwall is Boscarne, a few kilometres west of Bodmin. At Cotton Wood, the downstream limit of the workings, the tin gravel was '15 feet deep, and its bottom was 50 feet below the surface of the ground' on the south side of the valley.[143] Much of this rich deposit derived from the small tributary draining the tin stockwork area around Mulberry in Lanivet parish. The well wooded valley below the Roman camp at Nanstallon (Tregear of 19th century writers) still exhibits a flood plain of rough ground and pools, clear signs of former tin working. The valley did not escape the attentions of bronze age tinners, who must be commended for having utilized all the county's deepest and most important deposits of alluvial tin.

There exists in the museum at Bodmin a socketed 'celt' (Fig. 116):[144]

> About ten days since [12 Oct. 1847] there was found in an ancient tin stream work at Boscarne . . . a wedge-shaped, or rather chisel-shaped, hollow piece of copper about 5 inches long, and from an inch to 1½ inch broad. The

copper is soft, easily cut with a knife. It is supposed to have been a mining tool, used by the 'old men', before the use of iron. Near the spot where this relic was found there was discovered some few months since an ancient wooden shovel.

Information in Bodmin Museum gives the find spot as 'opposite Cotton Wood', 1·6 km downstream from Nanstallon camp. The implement is described nowadays as a socketed chisel, of middle bronze age date (*c.* 1100–1000 BC). The only others known are from Torquay in Devon, Rymer Point in Suffolk, Soham in Cambridgeshire, and Kergoff Noyal in Brittany.[145] The Boscarne example is unique in possessing a small hole deliberately formed in the casting seams at the time of manufacture, presumably for a pin to fasten the wooden handle.

Also in Bodmin Museum are a socketed, side-looped spearhead and a tanged knife (Fig. 117), both from Boscarne and both presented to the Museum by Richard Flamank in 1843. A label in the spearhead's socket states that it was found in 1840. The find circumstances are not recorded, but they probably came from the streamworks of which Flamank was the proprietor. Both objects are contemporary with the socketed chisel. The spearhead is characteristic of the middle bronze age, earlier examples having more sharply angled blades. The tanged knife or dagger is rare, but others are known from Reach Fen, Cambridgeshire, and from the River Thames.

The Roman camp at Nanstallon was constructed on that site partly because of its central position in the county commanding easy routes inland up both the Camel and Fowey rivers. More pertinent still is its position overlooking the most important tin stream in east Cornwall, and the closest overland to the fully Romanized province that extended barely west of Exeter. In spite of the large number of Roman finds in Cornwall, they constitute little more than a veneer over a county left very much to look after its own affairs.

Nanstallon itself was a fort the right size for a *cohors quingenaria equitata* (six centuries of infantry and four *turmae* or cavalry troops).[146] Built soon after the Claudian invasion and Vespasian's thrust into the south-west with his II Augusta legion, the camp was systematically dismantled 'during the reign of Vespasian [68–79] or, at the latest, very soon afterwards', leaving the presumably peaceful natives to attend to their own mining interests.

From room 1 in barrack No. 4 is a crucible containing a slag rich in silver with minor impurities of copper, iron, calcium, and magnesium. With it was a bronze weight inscribed 'S' (*semis* or 14 g) for a weighing-scales, suggesting that the soldiers carried out small-scale repairs to their uniforms or other equipment. The silver might have been from Cornish argentiferous galena, but this cannot be proved, and the Romans may well have imported it from the Mendips where lead mining was under their control in the late '40s. What is beyond doubt is that the Romans supervised, or merely encouraged, the exploitation of tin in the valley below the camp during the 30 or so years of its existence.

Nanstallon was recognized as Roman *c.* 1820 at a time when the streamworks at Boscarne was operational. Roman finds were made at both places, and it is now not easy to be certain which objects came from the camp and which from the tin stream. There is also confusion over the extent of the tin-streaming operations. According to the anonymous 'Stannator'

in 1828,[147] whose name should token a familiarity with tin working, finds were made in the valley flowing from the tin-bearing ground around Mulberry through Ruthern Bridge to join the Camel at Cotton Wood west of Boscarne.

> In the stream work at Ruthern Bridge, the workmen have found spades made of the wood of oak or holly; they discovered likewise near the same spot, what is commonly called a Jew's house, where there are furnaces ... resembling a straw bee-hive, and will contain about three gallons of water; of these furnaces there are six in a line, closely set together, and the whole in the vicinity of wood, where they produced charcoal to reduce the tin into its metallic state...

After further discussion about the Jews and the Phoenicians, he continues:

> I have now only to add that several coins of Vespasian have been found in the streamwork.

Streaming certainly took place at Ruthern Bridge in the time of Henry VIII,[148] but it is most likely that Stannator's account really refers to Boscarne, or at least the section at Cotton Wood which is less than 1·6 km from Ruthern Bridge. That finds came from Boscarne at this time is supported by the following account:[149]

> A few days since [*c.* 10 Apr. 1822] was found near Boscarne ... a gold fish hook, size No. 3, in the bed of an old river where some men were working for tin, and not far from the same spot were taken up several Roman coins, of the reigns of Vespatian [sic], and some of the later Emperors etc. The whole are in the possession of Richard Flamank Esq. of Bodmin, the land owner.

The fish hook is a mystery. Size No. 3 is meaningless, since in 1822 there were no standard sizes, which depended on the whim of the manufacturer (Wheeler, pers. comm.). However, there exists in Bodmin Museum the pin part of a Roman bronze fibula, which looks very like a fish hook and may be the object in question. Two Roman coins believed to be from Boscarne or from the camp at Nanstallon are most likely to be from the streamworks, one of them being a coin of one 'of the later Emperors' long after the camp had been abandoned. The coins are –

> As of Vespasian (69–79 AD); bronze, 24 mm diam.
> Ob. [IMP CAE] SAR VESPASIAN AVG COS [III ?]
> Rev. S C with legend uncertain. It shews, perhaps, Concordia holding corn-ears and cornucopiae with an altar in front.

> Billon or bronze tetradrachm of Claudius II (268–270 AD).
> Obv. AVΓ Κ ΚΛΑVΔΙΟC CEB (i.e. IMP CLAVDIVS AVG).
> Rev. Eagle standing right, head left, wreath in beak.
> Date LΓ = 269–270.

There also exists in Bodmin Museum a tin ingot weighing 3·2 kg when found, which is described as

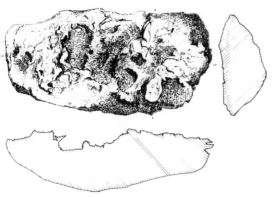

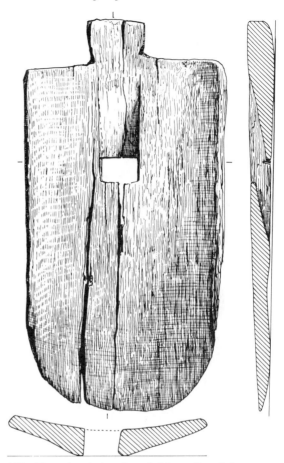

118   Tin ingot from Boscarne, much corroded on upper surface; weight 2·7 kg (heavier when found); length 20·3 cm (courtesy Bodmin Museum)

coming from Boscarne (Fig. 118). The find circumstances are not recorded, but it may well be from the 'Jew's house' recorded by Stannator in 1828. The ingot was not mentioned by Iago,[150] who made a valiant attempt at sorting out which finds in Bodmin Museum came from Nanstallon Roman camp and which came from the tin stream at Boscarne. Iago mentioned a coin of Trajan (98–117) found 'across the ford' on the Boscarne side of the river before the construction of the bridge there in 1830. Maclean also referred to a 'brass ounce of Trajan' in excellent condition (doubtless a sestertius), as well as coins of Vespasian.[151]

Iago also mentioned, but did not illustrate, a small bronze pendant and a bronze 'tack-stud' from 'Mr Flamank's stream work at Boscarne Moor'.

About 15 sherds of pottery are listed by Iago as from either Nanstallon or Boscarne, but not all of the latter are Roman. The base of a pot and a small fragment of internally glazed ware are 16th or 17th century and roughly contemporary with a fine large pitcher of micaceous earthenware, now in Truro museum, and presented by Iago in 1881–2; when found it was said to have been full of tin. The only important pottery from the streamworks is two sherds of samian, the larger one now lost and the smaller still in Bodmin museum (Fig. 119), found at Boscarne *c.* 1840. The larger sherd

119   Sherds of Samian pottery from Boscarne tin stream (drawing from Iago, 1890, plate V)

120   Tinner's oak shovel, probably mediaeval, from Boscarne tin stream; length 31·7 cm (courtesy Bodmin Museum)

appears to show a hare behind the forked tail of a swallow; in fact the 'tail' is the elongated jaw of a dog, part of a well known series of hunting scenes on south Gaulish pottery. The smaller fragment was identified[152] as part of a Dragendorff type 29 bowl of early Flavian ware from south Gaul, probably from La Graufesenque, and contemporary with the occupation of Nanstallon camp.

One piece of black glass, supposedly Roman (lost by 1890), formerly in Bodmin Museum, had been found with Roman pottery at Boscarne when the deposit was first worked by Lander in 1849.

Stone querns surviving at Bodmin Museum were 'discovered in excavating at Boscarne', but these are probably late, along with ancient clay pipes and shoes, which Iago noted were found at a depth of 5·5 m at Boscarne.

More than one wooden shovel survives from Boscarne. Apart from the mediaeval type in Bodmin Museum (Fig. 120), two of more oval shape were

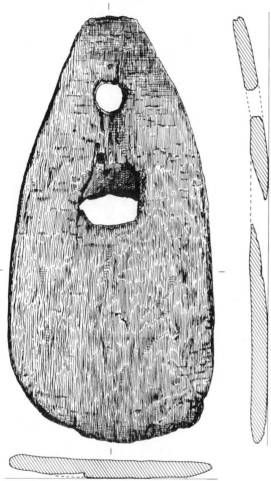

121   Tinner's oak shovel from Boscarne tin stream; ^{14}C dated to between AD 635 and 1045 (1140 ± 100 BP); length 34·3 cm (courtesy County Museum, Truro)

presented to Truro Museum in 1870 (Fig. 121): one has now been ^{14}C dated to between AD 635 and 1045. The importance of this is stressed in Chapter 27.

PENWITH

The account by Diodorus Siculus of 'Ictis' and 'Belerion' leaves the modern reader with the impression that the only ancient Cornish port was St Michael's Mount and the only tin workings were in its vicinity. The large number of finds from the tin streams of central Cornwall dispel that illusion. Were it left to the archaeological record, the far west of Cornwall would appear comparatively unimportant, for it is in the perverse nature of things that proof of tin streaming in antiquity in the Hundred of Penwith is scarce. This is particularly true of the Land's End

peninsula where the only recorded datable find is a coin of Vespasian (AD 69–79) 'in a stream work at Buryan'.[153]

The one important tin working here was on the eastern border of the parish at Bojewans where 0·6–2·7 m of tin ground lay beneath 2·4–6 m of overburden. There were also workings at Tregadgwith in the valley which joined the Lamorna stream below Bojewans.[154] Borlase drew a finger ring, bearing an intaglio head, found in a tin stream near Penzance in 1823. It is clearly the same object later described as coming from St Buryan,[155] probably from the same workings, but it is mediaeval and dated to the 15th century by Whitley.[156] Streaming was attempted in other south-facing valleys west towards Land's End, but 'the proceeds have not sufficed to pay the workmen'.[157] This is hardly surprising since the valleys lie south of the St Just metalliferous lodes. The Bojewans tin derived from lodes in Sancreed.

It is possible that the early tinners in West Penwith concentrated their efforts on the lodes outcropping in the cliffs of the north coast between Porth Nanven, south of Cape Cornwall, and Porthmeor Cove in Zennor. The only other parts of the county where this geological arrangement occurred is between St Agnes and Perranporth, and in St Austell Bay. However, in the latter area the lodes tend to run parallel with the coasts so that outcropping lodes in the cliffs are fewer. Only in St Just are they numerous. Reid and Flett found 'shallow trenches and mines worked by "the old men",'[158] though without accompanying evidence of age. The possibility of early lode mining has already been discussed, but it is worth adding here that Dines wrote of St Just that an 'adit driven on a lode from sea level will unwater some 300 to 400 ft of backs'.[159] This would permit a considerable tonnage of tin and copper to be worked without any problem of drainage.

The valleys on the north side of West Penwith are all short. Most have probably been tried for stream tin. There remain some unworked deposits which are attracting attention now, but work had ceased in most of the valley alluvials well before Henwood published his paper on 'Detrital Tin Ores' in 1874. One family was still making a living after 70 years working at Bosworlas, 1·6 km south-east of St Just churchtown. The deposits here were easily worked. Henwood recorded no more than 1 m of top soil and a few centimetres of gravel resting on 0·76 m of tin ground containing more or less rounded fragments of tin-rich veinstone and pure cassiterite.

Carne[160] spoke of wood-tin and resinous tin found in workings at Pillianeath [Pillinath] and Pemeder, *c.* 2·5 km south of St Just Church in the Nanquidno valley. He noted that an ingot of corroded 'Jew's house tin or Jew's bowl' had been found here in the workings which 'have not been wrought for perhaps upwards of fifty years'. The ingot could be prehistoric, but is just as

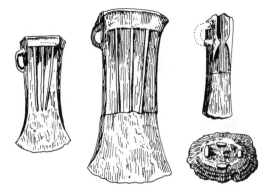

122 Part of late bronze age hoard from Kenidjack, St Just, (redrawn from Borlase, 1879). The two socketed axes are now in Cambridge; the winged palstave (3) and the casting jet (4) are now in Truro Museum
Note on the casting jet (4), the four white oval spots show the positions of the channels along which the molten metal was poured; four are characteristic of axes made in south-west England and south Wales

likely to be mediaeval. The same ingot was mentioned by Le Grice in 1846, who added that another piece of Jews' House tin had been found 'in the centre of a barrow' at Bossuliack Croft [Bosilliack] near Lanyon in Madron,[161] which may be proof of bronze age smelting.

Better evidence comes from Kenidjack Castle, 1·6 km north-west of St Just where 137 m outside the wall of the promontory fort were the foundations of a small hut. In it were found two socketed axes of the so-called South Wales type, a 'winged' palstave, 20 or 30 pieces of copper, a casting-jet, and a piece of smelted tin, unfortunately long-since lost (Fig. 122). The axes contained 17% tin and the palstave only 0·5% less. Of the pieces of pure copper, Borlase noted that a few showed 'the form of the stone bowl into which the metal had been made to run'.[162] Collins, who analysed the finds, found traces of iron, tin, and silica in the copper, typical impurities in copper won from Cornish lodes.

More 'fused tin' was found by Borlase in 1863 when he excavated at the Romano-British settlement at Carn Euny in Sancreed, with samian pottery, a spindle-whorl, various bones and iron objects in the 'fogou' or subterranean passageway.[163] One of the earliest recorded Roman coin hoards came from the heart of the St Just mining region. In or before 1737 about 100 copper coins, some clearly identifiable as Antoninus Pius (AD 138–61) were dug up in a field close to the 'fogou' at Boscaswell.[164] More important, however, is the evidence for the manufacture of tin or pewter dishes in St Just in the later Roman period. Whether the tin was won from lodes or alluvial deposits, it was probably of local origin rather than an import from elsewhere in the county. The Revd William Borlase

illustrated two stone 'bowls' found at Leswyn (now Leswidden) at the head of the Nanven valley less than two miles south-east of St Just church (Fig. 123).[165]

Stone bowls are a common feature of Romano-British sites in Cornwall, but what Borlase did not realize is that the objects are moulds. He presented them to the Ashmolean Museum in Oxford where their true nature was recently revealed by Brown.[166] When the supposed bowls were placed together they fitted neatly, leaving a space between them for molten metal. The smaller mould piece has a channel cut across the rim, above and below, through which the metal was poured. Another stone originally fitted inside the smaller of the two allowing two dishes to be cast at the same time, while the underside of the larger stone also served as the base of yet a third mould, probably for a shallow dish.

It is not uncommon for moulds to contain several items in the same block. In Cornwall there are two socketed axes on opposite sides of a 'greenstone' block from Gwithian (now in Truro Museum) of late bronze age date, while Brown cites several Roman examples from up-country. Borlase described the stone of the moulds as 'a particularly talky Moor-stone, or granite, called commonly Ludgvan Stone, from the parish which it is most plentifully found in'. De la Beche commented that Ludgvan granite was very micaceous and an esteemed, durable material.[167] Borlase's 'talky', talc-like, refers to the very fine white mica, and it is clearly the 'dove-coloured' granite described in his *Natural History of Cornwall*.[168] The quartz was fine-grained 3 mm or less, 'with a vast amount of silvery-talc [muscovite mica]' and regular spots of 'black cockle [tourmaline]'. In Borlase's own day the stone, commonly called 'Silver Stone', was 'much coveted for walling in ashlar-work, being tough, keeping a good edge, and working easy'. These qualities made it an admirable choice for a Roman mould, comparable to the moulds of Bath stone (oolite) found in excavations at Lansdown, Somerset, and displayed in the Roman Baths museum at Bath.

The rough castings from the mould would have been trimmed and polished to produce dishes similar to one from the hoard of pewter found at Appleford in Berkshire. This would have been done on a lathe.[14] Dishes of the St Just type were fairly widespread in southern Britain, corresponding to type 4a listed by Peal.[169] They date to the late 3rd or 4th centuries when pewter was manufactured in at least three places along the Fosse Way: Camerton, Lansdown, and Nettleton Shrub.[170] Whether the St Just dishes were of pewter or pure tin is not known – none survives.

Contemporary pewter comes from St Erth, but before leaving West Penwith the furnace excavated by Leeds at Chûn Castle should be noted.[171] Leeds dated the Castle to the 2nd or 3rd century BC, but reexamination of the finds by Thomas suggests that the

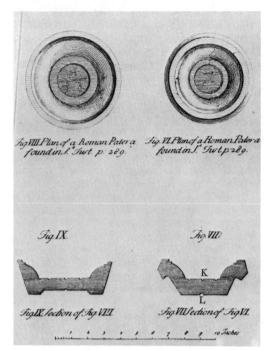

123*a* **Late Roman pewter moulds found at Leswidden,
St Just, in or before 1753 (Plate 25, Borlase, 1769)**

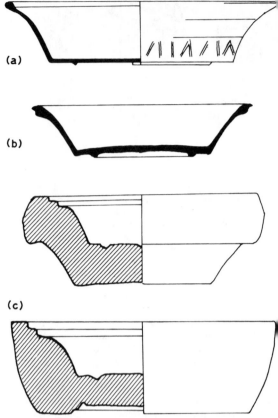

123*b* **Stone moulds from Leswidden; a pewter dish
from Appleford, Berkshire, showing a close
similarity to the reconstructed section (*b*) of a
bowl from the Leswidden moulds *c* (redrawn
from Brown, *CA*, *9*, 1970)**

furnace belonged to a post-Roman occupation of the site.[172] The furnace, in hut C, was associated with pottery bearing on its base the marks of chopped grass, a type common throughout post-Roman Cornwall until the 11th century, as well as sherds of amphorae imported during the years 500 to 700. The furnace was constructed of stone blocks and measured $3 \times 1 \cdot 5$ m. Unconvinced by Leeds' interpretation of its construction, the evidence was examined by Tylecote,[173] who suggested that it contained three small furnaces to which air was admitted by flues. However, recently he has doubted his own interpretation and is not entirely convinced of its metallurgical purpose (pers. comm. 1983).

An oval block of smelted tin weighing 4·9 kg and measuring $22 \times 16 \times 7$ cm (now in Truro Museum) was found on the inner face of the fort near the furnace. Like many ingots of tin it suffers badly from conversion of the metal to cassiterite and has lost some of its mass in recent years.

About the summer of 1910, Noall of St Ives watched the gradual removal of stone hedges and several ancient huts at Bussow, probably on the site of the reservoir then being constructed. Numerous sherds indicated a bronze age settlement whose inhabitants had been familiar with tin streaming. Not only were samples of stream tin found in a barrow, but there was also recovered a spindle whorl of smelted tin.[174] This

badly corroded object is now in Truro Museum. The only other Cornish example is from Trevelgue promontory fort near Newquay, and is also in the Museum. Tin spindle-whorls are also known from Glastonbury Lake Village and from Corbridge south of Hadrian's Wall.[175]

At the courtyard house settlement at Porthmeor, Zennor, smelted tin weighing 1 kg was found in house 2 in 1935. The original metal had been largely replaced by tin oxide. Another piece of tin was found in house 1, together with a piece of 'doubtful slag'.[176] These items do not appear to have been given to Truro Museum where other finds from the site are kept.

In 1756 the farmer at Bosence, St Erth, was driving his oxen from the field which contained the Roman camp when he noticed the foot of one of them sink deeper than normal. Goaded by the possibility of hidden treasure, the farmer dug deeper and soon discovered[177]

124   *a* 'Tin' jug and *b* 'tin' dish from Bosence; *c* broken 'tin' jug from Caerhays buried in or soon after AD 282 in Old Park Wood (Haverfield, 1924)

a perpendicular pit, circular, of two feet and half diameter. Digging to a depth of 18 feet, there he found a Roman *patera* (Fig. I and II ibid.); about 6 feet deeper the jug Fig. III., near by among the rubbish the stone, Fig. IV.; then a small mill-stone about 18 ins. diameter. Digging further still they found another *patera* with two handles, in other particulars much of the shape and size of Fig. II. Intermixed were found many pieces of leather, shreds of worn-out shoes. Having sunk to the depth of 36 feet, they found the bottom of the pit concave, like that of a dish or bowl: there was a sensible moisture and wet clay in all parts of the pit, in each side there were holes at due distances capable of admitting a human foot, so that persons might descend and ascend.

Most writers have followed Borlase in assuming the structure was a well. Not so Piggott, who interpreted the shaft as a ritual one containing votive offerings, similar to German examples over 30 m deep.[178]

The jug, or flagon, is very like that from Caerhays already mentioned (*see* Fig. 124). It is clearly late Roman, held over 5 l and weighed 3·4 kg. Together with one of the *patera* it was given by Borlase to the Ashmolean Museum in Oxford, where recent analysis found the flagon to be 85% tin, with 10% copper, only 3% lead and 2% iron. This percentage of tin is not unusual for Roman pewter, but the amount of copper appears to be unique and strongly suggests the use of a locally won, copper-rich tin ore, quite possibly from an open-cast working on a lode. Bosence is close to Godolphin where evidence has already been seen of probable lode-working in the bronze age.

The *patera*, or dish, was also analysed, showing a more normal composition: $85 \pm 5\%$Sn; $15 \pm 5\%$Pb; less than $0·5\%$Cu. The dish could, therefore, be the product of an up-country pewter manufacturer, perhaps in Somerset. The dish is flat-bottomed, *c.* 15 cm diameter at the top. Scratched on the inside of the base is a large letter R with the surrounding legend: *Aelius Modestus deo Marti* (Aelius Modestus [dedicates it] to the God Mars). The meaning of the letter R is not known.

The inscription lends support to Piggott's theory, for here most certainly is a shaft 'communicating with other-world deities . . . in which objects rendered holy by sacral use, and the bones or ashes of sacrifices, were buried in the consecrated area'.

There is proof of tin working back to the bronze age near 'Ictis', 6 km south-west of Bosence. Pocock, who visited St Michael's Mount on 4 October 1750, described a vein of tin on the Mount (several still exist there) 'worked some time ago, and a very rich ore was dug out of it. Mr Borlace [sic] showed me a celt or copper instrument of war, which, if I mistake not, he said was found here'.[179] Borlase makes no mention of such a find himself; Pocock was probably mistaken, confusing some axe shown him by Borlase with an account of a discovery reported by Leland and which Borlase would have known. Leland, who visited Cornwall about 1540, tells us that:[180]

> There was found of late Yeres syns Spere Heddes, Axis for Ware, and Swords of Coper, wrappid up in lynid scant perishid, nere the Mount in S. Hilaries paroch in Tynne Works.

Camden tells the same tale in *Britannia*,[181] placing the tin works 'at the foote of this mountain'. He added that similar finds were sometimes met with 'within the forest Hercinia in Germanie, and not long since in our Wales'. The description is vague, yet it can hardly refer to anything other than a hoard of the later bronze age. If 'rapiers' are substituted for 'swords', then all the objects mentioned by Leland have been described from other streamworks. The observation that the weapons were 'wrappid up in lynid' is important, for it confirms waterlogged conditions. The location and wetness fit one place – Marazion Marsh.

The Ponsandean stream, which drains Marazion Marsh, flows in a valley that has been worked for tin at least as far inland as Tregilliowe (or Tregilso, as Henwood called it),[182] where a mere 2 m of peat overlay 3 m of tin ground. At Marazion Marsh itself (*see* Fig. 125) the tin ground lay at a much greater depth. Henwood's cousin George published a section of the workings in 1855.[183] Working began here in 1828 after the Wheal Darlington copper mine recommenced work, the pumps draining off all the surface water of the marsh.

	Metres
Peat and grass roots	1·6
Alluvial soil	0·45
Pebbles and freshwater sand	?
Sea sand with shells and calcified seaweed	5·5
'Forest bed' – oak, holly, hazel, and large quantities of nuts; all trees lay with roots towards the sea	?
Alluvial soil with tree roots	0·76
Freshwater sand with a few shells	0·76
'Large trees', probably oaks; all trees lay with roots towards land	0·15
'Tin ground' mixed with usual pebbles and much extraneous matter requiring great care in separation, for the tin grains are very small. Sand occurred in regular laminae as if deposited by a series of tides and each containing sea shells	0·76

A considerable quantity of tin was recovered at this time, though the speculation was not a great success owing to squabbles between the adventurers and the spending of too little capital. The workings were aptly called Bog Mine, and would have been difficult to work in the bronze age, or indeed at subsequent periods, especially in the wetter iron age.

No prehistoric objects appear to have been discovered during the workings discussed by Henwood, but in 1849 Edmonds described a supposed Phoenician smelting house.[184] The river, recently diverted, flowed westwards 'to a considerable distance' and undermined the dunes that separated the Marsh from the sea. Its exact site is not now known. Here Edmonds found at a depth of 3·6–6 m 'the remains of ancient walls rudely built of unhewn stones mixed with clay, and near them great quantities of ashes, charcoal, and slag ... grains of tin being frequently imbedded in the slag. Some very ancient broken pottery, of rude manufacture, was also found, and much brick'.

None of these items survive, and the descriptions are too vague to give precise dates, though something comparatively modern is likely, were it not for the discovery 'within a few inches of one of the walls ... two fragments of a bronze vessel resting on a layer of charcoal'. The fragments, *c.* 15 × 10 cm with *c.* 12 mm gauge, came from a vessel 1 m in diameter, 'the mouth being bent back into a horizontal rim three-quarters of an inch broad'. If 'bent back' means bent outwards, the vessel sounds very like the cauldron from Broadwater, Luxulyan, made out of a single piece of bronze. A fragment was presented to Hunt at the Museum of Economic Geology in London, and it may yet lie unrecognized in the Victoria and Albert Museum where metalwork from the Geological Museum was sent in 1901. Hunt analysed 25 grains by weight (far less revealing than a simple sketch would have been):

	Grains
Copper	18.0
Tin	2·25
Iron	1·0
Loss	3·0
Earthy matter	0·75

Fortunately, other items have turned up at Marazion, including a large hoard of Roman coins in 1793, when Moyle drained *c.* 28 ha of the marsh. The labourers, in cutting open the drains, discovered an earthen pot containing nearly 1 000 coins, many badly corroded, but some sufficiently legible to recognize emperors who lived between AD 260 and 350,[185] that is from Gallienus (253–68) to Constantine II (337–61), all well represented in other Cornish hoards. The hoard does not prove tin streaming at Marazion, any more than the discovery in 1825 of a hoard of 'some

plan of the Bog streamwork

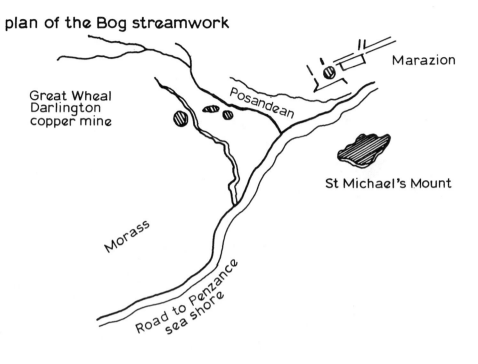

Great Wheal
Darlington
copper mine

Posandean

Marazion

St Michael's Mount

Morass

Road to Penzance
sea shore

section of the Bog tin stream

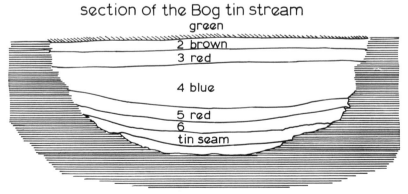

green

2 brown

3 red

4 blue

5 red

6

tin seam

125   Section and plan of bog tin streamworks at Marazion Marsh; the 'tin ground', marked as 'tin seam' on the section, was only 0·76 m thick (redrawn from frontispiece of George Henwood, 1855)

thousands' of minimi of Tetricus and Victorinus (*c.* AD 260) at Hayle causeway, proves tin streaming in the Hayle River. Taken in conjunction with the site at Bosence, and the milestones of Constantine I in St Hilary Church and of Postumus at Breage, the hoards serve to emphasize the importance the Romans placed in the Cornish tin fields from the late 3rd century.

Possible relics of tin working were found in 1882 'in draining at Marazion' Marsh. Borlase had in his collection two palstaves – one broken, the other nearly

126   Alluvial tin workings in the Hayle valley between St Erth church and Trewinnard; during these operations a stone axe was discovered in the tin ground at a depth of 3·6–4·2 m (photo J. H. Trounson, c. 1930)

perfect – found with others 'in an iron vessel'.[186] An iron vessel seems unlikely; perhaps it was bronze coated in bog iron ore, and it had probably gone before Borlase had a chance to see it. There is no record of the depth at which the items were found, but it was most probably shallow, putting them on a similar level to Edmonds' finds.

The Hayle River and its short but important tributary flowing north to join it at Relubbus drain mineralized areas on both sides of Godolphin Hill. At least 14 lodes are close enough to have supplied alluvial tin. Henwood recorded how the valley had been exploited near St Erth bridge, using a steam engine for drainage like the one employed at Marazion Marsh (Fig. 126). He considered the alluvials were neither rich nor extensive,[187] but they have been worked in the 20th century up to 1913, again in 1927–30, and more extensively between March 1942 and April 1945. According to Gregory:[188]

> A typical section would be surface to 1 foot, sandy soil; 1 foot to 3 feet, sandy fine gravel; 3 feet to bedrock at varying depths, coarse gravel and sand and occasional light boulders.

The depth of ground varied from 3 feet down to a maximum of 35 feet, the latter being near St Erth Church. The average being about 24 feet.

The attractive ponds near St Erth Church are the result of alluvial workings.

> Tin values were fairly evenly distributed throughout the gravel below the first 5 feet, though in several places higher values were observed in the bottom section of the drill holes [in prospecting]. It was interesting to observe that the tin was sub-angular in the immediate vicinity of the known lode series, and well water worn some distance away. It was this indication which probably led old prospectors to the locality of lodes and was known as shoding, a method still used by prospectors.

Drilling at Hayle causeway, the present limit of the estuary, gave no worthwhile tin content. Here it lay beneath 12 m of marine sands, a formation detectable upstream as far as Carbis Mill, half way to Relubbus.

> In the Relubbus section of the work, 84 000 cubic yards of material were removed yielding 30·25 tons of black tin 72 per cent Sn = 0·80 lb per cubic yard. In the St Erth section 171 300 cubic yards gave 55·45 tons of black tin 72 per cent Sn = 0·73 lb per cubic yard.

The width of the tin-bearing gravels varied: at St Erch church they were *c.* 90 m, at Carbis 45 m, only 27 higher up, widening again at Relubbus.

In or before early 1930, a stone axe was found 'in the tin gravel at a depth of 12 or 14 feet'. Capt. Taylor of East Pool, who was connected with the St Erch workings, thought it was flint. Others were less sure of the lithology and thought it more like basalt; certainly a 'greenstone' axe would be more appropriate to the area. Sadly it was taken to London by Hirsch of King William House, East Cheap, and never heard of again.[189] The only comparable artifact from a tin working is the polished stone axe found at La Villeder, Morbihan. Whatever the lithology of the St Erch axe, it is eloquent testimony of an early tin-streaming enterprise, comparable in age to the jet slider from the Pentewan valley, and a fitting conclusion to the catalogue of finds from Cornish tin streams.

Artifacts discovered in prehistoric and early historic Cornish tin workings

Found in open-cast working on tin lodes
'Many' middle bronze age palstaves at Godolphin

Found in tin streamworks

Locality	Early Bronze Age *c.* 2100–1500 BC	Middle Bronze Age *c.* 1500–800 BC	Late Bronze Age *c.* 800–500 BC	Iron Age *c.* 500 BC–AD 43	Romano–British AD 43–420	'Dark Ages' *c.* AD 420–1066
St Blazey – Par		bronze pin with amber setting* rapier and palstave*				
Pentewan valley & St Austell Moor	jet slider*	socketed chisel* socketed spearhead* and (?) rapier*		tankard*	[handled ingots]†	Saxon silver hoard*
Paramoor and Porthluney valley		rapier			fibula	penannular brooch and finger ring
River Fal (middle reaches)				Trenowth collar	Hallivick pewter bowl	
Carnon Valley	flat axe* 2 antler picks				coins* tin bowl	
St Columb		bulb-headed pin (Treloy?) [Blue Anchor‡ palstave]				
Red Moor, Lanlivery				La Tène fibula		
Goss, Tregoss and adjoining Moors		Loath-to-Depart palstaves (two)*	Broadwater cauldrons (two)*			penannular brooch near Lanivet*
Treloy, St Columb-Minor		'bronze age axes'*			tin bowl with cover* disc brooch* Roman coins, rings and brooches	

Locality	Early Bronze Age c. 2100–1500 BC	Middle Bronze Age c. 1500–800 BC	Late Bronze Age c. 800–500 BC	Iron Age c. 500 BC–AD 43	Romano–British AD 43–420	'Dark Ages' c. AD 420–1066
Lanherne, St Mawgan-in-Pydar		hoard with rapier, socketed palstave, axes, saw, and probable Breton axes*				
Bodmin Moor		Bolventor spearhead*			Parson's Park tin bowl	
Boscarne, Bodmin		socketed chisel, spearhead, 'knife'			coins samian ware and fibula	oak shovel
Land's End peninsula					Roman coin, St Buryan	
St Erth	stone axehead*					
Marazion Marsh		hoard of bronze age spears, axes, swords (rapiers?)				

* signifies the object was found on the tin ground, or at such a depth as to prove the same. Note that the lack of this symbol does not imply discovery at any shallower level. No object has been recorded as found on the surface or above the tin ground. The implication is that all artifacts were found on the tin ground

† Date uncertain; Continental Roman parallels known

‡ Probably from old tin working (not proven)

Finds from archaeological sites and other indirect proof of tin mining

Locality	Early Bronze Age c. 2100–1500 BC	Middle Bronze Age c. 1500–800 BC	Late Bronze Age c. 800–500 BC	Iron Age c. 500 BC–AD 43	Romano–British AD 43–420	'Dark Ages' c. AD 420–1066
Caerloggas Down, St Austell	cassiterite and tin slag					
Caerhays					coin hoard in 'tin' jug	
Red Moor, Lanlivery		Bodwen rapier mould				
St Columb Minor				Trevelgue tin spindle-whorl and metal working		
St Mawgan-in-Pydar					tin ingot from Nanskeval, metal working at Carloggas	
Trevisker, St Eval		cassiterite pebbles				
Bodmin area				torc with tin core 'near Bodmin'		
Land's End peninsula		palstaves and smelted tin at Kenidjack		smelted tin at Porthmeor tin-spindle whorl at Bussow	stone moulds for 'pewter' from Lewidden stream tin and smelted tin from Carn Euny	tin smelting at Chûn
St Erth					pewter jug and patera from Bosence	
Marazion Marsh		2 palstaves in iron (?) vessel smelting site (LBA?)			Roman coins	
Prah Sands					tin ingots	

NOTES AND REFERENCES

1   BORLASE, 1872, 5
2   LYSONS & LYSONS, 1814, **3**, ccxxv
3   ROWLANDS, 1976, **1**, 71, 74
4   Carlyon Deeds DD/CN/2437 at County Record Office, Truro
5   St Aubyn collection HA/7/22 and HA/7/30 at County Museum, Truro
6   BRITTON & BRAYLEY, 1801, 415
7   HERITY, 1969
8   SYMONS, 1877
9   PRYCE, 1778, 68
10 MILES, 1975, esp. 41 and Plate XII
11 BORLASE, *Ancient Cornwall*, 1871, **1**, 17
12 COLENSO, 1829
13 WINN, 1839
14 McINNES, 1968
15 BORLASE, 1871, **1**, 42
16 STOCKER, 1852
17 CORCORAN, 1952
18 WILSON, 1971, 58
19 RASHLEIGH, 1789; RASHLEIGH, 1794
20 ROGERS, 1867
21 HENWOOD, 1869, *JRIC* 'Chairman's Address', esp. xiii–xiv. The gold had a specific gravity of 16·52 and contained 90·12%Au, 9·05%Ag, 0·83% Si plus FeO
22 WILSON & BLUNT, 1961, who give a complete bibliography of published material on the hoard
23 HENCKEN, 1932, 262
24 Seventeen silver coins, 14 of Edward the Elder (879–924) and three of Athelstan (924–940) were found at Penhale, a mile north-west of Fowey, when a gas-main was being laid. Three coins are in the British Museum and the remainder are in the County Museum, Truro. The hoard is believed to belong to the period when Athelstan's invasion of Cornwall *c.* 929–30 resulted in the setting up of the Bishopric of St Germans in 931. When found the coins were believed by the workmen to be buttons or 'sheep dip tokens'. The Saxon mint at Launceston was much later, the earliest surviving coin being a unique silver penny of Ethelred II (978–1016)
25 WINN, 1839
26 THURSTON, 1930, 14–16
27 FLOWER, 1872
28 WALKER, 1962, 30
29 CLARK, 1947
30 BORLASE, 1758, 214
31 KLAPROTH, 1787, 12
32 POLWHELE, 1803, **4**, 133 footnote
33 ROWLANDS, 1976, **2**, 402
34 ANON., 1864a, 22
35 *The Gentleman's Magazine*, 6 Dec. 1787
36 RENNIE, 1957, 214 Fig. 4 No. 3

37 WHEELER & WHEELER, 1932, brooch No. 39
38 DICKENSON, 1982, *cf* brooch Fig. 5 No. 31
39 HAVERFIELD, 1900a
40 REID & TEALL, 1907, 60
41 BORLASE, 1758, 163
42 HAVERFIELD, 1924, 23, 35. The bowl passed into the possession of Philip Rashleigh of Menabilly who exhibited it at the Society of Antiquaries in 1807 (*JRIC* 1869, xi). It was figured in *Archaeologia*, 1812, **16**, Plate 9, and is briefly mentioned in *Proc. Soc. Antiq.*, 1870, **4**, 492–3
43 TOYNBEE, 1962, 176 and Plates 137–8
44 THOMAS, 1981, 88
45 GRAY, 1929
46 SALWAY, 1981, 635–6. For a good summary of the Roman mines see Gough, 1967 revised ed., Ch. 2, 19–47
47 CUNLIFFE, 1978, 10–11
48 GOODBURN, 1972, 23, 35 and Plate 15.2
49 BORLASE, 1871
50 The Rt. Hon. R. POLE CAREW, letter to Samuel Lysons 20 April 1807 printed in *Archaeologia*, 1812, **16**. The collar then belonged to Philip Rashleigh of Menabilly who died in 1811. The published plate was engraved in 1809 from a pen and ink sketch signed 'Fred.ᵏ Nash, del. 1807' and now in the Library of the Society of Antiquaries in London. Lysons (*History of Cornwall*, 1814, ccxxi) dated the discovery 1802. It is briefly mentioned in *Proc. Soc. Antiq.*, 1870, **4**, 493
51 MEGAW, 1967. Note that the place is normally spelled Trenowth and appears as such on Ordnance Survey maps
52 SAUNDERS, 1960–1, who notes that 'Half a bivalve bracelet mould was found in the filling of the ditch. The bracelet would have been a simple one, penannular with knobbed terminals, a type which has Iron Age parallels'
53 BORLASE, 1871
54 *Cornish Mining Development Association*, 32nd Ann. Rep. 1980
55 HENWOOD, 1855, 8–9
56 TONKIN, MS *History of Cornwall* (compiled 1702–1736) at the County Museum, Truro
57 PRYCE, 1778, 136
56 BRITTON & BRAYLEY, 1801, 438–9
59 MATON, 1797, **1**, 133–5; RAISTRICK, 1967, 30–1
60 COLLIER, 1791, MS *Diary* at Plymouth City Library
61 JENKIN, 1967 (13), 43
62 BORLASE, 1871, **1**, 4
63 BURGESS, 1974, 191
64 REVD JOHN WHITAKER in Gilbert, 1838, **4**, 167
65 WHITLEY, 1873, 454–5 which briefly mentions the 5in long figure exhibited at the Society. It was believed to be 14th century

66 HENWOOD, 1873, 206–8. At the time the skeleton was discovered the arsenic works occupied an old tin-smelting house 'not far below the village of Perranwell' (Gilbert 1838, **3**, 304), but when Henwood published his account the arsenic works were at Bissoe

67 KEITH, 1915, 35–7

68 BUCK, 1861, *Natural History Review*, **1**, 155

69 GILBERT, 1817, 209, 211

70 *West Briton*, 3 Nov. 1815

71 RICHARD SAINTHILL, letter dated 23 Dec. 1851, published in *Proc. Numismatic Soc.* 26 Feb. 1852, 12–14

72 HENWOOD, 1828, 60

73 ANDREW, 1936

74 PAINTER, 1977

75 SHELL, 1978

76 DE LA BECHE, 1839, 525

77 SYMONS, 1884, 166

78 THURNAM, 1871, 363, 453, 468 Fig. 170; SCARTH, 1858, does not illustrate the finds which were found together on a flat stone in the bottom of the central cist

79 BORLASE, 1871, **2**, 2

80 JENKIN, 1964 (8), 58

81 PENDERILL-CHURCH, 1982, *in litt.*

82 HENWOOD, 1873, 215–16

83 CAMM & HOSKING 1984

84 EVANS, 1881, 400

85 HENCKEN, 1932, 109, 166, and 292 wrongly placed under St Austell parish

86 GRIMES, 1951, 127, Fig. 41 No. 2

87 SMYTH, 1889, 'Presidential Address' *TRGSC* **11**, esp. 136

88 SKEWES, *Royal Cornwall Gazette*, 14 Dec. 1877

89 HARRIS *et al.*, 1977

90 ROWLANDS, 1976

91 HENWOOD, 1855, Lecture 2, 11 footnote

92 BOASE, 1832, 246–7

93 BARROW, 1908, 393

94 FLETT in Ussher *et al.*, 1909, 173; GREGORY, 1947, 24, notes that the irregular contour of the bottom 'prevented satisfactory dredging' on Goss Moor so that 'much of the values were not recoverable' in workings before the second world war. Some of the large pools south of the A30 road, a few kilometres east of Indian Queens, are the result of 20th century gravel extraction, though a steam-powered tin dredge worked here about the 1920s

95 BORLASE, 1871, **1**, 6

96 KLAPROTH, 1787, 13

97 WHITLEY, 1917

98 HENWOOD, 1828

99 'CORNUBIENSIS', 1795, *Gentleman's Magazine*, July, 561–2 and Plate III

100 HAWKES & SMITH, 1957

101 LEEDS, 1930

102 CURWEN, 1948

103 HAWKES & SMITH, 1957

104 BURGESS, 1980, who states that British cauldrons 'comfortably pre-date' Continental examples, those from Llyn-Fawr being 'old when deposited'

105 ANON., *Mining Journal*, 7 Nov. 1835

106 BARHAM, 1860, 'Presidential Address' *JRIC*, 16

107 HENCKEN, 1932, 201, 306; Nicholas Whitley 1873, also describes it as from 'Lanivet, Gossmoor', 3 m below the surface

108 FOWLER, 1963

109 DICKENSON, 1982. I am grateful for her comments on the Lanivet brooch

110 STEPHENS MS note book (no date, *c.* 1940) at County Museum, Truro, quoting from a lecture by C. K. C. Andrew. It is hoped that a report of the Trevelgue excavations will be published in the not too distant future

111 B. RICHARDSON, The Museum of London, pers. comm. 1982

112 MILES & MILES, 1973

113 COLLINGWOOD & RICHMOND, 1969 (ed.), 299; Reginald Smith of the British Museum, in a letter to Truro Museum (7 Mar. 1921) suggested the brooch was 3rd century AD because of the 'depth of the catch plate'

114 HENWOOD, 1828, 65

115 BORLASE, 1758, Plate XX

116 DE LA BECHE, 1839, 525. Some large 'antedeluvian bones', the largest 0·45 m in circumference and under 0·35 m long (perhaps from a whale), were reported to Dean Buckland by a Mr Punnett in a letter dated St Columb 19 Jan. 1828. The streamworks is not named, but Treloy is the only likely place. This is recorded in Sir W. Warington Smyth's 'Presidential Address', *TRGSC* 1889, 133

117 HENWOOD, 1873, 221

118 Revd F. V. JAGO, letter dated 20 May 1813 to Samuel Lysons, *Archaeologia*, 1814, **17**, 337–8. In 1814, when Lysons published *Cornwall* as *Magna Britannia*, **3**, he dated the hoard 1812 (page ccxx), an error perpetuated in later publications

119 WAY, 1847, 16

120 EVANS, 1881, 185

121 EVANS, 1881

122 THREIPLAND, 1965, 76 for analysis of the tin

123 HOGG, 1825, 75

124 T. J. TREVENNA, 1982, pers. comm.

125 HENDERSON MS 1916–17, **3**, 69

126 HAVERFIELD, 1900

127 HAVERFIELD, 1924, 10 and Fig. 11

128 WARNER, 1967

129 HAVERFIELD, 1894; BOON, 1971, 473–4, also concluded that the inscription could not possibly be

DDNN and favoured IENN which indicated 'perhaps the initials of some registration official'

130   APSIMON & GREENFIELD, 1972

131   CHOPE, 1967 (ed.), 18

132   GOVER, 1948

133   GILBERT, 1838, 24

134   PATTISON, 1849, 57; RODD, 1850, 58, describes a 'greenstone' ladle found at Berriow, North Hill, about 1833 and now in Truro Museum. With it were two stone moulds, *c.* 0·35 × 0·3 × 0·15 m deep with the sides tapering to a smaller base. They were probably mediaeval. Similar moulds are said to have been found near Higher Cusgarne, Gwennap, with the remains of a furnace and much charcoal, according to Francis' descriptive poem *Gwennap* (1845)

135   MALAN, 1888, esp. 350

136   PATTISON, 1847

137   HENWOOD, *Mining Journal*, 31 Oct. 1857

138   HENWOOD, 1860; *see also Royal Cornwall Gazette*, 20 Apr. 1860, 7

139   *JRIC*, 1864, 'Accessions List'. The torc was presented by Revd J. W. Murray

140   SMITH, 1925, 116, Fig. 126

141   THURNAM, 1857, *Proc. Soc. Antiq. London*, **4**, 148. The mould was presented to the Museum of Practical Geology in London. It is now lost but may be buried in the Victoria and Albert Museum

142   LEONARD & WHELAN, 1928

143   WHITLEY, 1908

144   *Royal Cornwall Gazette*, 22 Oct. 1847

145   ROWLANDS, 1976, 45, 351

146   FOX & RAVENHILL, 1972

147   'STANNATOR', 1828, *The Cornish Magazine*, 96–8

148   GILBERT, 1838, **4**, 340

149   *See also Gentleman's Magazine*, 1822 (Apr.). The supposed fish-hook is unlikely to have been made of gold. Base metals were frequently mistaken for something more valuable. Apart from the usual error of believing tin to be silver, there is the record of tin streamers working at Shallow Water (Blisland-Bolventor border) in 1891 finding a horse-shoe on the tin ground beneath 2 m of peat. All who saw it confidently believed it to be silver, but analysis at Carvedras Smelting Works in Truro proved it to be iron (Revd W. Iago, on a printed sheet with photograph at Truro Museum)

150   IAGO, 1890, esp. 214–22

151   MACLEAN, 1868, **1**, 114–15. The Revd John Wallis, who owned some of the finds from Nanstallon and Boscarne, wrote in *The Cornish Register*, 1847, 19, that coins of Vespasian came from Nanstallon, but this does not mean none came from Boscarne tin streamworks

152   C. JOHNS, British Museum, 7 Sept. 1982 (*in litt.*)

153   WHITLEY, 1875. In Truro Museum is a rich specimen of wood tin from the streamworks at Bojewan in *c.* 1865

154   HENWOOD, 1873, 196–7

155   CORNISH, 1884–5 in 'Meetings of the Society', *PZNHAS*, 103

156   WHITLEY, 1873, 454–5

157   HENWOOD, 1873, 196–7

158   REID & FLETT, 1907, 86–7

159   DINES, 1956, **1**, 60

160   CARNE, 1821

161   LE GRICE, 1846

162   BORLASE, 1879. Tin is not mentioned in this paper but is itemized in Borlase, 1882 MS, item 121

163   BORLASE, 1872, 260; he excavated the 'fogou' in 1863

164   BORLASE, 1769 (ed.), 300

165   BORLASE, 1769 (ed.), 310 and Plate XXV

166   BROWN, 1970

167   DE LA BECHE, 1839, 499

168   BORLASE, 1758, 99–100

169   PEAL, 1967

170   CAMBELL *et al.*, 1970, 30

171   LEEDS, 1927

172   THOMAS, 1956. Although the furnace may well be post-Roman, Leeds (1927) noted that the tin ingot was beneath the floor in the hut east of the furnace, on a lower floor associated with Glastonbury type pottery, so that the ingot probably is 3rd or 2nd century BC as Leeds supposed

173   TYLECOTE, 1962, 63–5 and Fig. 11

174   NOALL, 1929

175   HEDGES, 1964, 11

176   HIRST, 1937. Haematite from this site is discussed earlier in the present volume under evidence for lode mining in Cornwall

177   BORLASE, 1769 (ed.), 316 and Plate XXVIII

178   PIGGOTT, 1968, 72

179   POCOCK, 'Travels through England 1750', in Chope, 1967 (ed.), 195

180   CHOPE, 1967 (ed.), 27

181   CAMDEN, 1636 (ed.), 188

182   HENWOOD, 1873, 197

183   HENWOOD, 1855, 18–21

184   EDMONDS, 1849

185   HITCHENS & DREW, 1824, **2**, 331; ROWE, 1953, 228, dates the drainage of the Marsh

186   BORLASE, 1882 MS

187   HENWOOD, 1873, 198–9

188   GREGORY, 1947

189   OPIE, letter to Truro Museum, 29 Mar. 1930

26 Cornish tin ingots

Several ingots have already been discussed – those from St Martin's in the Isles of Scilly (*see* Fig. 47, Chapter 20), from Carnanton in St Mawgan-in-Pydar (Fig. 111, *above*), from Chûn in Morvah, and from Boscarne near Bodmin (Fig. 118, *above*). Others are conveniently dealt with in a single section.

Most ingots are undated and need not be prehistoric. Mediaeval mining and smelting were primitive with 'sluggish progress' (to use Lewis' words),[1] until the introduction of improved methods from Germany in the early 16th century. It would be unwise to assume that all ill-shaped ingots are premediaeval. The letter of the first Warden of the Stannaries, William de Wrotham, in 1198 refers to two smelting processes. The first, probably crude, took place at the mine, while the second for stannary taxation ('coinage') took place only at towns designated by the Warden. This method continued until the introduction of the single-process blowing houses in common use by the mid-14th century. De Wrotham's first smelting must have produced ingots indistinguishable from those familiar in the preceding millennia.

It is also certain that tin was smelted near tin streams for illicit sale to avoid the payment of taxes, though Hatcher warns against overestimating the extent of evasion.[2] Richard Earl of Cornwall derived enormous revenues from the Stannaries. Until the Charters of Edward I in 1305, which freed tinners from ordinary taxation, there would have been extra cause to smelt tin for sale untaxed. Tinners working the 'wastrel' or unenclosed land in both Devon and Cornwall had to acknowledge the land owners' right by payment of toll tin or lord's dish, which usually amounted to a fifteenth part. There was always a temptation to avoid this when times were hard – which they usually were. As recent a publication as Thomas Oliver's *Autobiography of a Cornish Miner* (Camborne, 1914) relates that when he was a lad of about 12 in 1842 working at the stamps at Germoe, 'we had very hard times, everything was very dear, and the working people were half starved'. He earned only ten shillings a month and his diet was so inadequate that 'sometimes I was so feeble that I could scarcely crawl along'. Conditions at the end of the 16th century were no better. Carew shrewdly observed[3]

that the parishes where tin is wrought rest in a meaner plight of wealth than those which want this damageable commodity, and that as by abandoning this trade they amend, so by reviving the same they decay again, whereas husbandry yieldeth that certain gain in a mediocrity which tin works rather promise than perform in a large measure.

The sale of uncoined tin was prohibited by the Devon Convocation in 1552 and by the Cornish Convocation in 1558.[4] The 'Laws of the Stannaries of Cornwall' drawn up at Truro in the second year of Queen Anne's reign state that anyone 'conveying, buying or selling of tin uncoined' forfeited the tin and paid its full value as a fine. References to convictions are hard to find, but in Devon, Finberg recorded how one Henry Wymmeslond, having admitted to selling to Agnes Staunton, a widow of Buckland Monachorum '214 pounds of corrupt and deceitful tin to the great prejudice of the King's people and to the damage of the said Agnes', was consigned to Lydford Castle. The dungeons for the Stannary of Tavistock were here, of evil reputation and described by an inmate of 1512 as 'one of the most annoious, contagious, and detestable places wythin this realme'.[5]

INVENTORY OF CORNISH TIN INGOTS

(For locations *see* Map 26).

Lostwithiel   Hext in 1891[6] refers to a Jew's house discovered 'some years ago' on Shirehall Moor just south of the town at the head of the estuary. Among the finds were 'many ingots of tin' covered with the cinders of the fuel used to reduce the metal. None is known to survive, but they were probably mediaeval or later when Lostwithiel was an important coinage town. In 1305, 204 282 kg tin were coined, dropping to *c.* 31 752 kg at the end of the 16th century.[7]

Fowey   Although the Fowey ingots are later than the period principally dealt with in this book, they are usefully included for comparison. In 1898, five ingots were dredged up from 2·5 m below the bed of the estuary in mid-stream off No. 2 Jetty at Fowey (Fig. 127). All measured 63 × 28 cm, varying in depth to give weights recorded as 149, 116, 146, 144, and 167 kg, giving a total weight of 722 kg. Each appears to have been cast in a rough-hewn granite mould with the

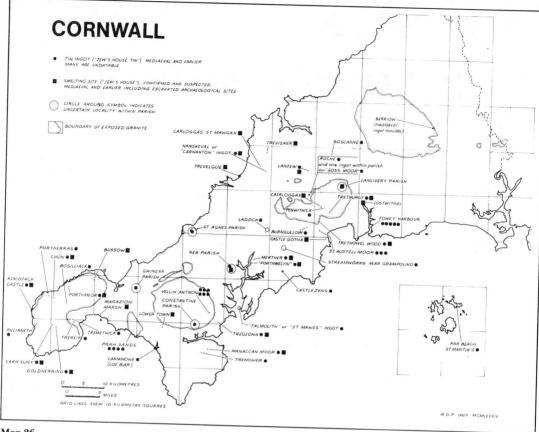

Map 26

sides tapering towards the base for ease of extraction. The Harbour Board offered them to the smelters Williams, Harvey & Co. of Hayle.[8] Only one was saved from the furnace and was presented to Truro Museum through Richard Pearce in 1907. It bears a hole for convenient lifting with a block and tackle, while the upper surface has been stamped with several marks, the clearest of which represent a cross. This may well be a blowing house mark comparable to others known to be in use in 1663.[9]

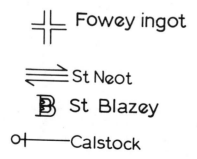

The well-known 'Lamb and Flag' stamp appeared in the 18th century and seems to have been used first at Treloweth, St Erth, a smelting house constructed for the use of coal in 1715. Gilbert[10] relates how his great-uncle, Henry Davies:

> contributed to the building, and the crest of his arms, a lamb carrying a flag, was adopted as a mark to distinguish the slabs of this house; all the different smelting and blowing houses having always used specific marks. The crest had, I presume, been originally taken in allusion to the Welsh and Cornish sound, at least of his name; *davas* being Cornish for a sheep, or perhaps a shepherd. This mark, however, conveyed to the minds of persons in Catholic countries some idea of consecration, and procured a preference for the Lamb Tin, although it never claimed to have the slightest superiority; and finally, all the other houses have taken the same, or similar marks.

127 **Fowey ingot, 1 of 5 recovered from the middle of the estuary in 1898, which could belong to a cargo lost in 1485; upper surface 57·5 × 28·5 cm, lower surface 50·5 × 22 cm, depth *c.* 13·5 cm (courtesy Truro Museum)**

Tradition maintains that the other popular mark of the 18th and 19th centuries – the 'Pelican in her Piety' – was stamped on tin sent to non-Christian countries.

At the time of their discovery, the Fowey ingots were assumed to be 14th or 15th century, 'if not earlier'. It is tempting to assign them to an ill-fated cargo of 1485, which would mean that a large number of similar ingots still await recovery from the same place. The full story is quoted by Gardiner.[11]

Chancellor John, bishop of Worcester
Date 22 August 1485–6 March 1486
Petitioners Roger Werth and Walter Yorke, citizens of the city of Exeter.
Complaint The petitioners had recently loaded a boat of Polruan (John Cornelys, master) in the water of Fowey with 61 pieces of Cornish tin; and it had been lying there, still loaded, for six weeks or more in great uncertainty and danger, all because of the chicanery of John Menheneke, mayor of the town of Lostwithiel, and Luke Fryse. The

128   Late Roman tin ingot from Trethurgy, St Austell; weight 7 kg (additional fragment weighed 5·5 kg), length 22·7 cm, width 16 cm (now in County Museum, Truro)

latter had brought a plaint before the same mayor and burgesses of that town, and they had arrested the boat and tin and were still holding them under arrest, not allowing the petitioners to have delivery of their own goods by finding sufficient security to the court, nor continuing the proceedings against them according to the law, but seeking, unfairly and unlawfully by all the devices that they could, to delay the boat and tin until it was perished and overturned by bad weather, to the utter undoing of the petitioners, contrary to all good law, reason and conscience without the chancellor's goodwill in this behalf.

Fowey was the principal port for the export of tin at this time – 65 t in 1498.[12]

The Fowey ingots resemble but are lighter than those described by Hawkins[13] from a Florentine MS *La Practica della Mercatura* written between 1332 and 1345 (printed in 1766) by Francisco Balducci. It tells how tin imported from Cornwall was 'in large slabs, of a long square form, each slab weighing about one cantaro and a third Barbaresque weight of Majorca, where and at Venice they make bundles of tin rods, bound together with rods of tin'. A *cantaro* of tin equalled 140 lb of Pisa, 145 of Genoa, and a cwt of tin of 112 lb of London, roughly equal to 120–2 lb at Bruges.

Lanlivery   In or before 1855 a tinner called John Hoare found 'at Lanlivery', probably at or close to the workings on Red Moor, the remains of a 'Jews'-house' smelting site together with specimens of the ore and

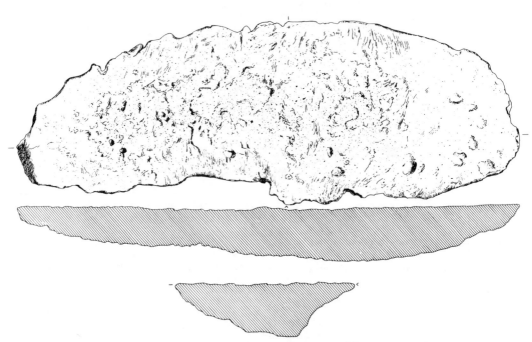

129   Tin ingot dug up in a garden at Penwithick Stents, St Austell; weight 7·7 kg, 41·8 × 14·7 × 4·4 cm

130   Outline and section of tin ingot found at Burngullow, St Mewan, in 1858; weight 33·8 kg, 50 × 28 × 7 cm

refined tin.[14] It may have been an ingot from this find that was sent to Hunt at the Royal School of Mines, who presented it to the Museum of Practical Geology in London. The ingot weighed 21 kg.[15]

Trethurgy, St Austell   Trethurgy was a small, isolated earthwork completely excavated in 1972–3 by Henrietta and Trevor Miles.[16] The bulk of the pottery was late Roman, providing a tentative occupation period between AD 250 and 550. Associated with Romano-Cornish wheel-made pottery within 'structure U' (probably not a house) was a roughly oval, plano-convex tin ingot (Fig. 128), typically much corroded, approximately 30 × 21 × 7 cm thick, and not dissimilar from that already described from Chûn castle of Romano-British or post-Roman date. With additional fragments it weighed 12·5 kg.

Penwithick, St Austell   Only 2·3 km north-west of Trethurgy is Penwithick. A 'pasty'-shaped ingot dug up by Mr Stephens in his garden (Fig. 129) was presented to Truro Museum in 1967. No datable objects were found with it. The ingot weighs 7·7 kg.

Trethowel Wood, St Austell   An irregularly shaped mass of smelted tin, with pebbles of stream tin and charcoal embedded in its base, was found immediately below the surface in Trethowel Wood which adjoins the St Austell River 1·6 km north of the parish church. The ingot weighed 36·3 kg and was sold about 1870 to Carvedras Smelting Co., Truro, for £3.6s.5d.[17] In 1922 a mass of tin cut along one edge from a larger piece, weighing 14 kg and with 'embedded pebbles of stream tin and fragments of charcoal' was presented to Truro Museum by Miss Daubuz, descendant of the smelting house owners.[18] Through decomposition the ingot has lost over 0·5 kg in weight since then. It may well be part of the Trethowel ingot, but this cannot be proved.

St Austell   In 1825 Thomas Hogg[19] wrote of two blocks of tin, each weighing nearly 11·7 kg, found 'in a mine near St Austell ... several years ago, about 80 feet under the ground'. This must be a garbled account of the handled ingots from St Austell Moor, 2·4 m below the surface, described in the finds from the Pentewan valley.

St Austell Moor   The handled ingots from this tin stream south of the town were described by Borlase in his *Natural History of Cornwall* (1758). He later noted that streamers working here in May 1765 found in the tin ground, only 1·5 m below the surface, 'a large Cake of Tin-ore weighing about six pounds, irregular in shape, cracked or jagged at the edges'.[20] Borlase assumed it was a natural lump containing native tin, though it was clearly an ingot much corroded with the outer layers converting to tin oxide. Part of the specimen was deposited 'in the Desk of Cornish Fossils at the Museum in Oxford [the Ashmolean]' and part 'in the Museum of the Royal Society, London'.

Burngullow, St Mewan   In the museum at Geevor Mine, St Just, is a plano-convex tin ingot (Fig. 130) loaned by the Royal Geological Society of Cornwall. It measures c. 50·5 × 28·6 × 7 cm in the middle, tapering to the edges, and of rough shape. It was found in 1858 when building the Cornwall Railway – later the GWR mainline. This is the same ingot which Francis Michell of Calenick told Henwood had been found 'some years ago ... at Burngullow in St Mewan' and sold to the Calenick Smelting Works.[21] It weighs 33·8 kg.

Goss Moor, Roche   The discovery of a block of tin of 'peculiar shape' in 1835 has already been mentioned. Another was found with 'a singular kind of ancient shovel' somewhere in the same parish about 1772. It weighed c. 9 kg and 'from its appearance seemed to be exceedingly ancient'.[22] It was found c. 1·2 m below the surface by tinners searching for ore. The description is very vague – 'about three inches thick, and its width and length were in proportion' – though it would seem to have been fairly regular, suggesting a mediaeval or later date. The Rashleigh family had so many mining concerns in this area that this ingot could have been one of those seen at Menabilly by Gregor in 1817.

St Wenn, or Withiel   Evans (1881) mentions a tin ingot found 'in a small stream at St Wenn' (Fig. 131). The ingot of typical plano-convex shape, 20–2 cm in diameter with a maximum thickness of 10 cm, was formerly in the Salisbury and South Wiltshire Museum. In 1932 it was described in a letter now at Truro Museum as crumbling and in a poor state of preservation. Perhaps it disintegrated completely, for it is no longer in the Salisbury collection.[23] The ingot may well be the one described[24] as having been found

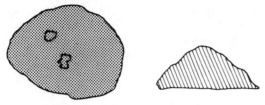

131 Outline and section of lost tin ingot, formerly in Salisbury Museum; perhaps the one from Lanjew, Withiel, 1826; *c.* 23 × 20 × 4 cm

in Withiel parish at a 'Jews' House' smelting site on Lanjew farm on the parish border, 1·6 km east of St Wenn church.

Castlezens, Veryan   In 1832 Hawkins reported the discovery 'made a few years ago, at the distance of about a furlong from Castle Zen . . . of a block of tin of a singular form, which has on it an inscription in Roman letters'.[25] Unfortunately the ingot no longer exists, so it is impossible to tell whether it was Roman, or much later like the ingot from Trereife described below.

Grampound, Creed   In the *Supplement* to his *Natural History of Cornwall*[26] Borlase wrote that Jonathan Crowle, a tinner, found on 17 July 1765 an ingot weighing 5–5·5 kg 'in a Stream-work near the borough of Granpont'. Borlase thought the Grampound ingot may have contained native tin like the ingot he also described from St Austell Moor, though he was unsure. It was much corroded and converting to tin oxide. The discovery was probably made near Golden Mill on the border of Creed parish where tin streaming began in 1753. Borlase himself noted that 'gold was in such plenty in this tin' that nuggets as large as a walnut were found. The same stretch of valley was being worked in 1796 when the *Sherborne Mercury* of 21 November advertised for sale a share in 'Tregony Bold Adventure Stream Work', an 'extensive sett now in full working'.

Ladock   The discovery of gold and stream tin in the valley below Ladock church in the early 19th century has already been mentioned. It was probably about this time that the ingot was found: 'a rude smelted block of tin, supplied by G. N. Simmons, found in Ladock, near Truro, and supposed to have been smelted when the Phoenicians traded to Cornwall for tin'. The block was exhibited in the Hyde Park exhibition of 1851.[27]

Merther   Revd F. Webber of Merther wrote to Dr Charles Barham on 6 November 1843 concerning discoveries made after he had reclaimed a piece of waste ground near his house.[28] The labourers dug to a depth of 0·6–0·9 m and found, near a spring of water, an area *c.* 3 m across filled with 'a black rubbish evidently

produced by the action of fire'. Webber also found 'several scoriae of great comparative weight, and also large masses of what appeared to be unwashed tin'. Specimens of the scoriae (slag) that he presented to Truro Museum no longer exist. He also noted that in the adjoining parish of St Michael Penkivel another 'Jews house' had existed at the head of 'Porthmelyn Creek, in the Tresillian branch, just below the site of the old manor house of the Carminows'. That puts it below Fentongollan at the head of what is now Merther Pond and no longer a creek.

St Agnes   An ingot of 3·5 kg from somewhere within the parish was sent to Hunt, Keeper of the Mining Records at the Royal School of Mines. According to Henwood[29] it was presented to the Museum of Practical Geology in London before 1874.

Kea   The Revd William Gregor, a mineralogist most famous as the discoverer of titanium, was interested in the decomposition of tin ingots. In a paper on the subject dated 7 September 1816,[30] he analysed an ingot 'presented to me many years ago by John Michell, of St Austle'. It had been found 'amongst other lumps [of tin?] . . . under the surface of a low and boggy ground in the parish of Kea; the tin was accompanied with a stratum of charred wood or charcoal', and supposedly came from an old smelting site. The exact location is not known, but it could represent the smelting of ore from the Carnon valley. Gregor does not describe the ingot's shape or size. However, it could be the ingot of 145 g from 'near Truro', examined by Tennant and Davies, and in 1874 at the British Museum.[31]

Gregor found the decomposition product to be 'Muriate of Tin', a compound he also found in an ingot from 'Pilianeath', St Just-in-Penwith. Henwood conjectured that the muriatic acid, an impure hydrochloric acid also known as spirits of salt, may have come from using dried seaweed, a substance still employed in his day as a fuel on the Isles of Scilly. Other masses of 'Jews House Tin' were known to Gregor. Unfortunately he gave no details beyond mentioning in a postscript, dated 12 February 1817, that two specimens were in Rashleigh's collection at Menabilly.

Antron, Mabe   In April 1913 at least six 'rough basin-shaped' pieces of tin were found *c.* 45 cm below the surface when Mr Harris was laying a drain in a field adjoining Mr Bernard's farmhouse at Vellin Antron. They were randomly placed. Two samples gave 96·88% tin with traces of iron; they were partially converted to tin oxide on the surface. The locality was locally known as Gravel Hill or Tin Pit Hill.[32] All the ingots were dispersed. Nine small fragments 'pinched' from the site were presented to Truro Museum by Walter Rogers, as was an ingot acquired by G. B.

Pearse of Hayle; unfortunately none of these now exists in the Museum's collection. Two ingots were bought by Capt. Henderson – the going price was five guineas each – while a third went to Sir William Sergeant of Lanivet, a noted mineral collector. One block weighed 4 kg and the total was between 22 and 27 kg. Fortunately one of the ingots survives in the museum in Helston; it is oval, 10 × 13 × *c.* 5 cm thick.

Constantine (near Falmouth) An ingot weighing 1·8 kg from this parish was sent to Hunt of the Royal School of Mines. In his usual practice he passed it on to the Museum of Practical Geology sometime before 1874.[33]

Tregedna, Budock Near Tregedna, 6–9 m of 'vegetable mould and hardened silt' overlay a very poor and thin deposit of tin ground resting on the local 'killas' (clay-slate). This must have proved profitable to work at some time for 'a mass of *Jew's-house-tin*, three or four pounds weight, was found among the remains of an ancient furnace near the well at Tregedna'.[34]

St Mawes The most enigmatic Cornish tin ingot, which has occasioned the outpouring of varied theories ranging from the fanciful to the well reasoned, is the St Mawes ingot (Fig. 132), sometimes less correctly referred to as the Falmouth ingot. Another resembling the letter H is said to have been found at Newton Moor, Condurrow, Camborne, in the late 19th century (S. A. Opie, *Old Cornwall*, 1932, **2**, (3), 44), but the report is so vague as to be valueless. Two others dredged up *c.* 1980 off Tyre and Sidon with pottery dated to 600 and 700 BC (R. S. Manx, pers. comm., 1985) cannot be confirmed.

Together with the Rillaton gold cup the St Mawes ingot is the most famous of all objects found in the county, yet when dredged up it attracted little attention and did not achieve international fame until after it had been published in 1863 by the Director of the Ordnance Survey, Col. Sir Henry James.[35] With typical Victorian inexactitude, he described its discovery as 'about forty years ago' (*c.* 1823), a date repeated by several later authors. However, the ingot had been published in 1838 by Penaluna who gave an earlier date for its discovery, and there is no reason why Penaluna's account should not be accepted:[36]

In the year 1812, as some bargemen were employed in taking up sand for manure in deep water off St Mawes, they discovered in their dredge something which appeared to be an ancient block of tin about one hundred and a half in weight, of the purest metal. It was much corroded, and so far as its appearance could furnish evidence, it seemed to have been originally cast in sand. From its having no mark which denoted the duty paid to the Duke of Cornwall, the inference is fair that it must have been deposited in the water before that event took place, but

how long is impossible to say. It is conjectured, however, that this piece of tin must have been manufactured in the days when the Phenecians [sic] visited the Cornish market for this article. Admitting this conjecture to be correct, this block of tin must have slipped into the water while putting on board a ship for exportation. This singular relic of antiquity is still preserved in Truro.

A cast of the ingot at the Royal School of Mines in London bore a label stating that one of the arms of the ingot had been 'cut off by the bargeman who found it, under the impression that it was silver'.[37] This is only a half-truth. Silver he may have thought it, but to prove its worth the ingot was taken to the Calenick Tin Smelting House near Truro for assay. In 1812 the works was owned by Ralph Allen Daniell, and the assay was made by his son Thomas, for the flat surface left where the arm is cut off is stamped with his initials TD. It is worth noting here that some modern archaeologists have assumed the arm was cut off in antiquity as a prehistoric equivalent to the more recent 'coinage'. Daniell's forethought in leaving his signature is to be commended. Unhappily, Thomas Daniell, who took over the works in 1823, was a poor business man. In 1828 his works were taken over and he was forced to reside in Boulogne to escape financial difficulties.

Daniell was a founder member of the Royal Institution of Cornwall with an interest in antiquities and, probably at the time his works were taken over, he presented the ingot to the Institution's Museum in Truro. Until then it had remained in obscurity, for in 1828 Richard Thomas,[38] a well known engineer and surveyor, wrote at length about the tin trade and argued that the Black Rock at the entrance to Falmouth Harbour could not possibly be the Ictis of Diodorus Siculus. Thomas made no mention of the St Mawes ingot, which he surely would have done had he been aware of its existence.

The find spot of the ingot is known with reasonable accuracy. Sand had been dredged from Carrick Roads since the 17th century:[39]

In Falmouth haven near St Mause [Mawes] Castle there is a sort of sand or rather coralline, that lies a foot under the ouse, which once being removed and the bed opened, this sand is taken up by a dredge and is used about Truroe, Probus &c...

The sand was extensively used in the early 19th century, as detailed by Worgan in 1811[40] and Thomas.[38] It is still worked commercially on a small scale by a Truro firm and has been sold under the trade name of Seagold. The main deposit lies on the 'St Mawes' or 'Vilt' Bank where the sea-bed is covered with up to 15 cm of living spherical rhodoliths, beneath which is mud containing dead maërl of the calcified seaweed *Lithothamnion calcareum*. It is rare in Britain with the main Cornish deposit lying 0·8–1·2 km north of Castle Point, St Mawes.[41]

132   **St Mawes ingot dredged up in 1812; max. length 91·3 cm, weight *c.* 72 kg (photo County Museum, Truro)**

The ingot may have been exported from a port on the Fal estuary, as tin has been to the present day, but it could have come from any part of the county. Whatever its port of origin, a likely explanation of its loss is that it had been on board a vessel, perhaps off the dreaded Manacles Rocks, when contrary winds forced it to run to shelter. A well laden ship, perhaps with a shifting cargo, unable to reach the quiet waters of Carrick Roads and the shelter of Pendennis Point foundered on the St Mawes side.

The ingot has generally been credited with great antiquity. When found it was said to contain crystals of cassiterite. This need not imply a great age for they could have been the black tin oxide romarchite (SnO) or, less likely, the white hydroromarchite ($Sn_3O_2(OH)_2$), minerals first described in 1971 after their discovery as crusts on pannikins (small metal cups) lost from an overturned canoe between 1801 and 1821 at Boundary Falls on the Winnipeg River, Ontario.[42] No more is known about the St Mawes ingot today than Penaluna knew in 1838, and his caution about its age should be echoed. Recent opinion suggests that it is unlikely to be earlier than mediaeval. Weighing nearly 72 kg – enough to make over a tonne of 10% tin bronze – it would have been beyond the purchasing power of any bronze age or iron age smith (P. J. Northover, pers. comm., 1982).

It has been doubted whether such a large ingot could have been smelted in antiquity, remembering that Agricola in 1556 tells how in Spain 'Some of the Lusitanians melt tin-stone in small furnaces, employing leather bellows hooped around with iron and drawn in and out like an accordion'. These were so inefficient that 'the smelter is not able during a whole day to melt much more than half a *centumpodium* of tin',[43] that is 50 *librae*, a little over 16 kg. Some Cornish ingots are heavier. The one from Trethowel Wood, St Austell, was 36 kg when found, but its date is not known. If the St Mawes ingot is mediaeval, its weight suggests the period of Edward I (1272–1307), for later ingots were heavier.[44]

Year	1305	1577	1587	1597	1604
Weight/kg	57	140	147	152	157

Manaccan An ingot of *c*. 0·2 kg from somewhere within the parish, was examined by Tennant and Davies; in 1874 it was in the British Museum. A fragment of tin from Manaccan Moor was formerly in the collection of John Jope Rogers at Penrose, Helston. It had been found by R. J. Cunnack together with fragments which 'had every appearance of slag'.[45]

Trenower, St Martin-in-Meneage A mass of 'Jew's-house-tin' was discovered here by Mr Cuttance and presented to Truro Museum in 1862–3. The ingot no

longer survives, but when examined by Henwood it weighed 0·56 kg.[46]

Carminowe, St Mawgan-in-Meneage In October 1862 Rogers found the remains of what seemed to be two ovens built of clay–slate stones, at the bottom of the slope of the hill where it joins the southern end of Loe Bar.[47] There were no finds to help date the structure, which Rogers thought might have been Roman, since pottery of that period had been found in October 1860 only 140 m up the hill.[48] What the 'ovens' really were is not known. Rogers speculated that they may have been a smelting site, noting that tin had 'been found not many years ago within a few paces of the spot'. No other details are known.

The site may seem a long way from the centres of tin mining, though it is not because the River Cober, which flows through Wendron into Loe Pool, was worked for stream tin, while in 1720 it was proposed to dredge the Loe itself. According to the prospectus 'immense quantities [of tin] lodged in its bottom and continually washed into it, as is visible from those that are occasionally thrown out on the sea sands when the Land Floods break over the Bar into the Ocean ... Thirty Thousand sacks of Tin having been sometimes spewed out at one of these breakings'.[49] More than a little imagination there, especially as no further word was heard of the dredging enterprise. But there must be some truth in it, and tin which escaped the Loe to reach the sea may, in part, have been responsible for saleable quantities of cassiterite recovered from beaches as far south as Gunwalloe, 3 km south of Loe Bar.[50]

Cunnack recorded that in the centre of Lower Town, 1·6 km upstream from Helston, 'a few feet beneath the surface' were found 'accumulations of slag and carbonaceous matter, which by their yielding on the vanning shovel particles of metallic tin, are evidently the remains of smelting operations'. In the same area elvan boulders were frequently found 'with the surfaces indented in deep hollows, where the tinstone had been rudely pounded into powder', and comparable to that seen in the photograph of the mediaeval finds from Vorvas near St Ives (*see above*, Fig. 75). A mediaeval or later crazing mill is also described by Cunnack, found about 1852 at Trelubbas, Wendron.[51]

Gwinear Carne (1821)[52] and Le Grice[53] mention a tin ingot weighing 15 kg 'found in a hedge' somewhere in the parish. Le Grice comments with feeling that 'having been offered for sale at the Angarrack Smelting House, the Goth of a refiner put it at once into a ladle and melted it down'.

Praa (or Prah) Sands, Germoe In February 1974 gales removed a good deal of the beach and also the marine edge of the sand dunes, revealing a black,

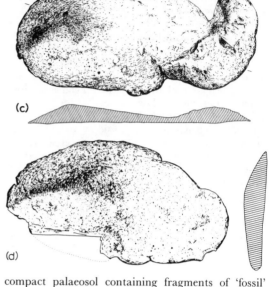

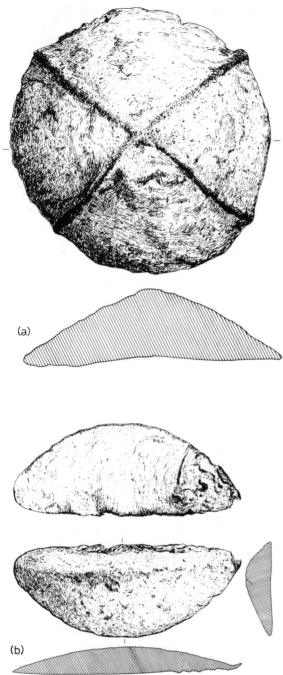

compact palaeosol containing fragments of 'fossil' wood up to 10 cm in thickness. Two ingots of tin were found by Mr Yelland 'at the bottom of the forest level', while two more were recovered by his children the following September (Fig. 133). Samples of the wood submitted to Harwell by Dr Keeley of the Dept. of the Environment, produced a ^{14}C date of 1290 ± 70 (*c.* AD 660). Whether the ingots belong to this period or are much older than the date of the forest's inundation by the dunes is not known, but it is tempting to regard the ingots as evidence of smelting in the post-Roman period. The ingots are in a splendid state of preservation and are now displayed in the County Museum, Truro. The tin had probably been poured into open stone moulds, perhaps of granite:

(*a*) A circular, plano-convex ingot weighing 3·8 kg bears a raised cross which must have been cut into the mould. The remaining ingots are all 'pasty' shaped.
(*b*) The most perfect example weighs 2 kg.
(*c*) When this ingot was poured the mould must have been tilted, for an irregular mass of tin spilled over the edge of one side. It weighs 2·06 kg.
(*d*) This small ingot had a piece cut out for analysis, which showed the ingots to be 99·5% tin with most of the remainder being iron.[54] It weighs only 744 g, and can only have been a little over 907 g when complete.

Goldherring, Sancreed   The annual excursion of the Penzance Natural History and Antiquarian Society on 28 August 1885 included in its itinerary a visit to the 'Jews' Smelting House' at Goldherring. An ingot of smelted tin had 'very recently' been discovered at the

a 3·8 kg, 18·8 × 4·6 cm *b* 1·87 kg, 25 × 10 × 3 cm *c* 1·9 kg, 26 × 12 × 2 cm *d* when complete *c.* 0·9 kg, 18·5 × 9·5 × 1·7 cm
133   Tin ingots, *c.* AD 600, from Prah Sands, Germoe

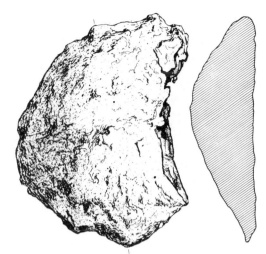

134 Tin ingot, probably from Land's End peninsula; weight 3·7 kg, dia. at section 18 cm (courtesy museum of the Royal Geological Society of Cornwall, Penzance)

spot and was shown to the party.[55] The ingot was not described and is now lost. The site includes a narrow strip of land leading down from the well – formerly an imposing structure – which was called in the 1838 Tithe Apportionment Schedule, *stennack* which signifies 'tin-bearing ground'. A Romano-British settlement adjoins the smelting house site, though whether the ingot is contemporary with it is not known. It could be much later.

The settlement was totally excavated in 1958–61 with indications that smelting had occurred in Romano-British times.[56] Site B, a rectangular hut set high at the east end of the oval settlement enclosure, had been re-used in more recent times, and at some time in its history it had certainly been a blowing house for smelting tin ore, perhaps in the mediaeval period. Pottery was a mixture of 3rd and 4th century sherds with some mediaeval. A sub-oval hearth, cut deep into the floor and contemporary with the early occupation of the site, was paved and contained a covered air-duct which led through the enclosure wall. The bottom 10 cm of the hearth were almost pure charcoal which proved to be gorse (*Ulex* species). The charcoal also filled the air-duct. No tin was found here, but three pebbles of stream tin were found – one in the courtyard at a 'fairly high level', and two in a pit at the south-east of the settlement.

St Just-in-Penwith Copeland Borlase had in his museum an ingot from Portherras, on the border of Morvah parish,[57] but nothing is known of the find circumstances. Carne (1821) mentioned an ingot weighing *c.* 2·2 kg from Pillianeth streamworks in the Nanquidno valley;[58] this could be the same one of 2·7 kg 'from St Just referred to by Le Grice'.[59] An ingot weighing 3·7 kg from an unknown locality, now in the Royal Geological Society's Museum at Penzance, probably came from the Land's End peninsula (Fig. 134).

135 Ingot of tin; 420 × 203 × 50 mm, weighing 12 kg, and 99·9% pure metal; found in or before 1845 at Trereife, Madron

Madron   As with St Just, several ingots have been found in this parish. In 1846 Le Grice wrote of the discovery 'only a few months ago' of a tin ingot in the centre of a barrow at Bossuliack [Bosilliack] Croft near Lanyon'.[60] It was free of decomposition except for the outer crust, which contained some chlorine – the muriate of tin of earlier writers. Flat stones 0·3 m below the ingot covered what may have been the primary burial cavity with ashes, if it is assumed that the structure had indeed been a bronze age burial.

In 1871 there was purchased for the Royal Institution of Cornwall's Museum at Truro a block of Jews-house tin found at 'Tremellack [Tremethick] Moor'.[61] It weighed 17·2 kg but suffered badly from corrosion and has lost several kilos since then. Of roughly oval shape, it measured 30 × 23 cm with a maximum thickness of 10 cm. Nothing is known of the find circumstances.

The most interesting ingot from the parish was found *c*. 1845 at Trereife when workmen removed a high bank of compact clay (Fig. 135). Within the bank was an inverted cone-shaped cavity, *c*. 0·9 m across the top and just as deep. Ashes lay on the flat stone of 0·3 m diameter at the bottom. A channel dug into the cavity on one side was assumed to have been the place for a bellows. The cavity had been filled with earth and rubble, on the top of which was a fine tin ingot weighing 13·4 kg. It is now in Penlee House Museum, Penzance. The ingot carries on one surface raised letters (EIC) and a cipher resembling an elaborate chi-rho. For a long time the ingot was believed to be Roman. However, archaeologists since the time of Haverfield[62] have accepted it as more recent since the marks are very similar to those of west country mediaeval merchants' marks,[63] and the letters could refer to the East India Company.

NOTES AND REFERENCES

1   LEWIS, 1908, 18
2   HATCHER, 1973, 85; he gives a good account (82–5) of the privations suffered by the tinners
3   HALLIDAY, 1953, 99
4   PENNINGTON, 1973, 143
5   FINBERG, 1949
6   HEXT, 1891, 155–6
7   MACLEAN, 1873
8   ROGERS, 1903
9   RIC/EN/147 Coinage Papers, 1663, MSS at County Museum, Truro
10 GILBERT, 1838, **1**, 365–6
11 GARDINER, 1976, 118 item C1/78/145
12 ROWSE, 1941, 73
13 HAWKINS, 1834, 129
14 HENWOOD, 1855, Lecture 2, 11 footnote
15 HENWOOD, 1873, 226 footnote
16 MILES & MILES, 1973
17 HENWOOD, 1873, 226 footnote
18 *JRIC*, 1923, 107 'Accessions for 1922'
19 HOGG, 1825, 75
20 *JRIC*, 1864, 25–6 *Supplement* to Borlase's *Natural History*
21 HENWOOD, 1873, 226 footnote
22 HITCHENS & DREW, 1824, **2**, 587
23 R. K. SAUNDERS *in litt.*, 1982
24 HENWOOD, 1873, 227 footnote, refers to *Gentleman's Magazine* 1826, 125. This may be an error as the reference cannot be found by the British Library
25 HAWKINS, 1832, 73–4
26 *JRIC*, 1864, 26
27 WAY, 1859, 69
28 REVD F. WEBBER, letter in *RRIC*, 1843, 38–9
29 HENWOOD, 1873, 252
30 GREGOR, 1816
31 HENWOOD, 1873, 252
32 HENDERSON, 1915
33 HENWOOD, 1873, 251
34 HENWOOD, 1873, 201–2 with footnote quoting Joshua Fox
35 JAMES, 1863
36 PENALUNA, 1838, **2**, 258
37 PETER, 1906, 237
38 THOMAS, *History of Falmouth*, 1828, 7–16
39 PYCROFT, 1856, 30, quoting 'The Improvement of Cornwall by Sea-Sand', *Phil. Trans.*, 1675, **10**, 293
40 WORGAN, 1811, 127
41 HARDIMAN *et al.*, 1976
42 ROBERTS, RAPP & WEBER, 1974, 524
43 HOOVER & HOOVER, 1950, 419–20
44 MACLEAN, 1873
45 HENWOOD, 1873, 251 footnote
46 *JRIC*, 1863, 18, under 'Donations'; HENWOOD, 1873, 203 footnote
47 ROGERS, 1863
48 ROGERS, 1861
49 JENKIN, 1962, **4**, 54
50 HUNT, 1887, 355
51 CUNNACK, 1867
52 CARNE, 1821
53 LE GRICE, 1846
54 BIEK, 1978, 'The archaeological iron and tin cycles', MS prepared for a Bonn Archaeometry Symposium
55 *PZNHAS*, 1885, 121 'Excursion notice'
56 GUTHRIE, 1969
57 BORLASE, 1882 MS, item 187
58 CARNE, 1821
59 LE GRICE, 1846
60 LE GRICE, 1846
61 *JRIC*, 1872, lxxxviii, under 'Donations'
62 HAVERFIELD, 1924, 24, 42
63 WORTH, 1891, for comparative Devonshire merchants' marks

27 Tin production in 'Dark Age' Cornwall

The present writer is indebted to George Rapp of the University of Minnesota for obtaining a ^{14}C date for the shovel from Boscarne (*see above* Fig 121). A Roman date was expected, but a result lying in the period AD 635–1045 is far more valuable and comes as a welcome addition to the small corpus of finds from Cornish tin streams datable to the 'Dark Ages' – to use an old-fashioned but appropriate term. There is the penannular brooch from a streamworks 'near Lanivet', another smaller one found with a finger-ring from the streamworks at Polmassick, and the knowledge that the Trewhiddle hoard had been deposited at a depth of 5 m in a streamworks which must have been in operation at the time. The tin ingots from Prah Sands were not recovered under ideal conditions, so their 7th century date has to be accepted on trust.

To this primary evidence can be added deductions from both general archaeological evidence and literary sources. In his study of Stannary Law – the law which dominated the Cornish tin industry from the 12th century – Pennington[1] showed that its origins were laid in pre-Norman Cornwall, possibly even before the Saxon conquest of the county in the previous century. The vague rules of tin bounding, which permitted streamers to work wherever they had a mind to 'without preliminary authorisation', had to develop in a sparsely populated Cornwall following the collapse of any Roman rule that might have been imposed earlier.

It could be argued that no Roman rule organized the tinners, so that the freedom of the tinners was a pre-Roman, even a bronze age right handed down from generation to generation. Whatever the truth of that, if mining had ceased in the post-Roman centuries and was restarted by the Normans or Anglo-Saxons, their laws would be expected to prevail. This is not the case. Stannary Law followed a unique set of rules, alien to the Normans. The antiquity of Stannary Law is emphasized by inherent differences that distinguish it from Anglo-Saxon mining law found, for example, in the lead districts of Wirksworth and High Peak in Derbyshire. It cannot be argued that tin mining was non-existent in the 11th century because it is not mentioned in the Domesday Survey of 1086. Rather, as Lewis pointed out,[2] tin was ignored in the Survey because for taxation purposes it was considered a royal property. The collection of taxes or 'farm tin' in the 12th century was designed to milk the profits of an old-established industry.

To be considered a 'royal property' the tin industry must have been lucrative, and there are occasional glimpses through ancient texts of Cornish tin traded over long distances. Hencken[3] is one of many authors who cites the story in the life of St John of Alexandria (died AD 614) of a ship's captain who, with his cargo of 20 000 bushels of corn, was blown westwards for 20 days and relieved a famine in Britain (*see* Appendix). Half his cargo was sold for coinage (*nominsmata*) and half for tin.[4] The story is roughly contemporary with Stephanos of Alexandria and his 'Brittanic metal' referred to in Chapter 1. These two references imply that there was nothing unusual about British – hence Cornish – tin reaching Alexandria. Nor should the miraculous element in St John's story be overlooked, for belief that the cargo was converted to silver on the voyage home is a sure sign that tin was obtainable with such ease that silver was a much more valuable commodity. Further evidence is given below for the diminished value of tin.

These stories belong to a period no more than a lifetime before Alexandria, once the second city of the Roman Empire, capitulated to the conquering Islamic hordes on 22 December 640. Trade with north-west Europe must have ceased then, if it had not already done so. Cornish tin failed to reach the Moslems, at least directly, for by the 9th century they looked east for their trade in goods which included tin, based on the trading centre of Kala in the mouth of the Tenasserim River (*see* Chapter 9).

In the 6th century there is abundant evidence of trade between the Mediterranean and western Britain. In this book in which a single product is examined in detail, it is all too easy to view all trade and contacts as

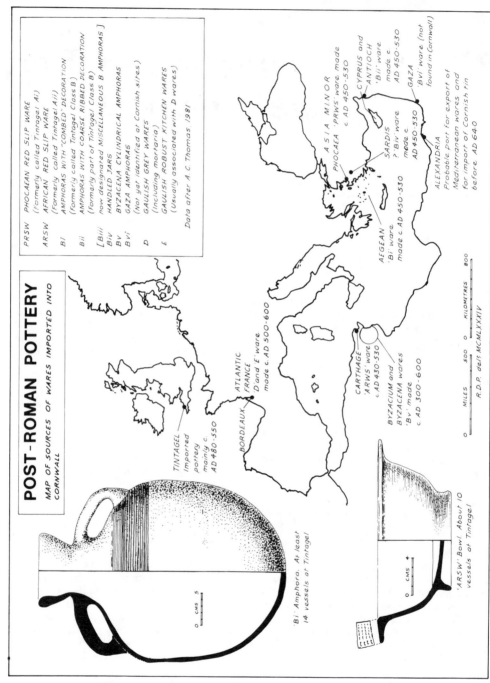

POST-ROMAN POTTERY

MAP OF SOURCES OF WARES IMPORTED INTO
CORNWALL

PRSW   PHOCAEAN RED SLIP WARE
          (Formerly called Tintagel Ai)
ARSW   AFRICAN RED SLIP WARE
          (Formerly called Tintagel Aii)
Bi   AMPHORAS WITH "COMBED DECORATION"
          (Formerly called Tintagel Class B)
Bii   AMPHORAS WITH COARSE RIBBED DECORATION
          (Formerly part of Tintagel Class B)
[Biii   now designated MISCELLANEOUS B AMPHORAS]
Biv   HANDLED JARS
Bv   BYZACENA CYLINDRICAL AMPHORAS
Bvi   GAZA AMPHORAS
          (Not yet identified at Cornish sites)
D   GAULISH GREY WARES
          (including mortaria)
E   GAULISH ROBUST KITCHEN WARES
          (Usually associated with D wares)

          Data after A.C.Thomas 1981

TINTAGEL
Imported
pottery
mainly c.
AD 480-550

BORDEAUX

ATLANTIC
FRANCE
D and E ware
made c. AD 500-600

CARTHAGE
'ARWS' ware
c. AD 430-530

BYZACIUM and
BYZACENA wares
'Bv' made
c. AD 300-600

ASIA MINOR
PHOCAEA 'PRWS' ware made
c. AD 450-530

SARDIS
? 'Biv' ware
made c.
AD 450-530

AEGEAN
'Bii' ware.
made c. AD 450-530

CYPRUS and
ANTIOCH
'Bii' ware
made c.
AD 450-530

GAZA
'Bvi' ware (not
found in Cornwall)

ALEXANDRIA
Probable port for export of
Mediterranean wares and
for import of Cornish tin
before AD 640

'Bi' Amphora. At least
14 vessels at Tintagel

0 CMS 5

'ARSW' Bowl. About 10
vessels at Tintagel

0 CMS +

0   MILES   500   0   KILOMETRES   800

R.D.P. del. MCMLXXXIV

Map 27

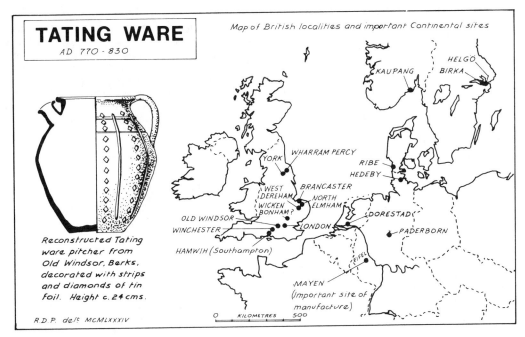

TATING WARE
AD 770 - 830

Map of British localities and important Continental sites

KAUPANG
HELGO
BIRKA

YORK
WHARRAM PERCY
RIBE
HEDEBY

WEST
DEREHAM
BRANCASTER
WICKEN
NORTH
BONHAM?
ELMHAM
OLD WINDSOR
DORESTAD
WINCHESTER
LONDON
PADERBORN
HAMWIH (Southampton)

EIFEL

MAYEN
(Important site of
manufacture)

O   KILOMETRES   500

Reconstructed Tating
ware pitcher from
Old Windsor, Berks,
decorated with strips
and diamonds of tin
foil. Height c. 24 cms.

R.D.P. delt MCMLXXXIV

Map 28

subservient to 'the Search for Ancient Tin'. However, as far as Cornwall is concerned it is difficult to envisage any alternative to tin that could have been the *raison d'être* of two-way international communications. Whatever products were exported from elsewhere in western Britain, and whatever link there may have been with the import of early Christian monasticism to western Britain, Cornish tin must have played a vital rôle. Why such a wealth of imported pottery should have ended up at Tintagel, far from the centres of tin streaming in Cornwall, is not clear. Nor is it appropriate here to delve into the enigmatic nature of the settlement of Tintagel, now seen as secular rather than monastic as formerly firmly believed.

All that is important here is to stress the archaeological evidence for a flourishing Cornish–eastern Mediterranean link, as strong in the 6th century as it had been in any previous period. The types of pottery – both North African and Phocaean Red 'A' Wares, and Aegean and Antioch (or neighbourhood) 'B' Wares of buff amphorae for wine or olive oil – have been intensively studied in recent years with their Cornish context admirably discussed by Thomas.[5] Details of the pottery need not be given here. Suffice it to say that the bulk of the Mediterranean wares reached Cornwall between about 480 and 550 from one or more of a number of possible ports, of which Alexandria is considered to be one of the most likely candidates.

Trade with France over the same period is exemplified by the 'D' wares from the Bordeaux area (500–600) and 'E' wares from 'Atlantic France' over a

similar period and perhaps as late as AD 700. In these later centuries it is in France and northern Europe that evidence of trade in tin is found.

The Norse sagas are silent on the subject of where tin was obtained, and rarely even mention the south-west of Britain. There is a Scandinavian tradition that an Abbey existed on Tresco, Scilly, in the late 10th century, and that Olaf Tryggvasson, later King of Norway, used the Isles of Scilly as a base from which he harried England in 989–93. The *Orkneyinga-saga* recounts the seizing of a vessel belonging to the monks of *Syllingar* (Scilly) *c.* AD 1150.[6] However, belief that Olaf Tryggvasson was converted to Christianity by a hermit on Scilly is dismissed by Jones[7] in favour of the version of the *Anglo-Saxon Chronicle* in which he was baptized shortly after an unsuccessful attack on London in 994.

The Norse were on good terms with the Cornish in the 9th century. The *Anglo-Saxon Chronicle* records that in 835 the Danes and the Britons of Cornwall united to combat Egbert of Wessex who, at Hingston Down near Callington, 'put to flight both Britons and Danes'. This appears to have been a Cornish attempt to retain their independence by enlisting the aid of foreigners with whom they had friendly trading contacts. After 926 when Athelstan banished the 'West Welsh' (Cornish) from Exeter and placed Cornwall firmly under the Saxon heel it is a different story, for in 997 the Danes were harrying in Cornwall as well as Wales and Devon.

In the early mediaeval period tin remained in constant demand for bronze and pewter. The Vikings were skilled metalworkers, as a glance at finds from such thriving settlements as Dublin and York proves. Viking iron spoons, sometimes with a bowl at each end, were at times coated with tin to give the appearance of silver. These are rare, but at the recent Coppergate excavations in York, three 10th century examples were found that may have been made in the local workshops.[8] Abroad, at the prosperous market and manufacturing centre of Hedeby at the neck of the Jutland peninsula were found moulds carved from antlers for the production of 'cheap' brooches of pewter.[9] There was no shortage of tin in Scandinavia to give inferior jewellery a silver glitter. In parts of Saxon England and Anglia poorly ornamented, flat disc-brooches, such as two from Long Wittenham in Berkshire, were similarly tinned.[10] Tin was also used in amounts of 1% or less in many Anglo-Saxon coins. Metcalf and Walker[11] refer to five dating from *c.* AD 740 containing over 5%, two in a Cambridge hoard containing 7% tin.

Some Saxon pottery was also decorated with tin foil. At Hamwih, the 8th century forerunner of Southampton, were found at least three sherds of 'Tating' ware. This distinctive pottery in the form of pitchers and small jars was usually decorated with tin foil applied in cross and lozenge shapes. Hodges[12] describes it as the finest pottery available in north-west Europe between *c.* 770 and 825. On the Continent it occurs as far east as Russia but was especially abundant at Scandinavian trading posts where imported wares were copied locally. Birka, on a small island west of Helgö in lake Mälaren, Sweden contained some of the best surviving examples, including one decorated with gold and tin foils (Fig. 136). Heavy mineral analysis of the pottery fabric has indicated the Eifel mountains of the middle Rhine as the major but not exclusive production centre. Dorested at the mouth of the Rhine produced over 50 vessels and must have been one of the towns from which Tating ware was traded. Dorested was captured by the Frisians *c.* AD 650, and it is the Frisians who are linked with British trade and with Cornwall, even though Tating ware itself in England is confined east of a line from Southampton to York.

There are historical references for believing that Cornish tin was well known on the Continent. The Abbé Raynal in 1782 mentioned 7th century fairs and markets established in France by King Dagobert to which Saxon traders came from England with lead and tin.[13] This trade increased. In the commercial treaty of 796, the first such treaty in English history, Charlemagne and Offa agreed to give protection to traders moving in both directions between the two countries.

In the following century Cornish tin reached much further afield. Gahlnbäck[14] in the Introduction to his study of the Moscow pewter industry described some of the early sources of tin, noting from customs regulations and the reports of travellers (without, unfortunately, expounding any details) that English tin was being imported into Russia through Germany and Austria as early as the 9th century.

In England the monk Aelfric (*c.* 955–1020) compiled a didactic *Colloquy*[15] between a master whose pupils were taught the elements of conversational Latin and men of various trades and callings. Aelfric makes the merchant say:

> I go aboard my ship with my goods, and I go over seas and sell my things, and buy precious things which are not produced in this country ... copper and tin, sulphur and glass and the like (*es et stagnum, sulfur et vitram et his similis*).

This wrongly implies that no tin was being mined in Cornwall. Whatever Aelfric may or may not have known about the Cornish industry, his words can best be interpreted as meaning that tin was transported to many parts of Britain by sea. For Aelfric in the Saxon south-east, Cornwall was a remote country, as foreign as Brittany, where the natives spoke their own language and had no love of their Saxon overlords.

Trade was considerable in Aelfric's time, with London the frequent resort of traders from every country between Norway and northern France. It was the focal point of two particularly important trade routes: one from Italy (and Italians were to dominate the later mediaeval Cornish tin trade) via the Rhine and Low Countries, and the other from Russia via the Baltic.[16] Cornish tin can hardly have been excluded from this trade. In the 7th and 8th centuries according to Stenton, and probably later according to archaeological evidence, trade with France and the Low Countries was largely in the hands of the Frisians. Much cloth sold abroad in this period under the name of 'Frisian' by Frisian merchants is believed to be of English origin.[17]

A distinctive form of pottery found on a scatter of sites in eastern England, but principally in Cornwall on the north coast from Mawgan Porth westwards, is the hand-made cooking pot distinguished by its two small bars of clay, each protected by a high lip from the heat of a fire over which it could be suspended. This 'bar-lug' or 'bar-lip' pottery appeared in Cornwall *c.* AD 800 and remained in use for several centuries. Continental examples have rounded bottoms, whereas the Cornish pots are flat-bottomed, and there are other minor differences. Nevertheless, it is generally assumed that the idea of 'bar-lug' pottery was introduced to Cornwall by Frisian merchants engaged in the tin trade, though this remains unproven.[18]

The rise of Christianity played its part in promoting a demand for Cornish tin. The tinner should take as his patron saint Paulinus, Bishop of Nola in Campania, who is credited with the introduction of church bells (hence campanile) early in the 5th century. Bells were

136   Tin-decorated Tating ware jug from grave 551, Birka, Sweden (published with permission of Statens Historiska Museer, Stockholm)

BAR-LIP POTTERY
c. AD 800-1100

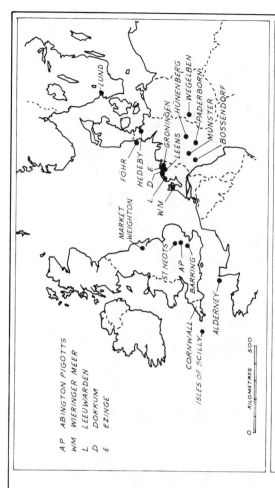

AP ABINGTON PIGOTTS
WM WIERINGER MEER
L LEEUWARDEN
D DOKKUM
E EZINGE

LUND
FÖHR
HEDEBY
L D E
W M
GRONINGEN
LEENS
HÜNENBERG
WEGELBEN
PADERBORN
MÜNSTER
BOSSENDORF
MARKET WEIGHTON
ST NEOTS
AP
BARKING
CORNWALL
ISLES OF SCILLY
ALDERNEY

KILOMETRES 500

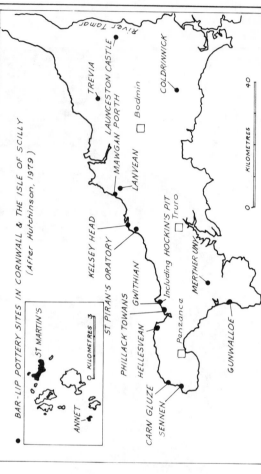

BAR-LIP POTTERY SITES IN CORNWALL & THE ISLE OF SCILLY
(After Hutchinson, 1979)

River Tamar
TREVIA
LAUNCESTON CASTLE
MAWGAN PORTH
Bodmin
COLDRINNICK
LANVEAN
KELSEY HEAD
ST PIRAN'S ORATORY
Including HOCKIN'S PIT
Truro
MERTHER UNY
PHILLACK TOWANS
GWITHIAN
HELLESVEAN
Penzance
CARN GLUZE
SENNEN
GUNWALLOE

KILOMETRES 40

ST MARTIN'S
ANNET
KILOMETRES 3

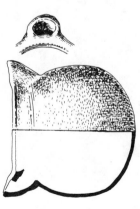

Reconstructed bar-lip pot from
Leeuwarden, Netherlands. Diam 24cms

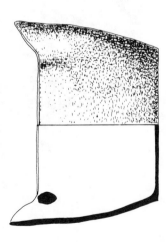

Reconstructed bar-lip pot from Hellesvean,
St Ives, Cornwall. Diam. 36 cms.
Only the Cornish examples have flat
bases.

R.D.P. delt. MCMLXXXIV

Map 29

used in Ireland in the time of Patrick (died 493), while an early reference to them on this side of the Irish Sea is in the time of Odoceus, Bishop of Llandâf following the death of St Teilo in 580, who removed the bells from his cathedral during a period of excommunication.[19]

There are increasing references to the use of bells before the time of Bede (673–735). St Patrick's bell, a hand-held *clogga* made from two plates of iron riveted together, was a type gradually replaced by bells of tin bronze. Price[20] writes of references to an unusual *tintinnabulum* around AD 800 which 'gave a joyful sound, as if struck by an angel', clearly not the dull sound of a *clogga* but a bell of cast bronze, a *campana* or *nola*. By AD 1000, not only the large religious houses held bells either hung or for hand-use; they were appearing in country churches everywhere.

All these fragments of information, in themselves sometimes insubstantial, taken together make an excellent case for the proposition that the Cornish tin industry in the 'Dark Ages' was healthy, thriving, and firmly set on the road that led in subsequent centuries to Cornwall's near monopoly of the world trade in tin.

NOTES AND REFERENCES

1   PENNINGTON, 1973, 9, 12, 72–3
2   LEWIS, 1908, 34
3   HENCKEN, 1932, 201
4   DAWES & BAYNES, 1948
5   THOMAS, 1981a; THOMAS, 1982
6   THOMAS, 1985, 212
7   JONES, 1975 (ed.), 132
8   HALL, 1984, 60 with Fig. 62
9   GRAHAM-CAMPBELL, 1979, 106; GRAHAM-CAMPBELL & KIDD, 1980, 101, 145
10 SMITH, 1923, 37
11 METCALF & WALKER, 1976
12 HODGES, 1981; HODGES, 1982
13 HATCHER, 1973, 16; he believes that Raynal had access to a now lost charter of Dagobert when he wrote his *Histoire philosophique et politique* in 1782 (**5**, 177)
14 GAHLNBÄCK, 1928, 4
15 GARMONSWAY, 1939
16 STENTON, 1947 (ed.), 533
17 POSTAN, 1984 (ed.), 209
18 DUNNING *et al.*, 1959, 48–9; BRUCE-MITFORD, 1956; HUTCHINSON, 1979
19 BRISCOE, 1899
20 PRICE, 1983, 78–94 for the introduction of bells into churches

Leaden seal, 55 mm dia., picked up in a field adjoining Lee Down near Bath, Somerset, in 1842. The inscription reads: S′ COMVИITΛTISSTΛИGИΛTORVMCORИVBIE (the seal of the community of tinners of Cornwall). The same design is on both faces. It is the form of seal attached to the Charter to the tinners of Cornwall signed by Edward I on 10 April 1305, 'having the print in it of one working with a spade in a tin work, and another with a pick'

Appendix

ST JOHN THE ALMSGIVER

Translated from the Greek by Dawes and Baynes, *Three Byzantine Saints*, 216–18, Mowbrays, London and Oxford, 1977 (first published 1948)

There was a foreign captain who had fallen upon evil days, he came to the blessed man and with many tears besought him to show mercy to him as he did to all others. So John directed that he should be given five pounds of gold. With these the captain went and bought a cargo, and no sooner had he gone on board than straightway, as it chanced, he suffered shipwreck outside the Pharos [famous lighthouse at entrance to Alexandria harbour], but he did not lose his ship. Then trusting to John's good will he again applied to him saying, 'Have mercy upon me as God had mercy upon the world'. The Patriarch said to him, 'Believe me, brother, if you had not mixed your own remaining monies with the money of the Church, you would not have been shipwrecked. For you had them from an evil source and thus the money coming from a good source was lost with it'. However he gave fresh instructions this time that ten pounds of gold were to be given him and he was not to mix other money with it. Again the captain bought a cargo and when he had sailed for one day a violent wind arose and he was hurled upon the land and lost everything, including the ship, and he and the crew barely escaped with their lives. After this from despair and destitution the captain decided to hang himself. But God, Who ever takes forethought for the salvation of men, revealed this to the most blessed Patriarch, who, hearing what had happened to the captain, sent him word to come to him without delay. The latter came before him with his head sprinkled with dust and his tunic torn and in disorder. When the Patriarch saw him in this guise he found fault with him and said, 'May the Lord be propitious unto you! Blessed be God! I believe His word that from to-day on you will not be wrecked again as long as you live. This disaster happened to you because you had acquired the ship itself, too, by unjust means'.

He immediately ordered that one of the ships belonging to the Holy Church of which he was head should be handed over to the captain, a swift sailer laden with twenty thousand bushels of corn. The captain, when he had received the ship, sailed away from Alexandria, and on his return he made a solemn statement to the following effect: 'We sailed for twenty days and nights, and owing to a violent wind we were unable to tell in what direction we were going either by the stars or by the coast. But the only thing we knew was that the steersman saw the Patriarch by his side holding the tiller and saying to him: "Fear not! You are sailing quite right." Then after the twentieth day we caught sight of the islands of Britain, and when we had landed we found a great famine raging there. Accordingly when we told the chief man of the town that we were laden with corn, he said, "God has brought you at the right moment. Choose as you wish, either one 'nomisma' for each bushel or a return freight of tin". And we chose half of each.' Then the story goes on to tell of a matter which to those who are ignorant of God's free gifts is either hard to believe or quite incredible, but to those who have experienced His marvellous works it is both credible and acceptable. 'Then we set sail again,' said the captain, 'and joyfully made once more for Alexandria, putting in on our way at Pentapolis [Cyrenaica].' The captain then took out some of the tin to sell – for he had an old business-friend there who asked for some – and he gave him a bag of about fifty pounds. The latter, wishing to sample it to see if it was of good quality, poured some into a brazier and found that it was silver of the finest quality. He thought that the captain was tempting him, so carried the bag to him and said, 'May God forgive you! Have you ever found me deceiving you that you tempt me by giving me silver instead of tin?' The captain was dumbfounded by his words and replied: 'Believe me, I thought it was tin! But if He who turned the water into wine has turned my tin into silver in answer to the Patriarch's prayers, that is nothing strange. However, that you may be satisfied, come down to the ship with me and look at the rest of the mass from which I gave you some.' So they went and discovered that the tin had been turned into the finest silver.

Mark ye, lovers of Christ, this miracle is not strange. For He Who multiplied the five loaves and at another time converted the waters of the Nile into blood, transformed a rod into a serpent, and changed fire into dew, easily accomplished this miracle, too, in order to enrich His servant and show mercy to the captain.

Bibliography

ABBREVIATIONS

Ant.J.   Antiquaries Journal
CA   Cornish Archaeology
IJNA   International Journal of Nautical Archaeology
JRIC/RRIC   Journal/Report of the Royal Institution of Cornwall
PPS   Proceedings of the Prehistoric Society
PZNHAS   Penzance Natural History and Antiquarian Society
QJGS   Quarterly Journal of the Geological Society of London
RCPS   Royal Cornwall Polytechnic Society
TRGSC   Transactions of the Royal Geological Society of Cornwall

F. D. ADAMS, *The Birth and Development of the Geological Sciences*, 1954 ed., Dover Publ., New York

W. F. ALBRIGHT, 'New light on the early history of Phoenician colonisation', *Bull. American Sch. Oriental Research*, 1941 (Oct.), **83**, 14–22

C. ALDRED, *The Egyptians*, 1971, Thames & Hudson

C. ALDRED, *Akenhaten: Pharaoh of Egypt*, 1972, London

E. M. EL ALFY, *The Mineral Resources of Egypt*, 1946, Cairo

J. C. ALLAN, *Considerations on the Antiquity of Mining in the Iberian Peninsula*, 1970, Royal Anthropological Institute of Great Britain and Ireland, Occasional Paper No. 27

F. R. ALLCHIN & N. HAMMOND (eds.), *The Archaeology of Afghanistan*, 1979, Academic Press

D. F. ALLEN, 'The Paul (Penzance) hoard of imitation Massiliote Drachms', *Numismatic Chronicle*, 7th ser. No. 1, 1961, 91–106

D. F. ALLEN, in Frere (ed.), 'The origins of coinage in Britain: a re-appraisal', *Problems of the Iron Age in Southern Britain*, 1961a, Council for British Archaeology

I. M. ALLEN, D. BRITON & H. H. COGHLAN, *Metallurgical Reports on British and Irish Bronze Age Implements and Weapons in the Pitt Rivers Museum*, 1970, OUP

J. ALLEN, *History of the Borough of Liskeard and its Vicinity*, 1856, London

C. K. C. ANDREW, 'A tin bowl found on Fowey Moor', *JRIC*, 1936, **24**, 331–4

ANON., 'Tin mines in France', *Mining Journal*, 1835 (10 Oct.), 53

ANON., 'Scientific summary', *JRIC*, 1864, **1**, 55

ANON., 'Trace elements in ores', *Royal School of Mines Report (1950–1953)*, 1954, London, 26

ANON., 'Possibilities for tin in Russia's industrial expansion', *Tin*, 1956 (Aug.), 163–6

ANON., 'Tin occurrences in the USSR', *Mining Journal*, 1958, 553

ANON., 'Sir George Turner's Some Memorialls towards a natural history of the Sylly Islands', *Scillonian Magazine*, No. 159, 1964, 154–7

ANON., *Tin*, 1974, Mineral Resources Consultative Committee, HMSO

ANON., 'Excavations of the mining and foundry sites of the Spring and Autumn Period and the Warring States Period at T'ung-lü-shan in Hupei Province', *Wen Wu* (Cultural Relics) No. 2, 1975, 1–18 (in Chinese)

ANON., 'Ingots from wrecked ship may help to solve ancient mystery', *Inst. Archaeo-Metallurgical Studies Newsletter*, No. 1, 1980, 1–2

A. M. APSIMON & E. GREENFIELD, 'The excavation of Bronze Age and Iron Age settlements at Trevisker, St Eval, Cornwall', *PPS*, 1972, **38**, 302–81

L. ARQUÉ, 'Les industries de l'étain en Franconie', *La Science Sociale suivante le méthode d'observation*, Ser. 6, 1906, Paris

A. ARRIBAS, *The Iberians*, 1964, Thames & Hudson

P. ASHBEE, 'The gold cups from Rillaton, Fritzdorf, and Eschenz', *CA*, 1977, **16**, 157–9

P. ASHBEE, 'The silver cup from Saint-Adrien, Côtes-du-Nord, Brittany', *CA*, 1979, **18**, 57–9

A. ASPINALL, S. E. WARREN, J. G. CRUMMETT & R. G. NEWTON, 'Neutron activation analysis of faience beads', *Archaeometry*, 1972, **14**, 27–40

W. BADCOCK, *Historical Sketches of St Ives and District*, 1896, St Ives

R. W. BAGLEY, 'The spread of Shang China', in *Treasures from the Bronze Age of China*, 1980, Metropolitan Museum of Art, New York

T. M. BAGNOLD, 'Tin mines of Banca &c', *Journal of Science*, 1820, 412–14

L. BARFIELD, *Northern Italy before Rome*, 1971, Thames & Hudson

T. F. BARHAM (1825), 'Some arguments in support of

the opinion that the Iktis of Diodorus Siculus is St Michael's Mount', *TRGSC*, 1828, **3**, 86–112

H. BARI & P. FLUCK, 'Incursions minéralogiques dans les Monts Métallifères Saxon', *Minéraux et Fossiles*, 1981, **74, 75**

S. BARING-GOULD, *A Book of Cornwall*, 1912 ed., Methuen

G. BARKER, 'Cultural and economic growth in the Bronze Age (Central Italy)', *PPS*, 1972, **38,** 170–208

G. BARKER, *Landscapes and Society: Prehistoric Central Italy*, 1981, Academic Press

H. BARKER & C. J. MACKEY, Item DM-29 in *American Journal of Science Radiocarbon Supplement*, 1959, 85–6

N. BARNARD, *Radiocarbon Dates and their Significance in the Chinese Archaeological Scene*, 1980, Australian National University, Canberra

N. BARNARD, 'Some observations on metal-winning and the societal requirements of early metal production in China', 1981, pre-print for the Moesgard Symposium on 'The Origin of Agriculture and Technology: West or East Asia?', held at the University of Aarhus, Denmark, 21–25 Nov. 1978

N. BARNARD, 'Further evidence to support the hypothesis of indigenous origins of metallurgy in ancient China', in D. N. Keightley (ed.), 1983, *The Origins of Chinese Civilization*, University of California Press

N. BARNARD & K. CHEUNG, 'Studies in Chinese archaeology 1980–1982: Reports on visits to mainland China, Taiwan, and the USA; Participation in conferences in these countries, and some notes and impressions', *Wen-hsüeh-she*, 1983, 359–64

N. BARNARD & T. SATŌ, *Metallurgical Remains of Ancient China*, 1975, Nichiōsha, Tokyo (English and Japanese text in Introduction)

G. BARROW, 'The high-level platforms of Bodmin Moor and their relation to the deposits of stream tin and Wolfram', *QJGS*, 1908, **64**, 384–400

G. BARROW & J. S. TEALL, *The Geology of the Isles of Scilly*, 1906, HMSO

D. B. BARTON, *Essays in Cornish Mining History*, 1970, **2**, Barton, Truro

P. M. BARTZ, *South Korea: a descriptive geography*, 1972, OUP

G. F. BASS, 'Cape Gelidonya: a bronze age shipwreck', *Trans. American Phil. Soc.*, 1967

G. F. BASS, *Archaeology under Water*, 1984, Penguin, London

G. F. BASS, D. A. FREY & C. PULAK, 'A Late Bronze Age shipwreck at Kas, Turkey', *IJNA*, 1970, **13**, 271–9

F. BAUDOT, 'Les mines d'étain de La Villeder', *Bull. Soc. Indust. Mines*, 1887, **1**, (3), 151–87

L. BAUMANN, 'Tin deposits of the Erzgebirge', *Trans. Inst. Mining & Metallurgy*, 1970 (10 May)

C. BECK, E. WILBUR, S. MERET, D. KOSSOVE & K. KERMANI, 'The infra-red spectra of amber and the identification of Baltic amber', *Archaeometry*, 1965, **8,** 96–109

S. BENTON, 'No tin from Kirrha in Phokis', *Antiquity*, 1964, **38**, 138

A. BERGEAT, 'Beitrage zur Kenntniss Erzlagerstätten', *Neues Jahrbuch Min.*, 1901, **1**, 135–56 (for an English summary *see Trans. Inst. Mag. Eng.*, 1901–2, **22**, 705–6)

J. F. BERGER, 'A mineralogical account of the Isle of Man', *Trans. Geol. Soc. London*, 1814, **2**, 58

L. BIEK, 'The archaeological iron and tin cycles', MS prepared for Proceedings of the Bonn Archaeometry Symposium, 1978

F. W. VON BISSING, 'On the occurrence of tin in Asia Minor and in the neighbourhood of Egypt', *J. Hellenic Studies*, 1932, **52**, 119

J. BOARDMAN, *The Greeks Overseas*, 1980 ed., Thames & Hudson

H. S. BOASE, 'Contributions towards a knowledge of the geology of Cornwall', *TRGSC*, 1832, **4**, 166–474

G. BOON, 'Aperçu sur la production des métaux non-fereux dans la Bretagne Romaine', *Apulum*, 1971, **9**, 454–503

G. BOON, 'A Greco-Roman anchor stock from North Wales', *Ant. J.*, 1977, **57**, (1), 10–30

W. BORLASE, *Observations on the Ancient and Present State of the Islands of Scilly*, 1756, Oxford

W. BORLASE, *The Natural History of Cornwall*, 1758, Oxford

W. BORLASE, *Supplement* to *The Natural History of Cornwall*, in *JRIC*, 1864

W. BORLASE, *Antiquities of Cornwall*, 1769 (2nd ed.), London

W. C. BORLASE, *Ancient Cornwall, a collection of drawings &c, Original and copied, illustrative of the Antiquities of that County*, collected and arranged by William C. Borlase FSA, Castle Horneck, March 1871, MS at County Museum, Truro. Most of the finds from tin streams are drawings copied 'chiefly by Revd Canon Rogers from a scrapbook, collected by J. J. Rogers Esqr in 1870', 1871

W. C. BORLASE, *Naenia Cornubiae: a descriptive essay, illustrative of the sepulchres and funereal customs of the early inhabitants of the County of Cornwall*, 1872, London & Truro

W. C. BORLASE, 'Archaeological discoveries in the parishes of St Just-in-Penwith and Sennen' *JRIC*, 1879, 190–212

W. C. BORLASE, MS *Catalogue*, in Truro Museum, of the objects in his museum at Lariggan, Penzance, 1882

W. C. BORLASE, *Tin-Mining in Spain Past and Present*, 1897, London

A. BOUSCARAS, 'Découvert d'une épave du premier âge du fer à Agade', *Rivista di Studi Liguri* 1964, 288–94

A. BOUSCARAS, 'L'épave des bronzes de Rochelongues', *Archéologie Sous-marine*, 1971, **39**, 68–73

J. BOUSQUET, 'Une monnaie d'or de Cyrène sur la côte nord de l'Amorique', *Annales de Bretagne*, 1961, **68**, 25–39

J. BOUSQUET, 'Deux monnais grèques: Massalia sestos', *Annales de Bretagne*, 1968, **75**, 277–9

J. BOUZEK, 'Syrian and Anatolian Bronze Age figurines in Europe', *PPS*, 1972, **38,** 156–64

R. L. BOWLEY, *The Fortunate Isles. A History of the Isles of Scilly*, 1949 (3rd ed.), Reading

V. BOYE, *Fund af Egekister fra Bronzealderen i Danmark*, 1896, Copenhagen

K. BRANIGAN, 'The Surbo bronzes, some observations', *PPS*, 1972, **38,** 276–85

K. BRANIGAN, *Aegean Metalwork of the Early and Middle Bronze Age*, 1974, Oxford

K. BRANIGAN, 'A Cypriot hook-tang weapon from Devon', *Proc. Devon Archaeol. Soc.*, 1983, **41,** 125–8

A. E. BRAY, *The Borders of the Tamar and Tavy*, 1879 (2nd ed.), London

B. BRENTJES, 'Ein elamitischer Streufund aus Soch Fergana (Usbekistan)', *Iran*, 1971, **9,** 155

J. BRIARD, *Les Dépôts Breton et l'Age du Bronze Atlantique*, 1965, Rennes

J. BRIARD, *The Bronze Age in Barbarian Europe*, 1979, Book Club Associates

C. S. BRIGGS, 'Double axe doubts', *Antiquity*, 1973, **47,** 318–20

C. S. BRIGGS, 'Double axe postscript', *Antiquity*, 1975, **49,** 63–4

C. S. BRIGGS, 'Notes on the distribution of some raw materials in later prehistoric Britain', in Burgess and Miket (eds.), *Settlement and Economy in the Third and Second Millennia BC*, British Archaeological Reports No. 33, 1976, 267–82

C. S. BRIGGS, 'Copper mining at Mount Gabriel, Co. Cork', *PPS*, 1983, **49,** 317–33

J. P. BRISCOE, 'Stories about bells', in William Andrew (ed.), *Ecclesiastical Curiosities*, 1899, London, 133–52

J. BRITTON & E. W. BRAYLEY, *A Topographical and Historical Description of the County of Cornwall*, 1801, London

C. G. BROWN & T. E. HUGO, 'Prehistoric and Romano-British finds from Mount Batten, Devon, 1979–1983', *Proc. Devon Archaeol. Soc.*, 1983, **41,** 69–74

P. D. C. BROWN, 'A Roman pewter mould from St Just, Penwith, Cornwall', *CA*, 1970, No. 9, 107–110

R. L. S. BRUCE-MITFORD, 'A Dark Age settlement at Mawgan Porth, Cornwall', in Bruce-Mitford (ed.), *Recent Archaeological Excavations in Britain*, 1956, Routledge & Kegan Paul

J. BRYANT, 'On the remains of an ancient crazing mill in the parish of Constantine', *JRIC*, 1882, **7,** 213–14

J. BULLER, *A Statistical Account of the Parish of Saint Just in Penwith*, 1842, Penzance

C. BURGESS, 'The Bronze Age', in Colin Renfrew (ed.), *British Prehistory: a new outline*, 1974, Duckworth, 165–233

C. BURGESS, 'The Bronze Age in Wales', in J. A. Taylor (ed.), *Culture and Environment in Prehistoric Wales*, British Archaeological Reports, British Series No. 76, 1980, 234–86

R. BURNARD, 'On the track of the "old men"', *Trans. Plymouth Inst.*, 1888, **10,** 95–112, 1889, 223–42

C. BURNEY, 'Tepe Yahya', *Antiquity*, 1975, **49,** 191–6

C. BURNEY, *From Village to Empire: an introduction to Near Eastern archaeology*, 1977, Phaidon

R. BURT, *John Taylor 1779–1863, mining entrepeneur and engineer*, 1973, Moorland Publ. Co.

J. B. BURY & R. MEIGGS, *A History of Greece*, 1978 (4th ed.), Macmillan

S. W. BUSHNELL, *Chinese Art*, 1921 ed., **1,** V & A Museum

E. R. CALEY, 'Results of an examination of fragments of corroded metal from the 1962 excavation at Snake Cave, Afghanistan', *Trans. American Phil. Soc.*, (New Ser.) 1972, **62,** 43–84

L. CAMBI, 'Problemi della Metallurgia Etrusca', *Studi Etruschi*, 1959, **27,** 415–32

W. CAMDEN, *Britannia*, 1636 ed., London

S. CAMM & K. F. G. HOSKING, 1984, 'The Stanniferous Placers of Cornwall, Southwest England', *Geol. Soc. Malaysia*, Bulletin 17. 323–56

J. CAMPBELL D. ELKINGTON, P. FOWLER & L. GRINSELL, *The Mendip Hills in Prehistoric and Roman Times*, 1970, Bristol

F. L. CANCRINUS, *Berchreibung der Bergwerke*, 1767, Frankfurt

J. CARNE, 'On the mineral production and the geology of the parish of St Just', *TRGSC*, 1821, **2,** 290–358

J. CARNE, 'An account of the discovery of some varieties of tin-ore in a vein, which have been considered peculiar to streams; with remarks on diluvial tin in general', *TRGSC*, 1830, **4,** 95–112

R. CARPENTER, *Beyond the Pillars of Heracles*, 1966, Delacorte Press

G. W. CARRIVEAN & G. HARBOTTLE, 'Ban Chiang pottery: thermoluminescence dating problems', *Antiquity*, 1983, **57,** 56–8

M. CARY, 'The Greeks and the ancient trade with the Atlantic', *J. Hellenic Studies*, 1924, **44,** 166–79

G. CASTALDO & G. STAMPANONI, *Memoria Illustrativa della Carta Mineraria d'Italia*, 1975, Carta Geologica d'Italia, **14**

C. CHAMPAUD, 'Notice sur trois types d'outiles Gallo-Romains retrouvés dans l'exploration minière d'Abbaretz (Loire Inférieure)', *Annales de Bretagne*, 1955, **62,** 293–9

C. CHAMPAUD, 'L'exploitation ancienne de cassitérite d'Abbaretz-Nozay (Loire Inférieure)', *Annales de Bretagne*, 1957, **64,** 46–96

C. S. CHARD, 'First radiocarbon dates from the USSR', *Arctic Anthropology*, 1962, **1,** 84–6

C. S. CHARD, *Northeast Asia in Prehistory*, 1974, University of Wisconsin Press

W. T. CHASE, 'Early Chinese black mirrors and pattern-etched weapons', *Ars Orientalis*, 1979, **11,** 215–58

T. CHÊNG, 'New light on Shang China', *Antiquity*, 1975, **49,** 25–32

E. N. ČERNYCH [CHERNYKH], 'Aibunar, a Balkan copper mine of the 4th millennium BC', *PPS*, 1978, **44,** 203–17

E. N. CHERNYKH, *Mining and Metallurgy in Ancient Bulgaria*, 1978, Sofia (in Russian). For an excellent English summary see T. A. P. Greeves' review in *PPS*, 1982, **48,** 538–42, reprinted from *J. Historical Metallurgy Soc.*, 1981, 45–9

E. N. CHERNYKH, 'The Constanta hoard and the problems of the Balkan-Caucasian links of the late Bronze Age', *Soviet Archaeology*, 1981, (1), 19–26 (in Russian with English summary)

V. G. CHILDE, *The Dawn of European Civilization*, 1939, London

V. G. CHILDE, 'Bronze dagger of Mycenaean type from Pelynt, Cornwall', *PPS*, 1951, **17,** 95

V. G. CHILDE, 'The socketed celt in Upper Eurasia', *University London Inst. Archaeology*, 10th Ann. Rep., 1954, 11–25

V. G. CHILDE, *The Prehistory of European Society*, 1958, Pelican, London

R. P. CHOPE, *Early Tours in Devon and Cornwall*, 1967 ed., David & Charles

D. M. CHURCHILL, 'The kitchen midden site at Westward Ho!, Devon, England: ecology, age, and relation to changes in land and sea level', *PPS*, 1965, **31,** 74–84

J. G. D. CLARK, 'Whales as an economic factor in prehistoric Europe', *Antiquity*, 1947, **21,** 84–104

J. G. D. CLARK, *Prehistoric Europe: the economic basis*, 1965 ed., Methuen

J. G. D. CLARK, *World Prehistory in New Perspective*, 1977, CUP

D. L. CLARKE, *Beaker Pottery of Great Britain and Ireland*, 2 vols., 1970, CUP

R. H. CLARKE, 'Quaternary sediments of south-east Devon,' *QJGS*, 1979, **125,** 277–318

G. COFFEY, *The Bronze Age of Ireland*, 1913, Dublin

H. H. COGHLAN, 'A note on Irish copper ores and metals', in H. H. Coghlan *et al.*, 1963

H. H. COGHLAN, *Notes on the Prehistoric Metallurgy of Copper and Bronze in the Old World*, 1975 (2nd ed.), OUP

H. H. COGHLAN, J. R. BUTLER & G. PARKER, *Ores and Metals*, 1963, Royal Anthropological Inst. Occ. Papers No. 17

J. W. COLENSO (1829), 'A description of the Happy-Union tin streamwork at Pentewan', *TRGSC*, 1832, **4,** 29–39

J. M. COLES, 'The Bronze Age in North Western Europe', in Wendorf & Close (eds.), *Advances in World Archaeology*, 1982, **1,** 265–321

J. M. COLES & A. F. HARDING, *The Bronze Age in Europe*, 1979, Methuen

R. G. COLLINGWOOD & I. RICHMOND, *The Archaeology of Roman Britain*, 1969 ed., Methuen

A. L. COLLINS, 'Fire-setting, the art of mining by fire', *Trans. Federated Inst. Mining Engineers*, 1893, 82–92

J. H. COLLINS, *On the Geology of Cornwall*, 1887, Geologists' Assoc. Excursion to Cornwall

J. H. COLLINS, 'The tin alluvials of the Goss and Tregoss Moors', *RCPS*, (New Ser.), 1909, **1,** 121–6

J. H. COLLINS, *Observations on the West of England Mining Region*, 1912, published as *TRGSC*, **14**

M. B. COLLINS, 'Tin production in the province of Yunnan, China', *Trans. Inst. Min. & Metall.*, 1909–10, **19,** 187–201, 208–11

D. COLLS, C. DOMERQUE, F. LAUBENHEIMER & B. LIOU, 'Les lingots d'étain de l'épave Port Vendre II', *Gallia*, 1975, **33,** (1)

D. COLLS, R. ETIENNE, R. LEQUÉMENT, B. LIOU & F. MAYET, 'L'épave Port-Vendre II et la commerce de la Bétique a l'époque de Claude', *Archaeonautica*, 1977, **1**

J. X. W. P. CORCORAN, 'Tankard and tankard handles in the British Early Iron Age', *PPS*, 1952, **18,** 85–102

A. COTTERELL, *The First Emperor of China*, 1981, Macmillan

J. COUCH, 'An account of some barrows in the parish of Pelynt, and the remains found on opening them', *RRIC*, 1845, Appendix VI, 33–7

T. COX, *Magna Britannia et Hibernia Antiqua et Nova*, 1720–33, London, Vol. 1, *Cornwal* [sic], 306–64

P. T. CRADDOCK, 'The copper alloys of the mediaeval Islamic world', *World Archaeology*, 1979, **11,** 68–79

P. T. CRADDOCK (ed.), *Scientific Studies in Early Mining and Extractive Metallurgy*, 1980, British Museum Occ. Paper No. 20

P. T. CRADDOCK, 'The composition of the non-ferrous metals from Tejada', in Rothenberg & Blanco-Freijeiro, *Studies in Ancient Mining and Metallurgy in South-West Spain*, 1981, London

H. CRAWFORD, 'The mechanics of the obsidian trade: a suggestion', *Antiquity*, 1978, **52,** 129–32

M. CRISTOFANI, *The Etruscans: a new investigation*, 1979 ed., Orbis, London

W. CULLICAN, *The First Merchant Venturers: The Ancient Levant in History and Commerce*, 1966, Thames & Hudson

B. CUNLIFFE, *The Roman Baths: a guide to the baths and Roman Museum*, 1978, Bath Archaeological Trust

B. CUNLIFFE, 'Ictis: is it here?' *Oxford J. Archaeol.*, 1983, **2,** 123–6

R. J. CUNNACK, 'Vestiges of ancient tin workings in the Loe Pool Valley', *Rep. Miners' Assoc. Cornwall & Devon*, 1867, **1,** 53–5

E. C. CURWEN, 'A bronze cauldron from Sompting, Sussex', *Ant. J.*, 1948, **28,** 157–63

J. H. DANA, *The System of Mineralogy*, 7th ed., 1951, **2,** John Wiley & Sons

G. DANIEL, *A Hundred Years of Archaeology*, 1950, Duckworth

G. DANIEL, *The Idea of Prehistory*, 1964, Penguin, London

A. DAUBRÉE, 'Kaolin de La Lizolle et d'Échassières, Allier', *Comptes-Rendus Acad. Sci.*, 1869, 1135–9

E. H. DAVIDSON, *Handbook of Cornish Geology*, 1926, Royal Geological Soc., Cornwall

E. H. DAVIES, revised 6th ed. of D. C. Davies' *A Treatise on Metalliferous Minerals and Mining*, 1901, London

G. M. DAVIES, *Tin Ores*, 1919, Imperial Institute Monograph

N. DE GARIS DAVIES, *The Tomb of Rekh-mi-Rē' at Thebes*, 1943, Metropolitan Museum of Art, New York

O. DAVIES, 'Two north Greek mining towns', *J. Helladic Studies*, 1929, **49**, 89–99

O. DAVIES, 'Ancient tin sources of western Europe', *Belfast Nat. Hist. & Phil. Soc. Proc.*, 1933, 44–50

O. DAVIES, *Roman Mines in Europe*, 1935, Oxford

L. DAVY, 'Sur l'ancienneté probable de l'exploitation de l'étain en Bretagne', *Comptes-Rendus Acad. Sci.*, 1897, 337–9

E. DAWES & N. H. BAYNES, *Three Byzantine Saints*, 1948, Mowbrays, London and Oxford

G. DE BEER, 'Iktin', *Geographical Journal*, 1960 (Jun.), **126**, (Reprinted in De Beer, 1962)

G. DE BEER, *Reflections of a Darwinian*, 1962, Nelson

J. DÉCHELETTE, *Manuel d'Archéologie Préhistorique*, 1924, 2 vols., Paris

H. T. DE LA BECHE, *Report on the Geology of Cornwall, Devon, and West Somerset*, 1839, London

M. VON DEWALL, 'The Tien culture of south-west China', *Antiquity*, 1967, **41**, 8–21

T. M. DICKENSON, 'Fowler's type G penannular brooches reconsidered', *Medieval Archaeology*, 1982, **26**, 41–68

H. G. DINES, *The Metalliferous Mining Region of South-West England*, 1956, HMSO

C. C. DOBSON, *Did Our Lord Visit Britain as they say in Cornwall and Somerset?*, 1936, Convent Publ. Co.

B. M. DONALD, 'Burchard Kranich (*c.* 1515–1578), miner and queen's physician, Cornish mining stamps, antimony and Frobisher's gold', *Annals of Science*, 1950, **6**, No. 3, 308–22

D. DUDLEY, 'Excavations on Nor'nour in the Isles of Scilly', *Arch. J.*, 1967, **124**, 1–64

F. W. DUNNING, W. MYKURA & D. SLATER (eds.), *Mineral Deposits of Europe*, 1982, Vol. 2 'South-east Europe', Mineral Soc. & Inst. Mining & Metallurgy

G. C. DUNNING, J. G. HURST, J. N. L. MYRES & F. TISCHLER, 'Anglo-Saxon pottery: a symposium', *Medieval Archaeology*, 1959, **3**, 1–78

R. EDMONDS, 'On the fragments of a bronze furnace (supposed to be Phoenician) discovered near St Michael's Mount – the *Iktin* of Diodorus Siculus', *PZNHAS*, 1849, **1**, 347–51

R. EDMONDS, 'On the name of Britain and the Phoenicians', *Trans. Devon. Assoc.*, 1871, **4**, 418–22

W. B. EMERY, *Archaic Egypt*, 1961, Penguin, London

A. EVANS, 'The prehistoric tombs at Knossos', *Archaeologia*, 1906, **59**, 391–562

A. M. EVANS, *An Introduction to Ore Geology*, 1980, Geoscience Texts, **2**, Blackwell Scientific Publications

J. EVANS, *Ancient Bronze Implements*, 1881, Longmans & Co.

B. M. FAGAN, 'Two hundred and four years of African archaeology', in J. D. Evans, B. Cunliffe & C. Renfrew (eds.), *Antiquity and Man: Essays in Honour of Glyn Daniel*, 1981, Thames & Hudson, 42–51

S. FAWNS, *Tin Deposits of the World*, 1905, Mining Journal

W. G. FEARNSIDE & O. M. B. BULMAN, *Geology in the Service of Man*, 1961 ed., Penguin, London

T. FIDDICK, *Dowsing, with an account of some original experiments*, 1913, Camborne

H. FIELD & E. PROSTOV, 'Tin deposits in the Caucasus', *Antiquity*, 1938, **12**, 314–45

H. P. R. FINBERG, 'The stannary of Tavistock', *Trans. Devon. Assoc.*, 1949, **81**, 155–84

I. M. FINLEY, *Aspects of Antiquity*, 1977 ed., Penguin, London

L. FLEURIOT & P. R. GIOT, 'Early Brittany', *Antiquity*, 1977, **51**, 106–16

W. H. FLOWER, 'On the bones of a whale found at Pentuan', *TRGSC*, 1872, **9**, 1878, 114–21

R. J. FORBES, *Studies in Ancient Technology*, 1964, **8**, E. J. Brill, Leiden

W. & B. FORMAN & J. POULÍK, *Prehistoric Art*, (n.d., *c.* 1950), Spring Books, London (the book concerns only Czechoslovakia)

E. J. FORSDYKE, *Catalogue of Vases in the British Museum*, 1925, vol. 1, London

E. FOWLER, 'Celtic metalwork of the fifth and sixth centuries AD: a re-appraisal', *Arch. J.*, 1963, **120**, 98–306

P. FOWLER & A. C. THOMAS, 'Lyonesse revisited: the early walls of Scilly', *Antiquity*, 1979, **53**, 175–89

A. FOX, 'Two Greek coins from Holne, South Devon', *Ant. J.*, 1950, **30**, 152–5

A. FOX, 'Excavations on Dean Moor in the Avon Valley, 1954–1956: the Late Bronze Age settlement', *Trans. Devon Assoc.*, 1957, **89**, 18–77

A. FOX, *South-West England, 3500 BC–AD 600*, 1973 ed., David & Charles

A. FOX & W. RAVENHILL, 'The Roman fort at Nanstallon, Cornwall', *Britannia*, 1972, **3**, 56–111

C. FOX, *The Personality of Britain: its influence on inhabitant and invader in prehistoric and early historic times*, 1952, National Museum of Wales, Cardiff

D. J. FOX, 'Tin mining in Spain and Portugal', in W. Fox (ed.), *A second Technical Conference on Tin*, Bancock, 1969, **1**, 225–65

S. M. FRANKENSTEIN, *The Impact of Phoenician and Greek expansion on the Early Iron Age Societies of Southern Iberia and Southwestern Germany*, 1979, unpublished PhD thesis, University of London, Institute of Archaeology

A. D. FRANKLIN, J. S. OLIN & T. A. WERTIME (eds.), *The Search for Ancient Tin*, 1978, (A seminar organized by Theodore A. Wertime and held at the Smithsonian Institution and the National Bureau of Standards, Washington DC, 14–15 March 1977), Smithsonian Institution, Washington

R. FRASER, *General View of the County of Cornwall*, 1794, Board of Agriculture

S. S. FRERE, *The Problems of the Iron Age in Southern Britain*, 1961, London

J. GAHLNBÄCK, *Russisches Zinn*, 1928, Leipzig (in spite of its title the book says nothing about Russian tin; *zinn* here signifies pewter and deals with the history of the Moscow pewter industry from the 17th century)

E. GALILI & N. SHMUELI, 'Israel' in 'Short Notes', *IJNA*, 1983, **12**, (2), 178

M. J. GALLAGHER, U. M. MICHIE, R. T. SMITH & L. HAYNES, 'New evidence of uranium and other mineralization in Scotland, *Trans. Inst. Min. & Metall.*, 1971, **80**, (B), 150–73

D. M. GARDINER, *A Calendar of Early Chancery Proceedings relating to West Country Shipping, 1388–1493*, 1976, Devon & Cornwall Record Society

J. GARLAND, 'Tin deposits in Galicia', *RCPS*, 1888, 56th Ann. Rep., 54–7

G. N. GARMONSWAY (ed.), *Aelfric's Colloquy*, 1939, Methuen Old English Library

M. S. GARSON, 'Younger granites of Egypt and associated mineralization', *Trans. Inst. Min. & Metall.*, 1977, **86**, (B), 161

S. GERRARD, 'Tin workings on Bodmin Moor', *The Trevithick Soc. Newsletter*, 1983, No. 43, 6–7

A. G. GIBSON, 'Cider mill, or mill for pulverizing the tin ore in ancient times', *Western Antiquary*, 1885, **4**, 238

C. S. GILBERT, *An Historical Survey of the County of Cornwall*, 1817, Plymouth & London

D. GILBERT, *The Parochial History of Cornwall*, 1838, London

A. GILMAN, 'Bronze Age dynamics in south-east Spain', *Dialectical Anthropology*, 1976, **1**, 307–19

M. GIMBUTAS, *Bronze Age Cultures in Central and Eastern Europe*, 1965, Mouton, The Hague

M. GIMBUTAS, *The Slavs*, 1971, Thames & Hudson

P. R. GIOT, J. BRIARD & L. PAGE, *Protohistoire de la Bretagne*, 1979, Rennes

R. GIRSHMAN, *Iran from the earliest times to the Islamic Conquest*, 1978 ed., Pelican, London

P. V. GLOB, *The Mound People: Danish Bronze-Age Man Preserved*, 1983, Paladin Books, London

R. GOODBURN, *The Roman Villa, Chedworth*, 1972, The National Trust

A. J. J. GOODE, 'The mode of intrusion of Cornish elvans', *Inst. Geol. Sciences Report* 73/7, 1973, HMSO

J. A. GORIS, *Étude sur les Colonies Marchandes Méridionales (Portugais, Espagnoles, Italiens) à Anvers de 1488 à 1567*, 1925, Louvain

G. GOSSÉ, 'Las minas y el arte minero de España en la antigüedad', *Ampurias*, 1942, **4**, 43–68

J. W. GOUGH, *The Mines of Mendip*, 1967 ed., David & Charles

M. DE GOUVENAIN, 'Sur la dissémination de l'étain et sur la présence du cobalt et de diverses substances dans les kaolins des Colettes et d'Echassières situés dans le département de l'Allier', *Comptes-Rendu Acad. Sci.*, 1874, **78**, 1032–4

J. E. GOVER, *The Place Names of Cornwall*, 1948, unpublished MS at the County Museum, Truro

W. GOWLAND, 'The early metallurgy of copper, tin, and iron in Europe as illustrated by ancient remains', *Archaeologia*, 1899, **56**, 267–322

W. GOWLAND, 'The metals in antiquity', *J. Royal Anthropological Inst.*, 1912, **42**, 235–87

F. GRAHAM, *Speculi Britanniae Pars. A Topographical & Historical description of Cornwall (by John Norden c. 1584)*, 1966, Newcastle-upon-Tyne

J. GRAHAM-CAMPBELL, *The Viking World*, 1979, Book Club Associates

J. GRAHAM-CAMPBELL & D. KIDD, *The Vikings*, 1980, British Museum

M. GRANT, *The Etruscans*, 1980, Weidenfeld & Nicholson

H. ST G. GRAY, 'Roman pewter, Meare', *Proc. Som. Archaeol. & Nat. Hist. Soc.*, 1929, **75**, 105–6

H. S. GREEN, A. H. V. SMITH, B. R. YOUNG & R. K. HARRISON, 'The Caergwrle bowl: its composition, geological source and archaeological significance', *Inst. Geol. Sciences Report*, 1980/81, 26–30

R. P GREG & W. G. LETTSOM, *Manual of Mineralogy of Great Britain and Ireland*, 1858, John van Voorst, London

W. GREGOR, (1816), 'Observations on a remarkable change which metallic tin undergoes, under peculiar circumstances, and on its partial conversion to muriate of tin', *TRGSC*, 1818, **1**, 51–9

J. S. GREGORY, *Russian Land: Soviet People*, 1968, Harrap

M. GREGORY, 'The St Erth valley, with particular reference to alluvial working during the recent war', *Trans. Cornish Inst. Engineers*, 1947, **2**, (2), 19–24

R. GRIFFITHS, *Report on the metallic mines of the Province of Leinster*, 1828, Royal Dublin Society

W. F. GRIMES, *The Prehistory of Wales*, 1951, National Museum of Wales, Cardiff

M. GUEDRAS, 'Discovery of tin in France', *Mining Journal*, 1904, (21 May)

M. GUIDO, *Sardinia*, 1963, Thames & Hudson

H. GÜNTHER-BUCHHOLZ, 'Erzhandel in zweten vorchristlichen Jahrstansend', *Prähistorische Zeitschrift*, 1959, **37**, 1–40

O. R. GURNEY, *The Hittites*, 1952, Pelican, London

A. GUTHRIE, 'Excavations of a settlement at Goldherring, Sancreed', *CA*, 1969, **8**, 5–39

R. HALL, *The Excavations at York: The Viking Dig*, 1984, The Bodley Head

F. HAMPL & H. KERCHLER, 'Das mittlebronzezeitliche Gräberfeld von Pitten in Niederöstreich', *Mitteilungen der Prähistorischen Kommission der Österreichischen Akademie der Wissenschaften*, 1981, **19, 20**

J. H. HAMPTON, 'Tin deposits of the state of Perak, Straits settlements', *Trans. Mining Assoc. & Inst. Cornwall*, 1887, 143–52

D. HARDEN, *The Phoenicians*, 1971, Penguin, London

P. A. HARDIMAN, M. S. ROLFE & I. C. WHITE, 'Lithothamnium studies off the south west coast of England', 1976, Shellfish & Benthos Committee, Fisheries Laboratory, Burnham-on-Crouch, Essex (duplicated type-script)

A. F. HARDING, 'Mycenaean Greece and Europe: the evidence of bronze tools and implements', *PPS*, 1975, **41**, 183–202

A. F. HARDING, 'Radiocarbon calibration and the chronology of the European Bronze Age', *Archeologické Rozhledy*, 1980, **32**, 178–86

A. F. HARDING & H. HUGHES-BROCK, 'Amber in the Mycenaean world', *Annals British School Athens*, 1974, **69**, 146–72

D. HARRIS, S. PEARCE, H. MILES & M. IRWIN, 'Bodwen, Lanlivery: a multi-period occupation', *CA*, 1977, **16**, 43–59

J. R. HARRIS, *Lexicographical Studies in Ancient Egyptian Minerals*, 1961, Deutsche Akademie de Wissenschaften zu Berlin Institut fur Orientforshung

R. J. HARRISON, 'A reconsideration of the Iberian background to Beaker metallurgy', *Palaeohistoria*, 1974, **16**, 63–105

A. HARTMANN, *Prähistorische Goldfunde aus Europa*, 1970, Berlin, Studien zu den Anfängen der Metallurgie, **3**

J. HATCHER, *English Tin Production and Trade before 1550*, 1973, Clarendon Press

E. DE HAUTPICK, 'Native tin and its origins', *Mining Journal*, 1912, 350 (this deals only with Russia)

F. HAVERFIELD, 'Roman inscriptions in Britain 1890–91', *Arch. J.*, 1894, **49**, 177–201 (read 1892)

F. HAVERFIELD, 'An inscribed Roman ingot of Cornish tin and Roman tin-mining in Cornwall', *Proc. Soc. Antiquaries*, 1900, (2nd Ser.), **18**, 117–23

F. HAVERFIELD, 'On a hoard of Roman coins found at Carhayes, Cornwall', *Numismatic Chronicle*, (3rd Ser.), 1900a, **20**, 209–17

F. HAVERFIELD, *Romano-British Cornwall*, 1924, part 5 of The Victoria County History, London

C. F. C. HAWKES, *The Prehistoric Foundations of Europe*, 1940, Methuen

C. F. C. HAWKES, 'Double axe testimonies', *Antiquity*, 1973, **47**, 206–12

C. F. C. HAWKES, *Pytheas: Europe and the Greek Explorers*, 8th J. L. Myres Memorial Lecture, 20 May 1975, Oxford

C. F. C. HAWKES & M. A. SMITH, 'On some buckets and cauldrons of the Bronze and Early Iron Ages', *Ant. J.*, 1957, **37**, 131–98

J. HAWKES, *The First Civilizations*, 1977 ed., Pelican, London

H. HAWKINS, *Through Cornwall with a Camera*, 1896, London

J. HAWKINS, 'On the intercourse which subsisted between Cornwall and the commercial states of Antiquity and on the state of the tin-trade during the Middle Ages', *TRGSC*, 1824, **3**, 113–35

J. HAWKINS, 'On the state of our tin mines at different periods, until the commencement of the eighteenth century', *TRGSC*, 1832, **4**, 70–94

J. F. HEALY, *Mining and Metallurgy in the Greek and Roman World*, 1978, Thames & Hudson

R. HEATH, *A Natural and Historical Account of the Islands of Scilly*, 1750, London

E. S. HEDGES, *Tin in Social and Economic History*, 1964, Edward Arnold

H. O'NEILL HENCKEN, *The Archaeology of Cornwall and Scilly*, 1932, Methuen

H. O'NEILL HENCKEN, 'Beitzsch and Knossos', *PPS*, 1952, **18**, 36–46

C. HENDERSON, MS *Antiquities* (of Cornwall), vol. 3 'Pydar and Powder Hundreds', 1916–17, at County Museum, Truro

J. S. HENDERSON, 'Notes on a discovery of "Jews' House" tin near Penryn in April 1913', *JRIC*, 1915, **20**, 91–2

G. HENWOOD, *Four Lectures of Geology and Mining*, 1855, Mining Journal, London

G. HENWOOD, 'Relics of ancient tinners', *Mining Journal*, 1860 (14 April)

W. J. HENWOOD, (1828), 'On some of the deposits of stream tin-ore in Cornwall, with remarks on the theory of that formation', *TRGSC*, 1832, **4**, 57–69

W. J. HENWOOD, 'On the metalliferous deposits of Cornwall and Devon', *TRGSC*, 1843, **5**, 1–386

W. J. HENWOOD, (1873), 'On the detrital tin-ore of Cornwall', *JRIC*, 1874, 191–254

M. HERITY, 'Early finds of Irish antiquities', *Ant. J.*, 1969, **49**, 21–54

G. HERMANN, 'Lapis lazuli: the early phases of its trade', *Iraq*, 1968, **30**, 21–54

F. & E. HESS, *Bibliography of the Geology and Mineralogy of Tin*, 1912, Smithsonian Institution

F. M. HEXT, *Memorials of Lostwithiel*, 1891, Truro

A. HEYWORTH & C. KIDSON, 'Sea-level changes in south-west England and Wales', *Proc. Geologists Assoc.*, 1982, **93**, 91–111

C. HIGHAM, 'Excavations at Ban Nadi, North-east Thailand, 1980–1981', *South-East Asian Studies Newsletter*, No. 5, 1981, 1–2

C. HIGHAM, 'Ban Nadi, North-east Thailand: further radiocarbon dates', *South-East Asian Studies Newsletter*, No. 7, 1982, 1–2

J. B. HILL, D. A. MACALISTER & J. S. FLETT, *Geology of Falmouth and Truro and of the Mining District of Camborne and Redruth*, 1906, HMSO

F. C. HIRST, 'Excavations at Porthmeor, Cornwall, 1933, 1934 and 1935', *JRIC*, 1937, **24**, Appendix II, 1–81

F. HITCHINS & S. DREW, *The History of Cornwall*, 1824, Helston

R. C. HOARE, *Ancient History of North and South Wiltshire*, 1812, London

R. HODGES, *The Hamwih Pottery*, 1981, Southampton

Archaeological Research Committee Report 2/CBA Research Report 37

R. HODGES, *Dark Age Economics: the origins of towns and trade AD 600–1000*, 1982, Duckworth, London

T. HOGG, *A Manual of Mineralogy*, 1825, Truro

B. W. HOLMAN, 'Heat-treatment as an agent in rock-breaking', *Trans. Inst. Min. Metall.* 1927, **36**, 219–44, 255–6, 259–62

H. C. & L. H. HOOVER, *Georgius Agricola, De Re Metallica* (1556), 1950, Dover Publ., New York (reprint of the translation first published by the *Mining Magazine*, London, 1912)

K. F. G. HOSKING, 'Oxidation phenomena in Cornish lodes', *TRGSC*, 1950, **18**, 120–45

K. F. G. HOSKING, 'Permo-carboniferous and later primary mineralisation of Cornwall and south-west Devon', in Hosking & Shrimpton (eds.), *Present Views of Some Aspects of the Geology of Cornwall and Devon*, 1964, Royal Geol. Soc. Cornwall, 201–45

K. F. G. HOSKING, *The Search for Tin Deposits*, 1974, Fourth World Conference on Tin, ITC, London

K. F. G. HOSKING, 'The native tin story', *Warta Geolog.* (formerly the *Geol. Soc. Malaysia Newsletter*) 1974a, **50**, 6–11

K. F. G. HOSKING, S. M. NAIK, R. G. BURN & P. ONG, 'A study of the distribution of tin, tungsten, arsenic and copper in the sediments, and of the total-heavy-metal in the water of the Menalhyl River, mid-Cornwall', *Camborne Sch. Mines Mag.*, 1962, **62**, 49–59

K. F. G. HOSKING & R. OBIAL, 'A preliminary study of the distribution of certain metals of economic interest in the sediments and waters of the Carrick Roads (West Cornwall) and of its feeder rivers', *Camborne Sch. Mines Mag.*, 1966, **66**, 17–37

G. F. HOURANI, *Arab Seafaring in the Indian Ocean in Ancient and Early Medieval Times*, 1951, Princeton University Press

T. N. HUFFMAN, 'Ancient mining and Zimbabwe', *J. South African Inst. Mining & Metallurgy*, 1974, **74**, 238–42

R. HUNT (ed.), *Ure's Dictionary of Arts, Manufactures and Mines*, 1875, 7th ed., London (3 vols., Vol. 4, *Supplement*, 1879)

R. HUNT, *British Mining*, 1887, Crosby Lockwood & Co, London

G. HUTCHINSON, 'The bar-lug pottery of Cornwall', *CA*, 1979, **18**, 81–104

W. IAGO, 'On some recent archaeological discoveries in Cornwall', *JRIC*, 1890, **10**, 185–262

IMM, *The Future of Non-Ferrous Mining in Great Britain and Ireland*, 1959, Inst. Mining & Metallurgy, Proceedings of a Symposium held 23–24 Sept. 1958

S. A. IMMERWAHR, 'The use of tin on Mycenaean vases', *Hesperia*, 1966, **35**, 381–96

ITC, *Statistical Yearbook, 1964*, 1964, ITC, London (a similar volume, differing only in recent statistics, was published in 1968)

ITC, *Tin Statistics, 1964–1974*, 1974, ITC, London

J. S. JACKSON, 'Bronze Age copper mines in Mount Gabriel, west County Cork, Ireland', *Archaeol. Austriaca*, 1968, **43**, 92–114

J. S. JACKSON, 'Metallic ores in Irish prehistory: copper and tin', in M. Ryan (ed.), 1978

J. S. JACKSON, 'Copper mining at Mount Gabriel, Co. Cork: Bronze Age bonanza or post-famine fiasco? A reply', *PPS*, 1984, **50**, 375–7

C. C. JAMES, 'Great Wheal Vor', *TRGSC*, 1945, **17**, 194–207

C. C. JAMES, *A History of the Parish of Gwennap*, 1949, Penzance

H. JAMES, *Note on the Block of Tin Dredged up in Falmouth Harbour*, 1863, Edward Stanford, London (an edition of 1872, with minor differences, was reprinted from *Arch. J.*, 1871, **28**, 196–203

J. JANECKA & M. STEMPROK, *Endogenous Tin Mineralization in the Bohemian Massif*, 1967, ITC, London

J. J. JANSSEN, *Commodity Prices from the Ramessid Period*, 1975, E. J. Brill, Leiden

A. K. H. JENKIN, *Cornwall and the Cornish*, 1933, Dent

A. K. H. JENKIN, *Mines and Miners of Cornwall*, 1961–70, Truro Bookshop, vols. 1–14, Federation of Old Cornwall Societies vols. 15–16; vols. referred to in this book are: 1962, **4**; 1963, **7**; 1964, **8**; 1964, **9**; 1967, **13**

J. JENSEN, *The Prehistory of Denmark*, 1982, Methuen

P. S. DE JESUS, 'Considerations on the occurrence and exploitation of tin sources in the ancient Near East', in Franklin *et al.* (eds.), 1978

M. JOLEAUD, 'L'ancienneté de d'exploitation de l'étain dans le nord-ouest de l'Espagne', *Anthropologie*, 1929, **39**, 134–6

G. JONES, *A History of the Vikings*, 1975 ed., Book Club Associates

W. R. JONES, *Tinfields of the World*, 1925, Mining Publications Ltd, London

B. JOVANOVIĆ & B. S. OTTAWAY, 'Copper mining and metallurgy in the Vinča group', *Antiquity*, 1976, **50**, 104–13

K. KANEKO (ed.), *Geology and Mineral Resources of Japan*, 1960 (2nd ed.), Geol. Surv. Japan

V. KARAGEORGHIS, *Salamis, Recent Discoveries in Cyprus*, 1969, McGraw-Hill

A. KEITH, *The Antiquity of Man*, 1915, Williams & Norgate

F. KELLER, *The Lake Dwellings of Switzerland and other parts of Europe*, 1878, Longmans, Green & Co

J. E. KIDDER, *Japan before Buddhism*, 1959, Thames & Hudson

J. KIM, 'Bronze artifacts in Korea and their cultural-historical significance', in R. J. Pearson (ed.), *The Traditional Culture and Society of Korea: Prehistory*, 1975, Centre of Korean Studies, University of Honolulu, 130–97

G. H. KINAHAN, 'Irish metal mining', *Proc. Royal Dublin Soc.*, 1886, **5**, 'Tin', 207

M. H. KLAPROTH, *Observations relative to the Mineralogical and Chemical History of the Fossils of Cornwall*, 1787, London

P. L. KOHL (ed.), *The Bronze Age Civilization of Central Asia*, 1981, New York

S. N. KORENEVSKY, 'Chemical composition of the bronze artifacts of the Tly burial ground', *Soviet Archaeology*, 1981, (3), 148–62 (in Russian with English summary)

U. S. KÜSSEL, 'Extractive metallurgy in Iron Age South Africa', *J. S. African Inst. Min. Metall.*, 1974, **74**, 246–9

A. LACROIX, *Minéralogie de la France et de ses Colonies*, 1901, 217–35, Paris (5 vols.)

L. R. LAING, *Coins and Archaeology*, 1969, Weidenfeld & Nicholson

B. B. LAL, 'The copper hoard culture of the Ganga Valley', *Antiquity*, 1972, **44**, 282–7

W. LAMB, *Excavations at Thermi in Lesbos*, 1936, CUP

C. C. LAMBERG-KARLOVSKY. 'Archaeology and metallurgy in prehistoric Afghanistan, India and Pakistan', *American Anthropologist*, 1967, **69**, 145–62

C. C. LAMBERG-KARLOVSKY, 'The Proto-Elamites on the Iranian plateau', *Antiquity*, 1978, **52**, 114–20

C. C. & M. LAMBERG-KARLOVSKY, 'An early city in Iran', *Scientific American*, 1971, **224**, No. 6, 102–11

G. W. LAMPLUGH, *The Geology of the Isle of Man*, 1900, HMSO

R. LATHAM, *The Travels of Marco Polo*, 1958, Penguin, London

L. DE LAUNEY, 'Une mine de lithine en France', *La Nature* (2nd Ser.), 1901, **29**, 43–4

E. T. LEEDS, 'Excavations at Chûn Castle in Penwith, Cornwall', *Archaeologia*, 1927, **76**, 205–37

E. T. LEEDS, 'A bronze cauldron from the River Cherwell, Oxfordshire, with notes on cauldrons and other bronze vessels of allied types', *Archaeologia*, 1930, **80**, 1–36

C. V. LE GRICE, 'Notice of an ancient smelting place of tin, generally called a Jews' House, lately discovered on the estate of Trereife, near Penzance', *TRGSC*, 1846, **6**, 43–6

A. G. G. LEONARD & P. F. WHELAN, (1928), 'Spectrographic analyses of Irish ring money and of a metallic alloy found in commercial calcium carbide', *Scientific Proc. Royal Dublin Soc.*, 1929, **19**, 55–62

G. R. LEWIS, *The Stannaries. A Study of the Medieval Tin Miners of Cornwall and Devon*, 1908, Harvard University (reprinted 1965, D. B. Barton, Truro)

H. G. LIDDEL & R. SCOTT, *A Greek-English Lexicon*, 1864, Oxford

F. DE LIMUR, 'Recherches sur les gisements probables des matières constitutives de certains objets en pierre trouvés dans les monuments mégalithiques du Morbihan', *Bull. Soc. Polymathiques du Morbihan*, 1893, **71**, esp. 81–7

G. LIPSCOMB, *A Journey into Cornwall*, 1799, Warwick

B. A. LITVINSKII, 'Towards a history of tin-mining in Uzbekistan', *Proc. Middle Asian State University*, 1950, **11**, 56–68 (in Russian)

A. LUCAS, 'Notes on the early history of tin and bronze', *J. Egyptian Archaeol.*, 1928, **14**, 97–108

A. LUCAS, *Ancient Egyptian Materials and Industries*, 1948, London

B. LYONNET, 'Découverte des sites de l'âge du bronze dans le N.E. de l'Afghanistan: leurs rapports avec la civilisation de l'Indus', *Annali Instituto Orientali di Napoli*, 1977, **37**, 19–35

D. & S. LYSONS, *Magna Britannia*, 1814, **3** *Cornwall*, London

S. MACADAM & J. A. SMITH, 'Notice of bronze celts or axe heads, which have apparently been tinned, also of bronze weapons and armlets found along with portions of metallic tin near Elgin in 1868', *Proc. Soc. Ant. Scotland*, 1872, **9**, 428–43

J. MACLEAN, *Deanery of Trigg Minor*, 1868, London and Bodmin

J. MACLEAN, 'The tin trade of Cornwall in the reigns of Elizabeth and James compared with that of Edward I', *JRIC*, No. 15, 1874 (Apr.), 187–90

E. F. MACNAMARA, 'A group of bronzes from Surbo: new evidence for Aegean contacts with Apulia during Mycenaean IIIB and C', *PPS*, 1970, **36**, 241–60

E. F. MACNAMARA, 'A note on the Aegean sword-hilt in Truro Museum', *CA*, 1973, **12**, 19–23

J. MACTEAR, *Trans. Inst. Min. & Metall.*, 1895, **3**, 2–39. The story is also quoted in O. G. S. Crawford, 'Iranian tin', *Antiquity*, 1940, **14**, 195–7

I. MCINNES, 'Jet sliders in late Neolithic Britain', in Coles & Simpson (eds.), *Studies in Ancient Europe, Essays presented to Stuart Piggott*, 1968, Leicester University Press, 133–40

R. MADDIN, T. S. WHEELER & J. MUHLY, 'Tin in the ancient Near East: old questions and new finds', *Expedition*, 1977, **19**, 35–47

A. MAJENDIE, 'A sketch of the geology of the Lizard district', *TRGSC*, 1818, **1**, 32–7

J. MAJER, *The [tin] Mines in Woods in the Bohemian-Saxon Border Region in the 16th and early 17th centuries*, 1965, Narodni Technicke Muzeum v Praze, Praha (in Czech with English summary)

J. MAJER, 'Der Zinnerzbergbau im sachsisch-bohmischen Grenzgebiet des westlichen Erzgebirges wahrend des 16. und fruhen 17. Jahrhunderts,' *Abhandlungen des Staatlichen Museums für Mineralogie und Geologie zu Dresden*, Leipzig, 1980, **30**, 153–213

P. H. MALAN, 'Notes on the neighbourhood of Brown Willy', *JRIC*, 1888, **9**, 314–52

E. MALLARD, 'Note sur les gisements stannifères du Limousin et de la Marche et sur quelques anciennes fouilles qui paraissent s'y rattacher', *Annales des Mines* (6th Ser.), 1866, **10**, 312–52

M. MANÈS, 'Mémoire sur les mines d'étain de Saxe', *Annales des Mines* (3rd Ser.), 1823–4, **8**, 499–595, 837–

86; **9,** 281–304, 463–76, 625–56 (Fire-setting is dealt with particularly in **9,** 292–4, and tin streaming in **9,** 653–6)

V. MARKOTIC (ed.), *Ancient Europe and the Mediterranean,* 1977, Arris & Phillips Ltd

L. MARSILLE, 'Les dépôts de l'Age du Bronze dans le Morbihan', *Bull. Soc. Polymathique du Morbihan,* 1913, 49–109

R. J. MASON, 'Background to the Transvaal Iron Age', *J. S. African Inst. Min. Metall.,* 1974, **74,** 211–16

H. J. L. J. MASSÉ, *A Short History and Description of the Church and Abbey of Mont S. Michel,* 1902, London

V. A. MASSON & V. I. SARIANIDI, *Central Asia: Turkmenia before the Achaemenids,* 1972, Thames & Hudson

P. M. MATERIKOV, 'Tin', chapter in vol. 3 of V. I. Smirnov (ed.), *Ores of the Soviet Union,* 1977, Pitman, 229–94

W. G. MATON, *Observations of the Western Counties of England in 1794 and 1796* (vol. 1), in Chope, 1967 ed., 233–78

I. S. MAXWELL, 'The location of Ictis', *JRIC* (New Ser.), 1972, **4,** 293–319

J. V. S. MEGAW, 'The Vix burial', *Antiquity,* 1966, **40,** 38–44

J. MELLAART, *The Archaeology of Ancient Turkey,* 1978, London

R. E. H. MELLOR, *Geography of the USSR,* 1966, Macmillan

R. J. MERCER, 'Metal arrow-heads of the European Bronze and Early Iron Ages', *PPS,* 1970, **36,** 171–213

J. R. MERRIFIELD, 'Modern carbonate marine sands in estuaries of south-west England', *Geol. Mag.,* 1982, **119,** 567–80

N. J. VAN DE MERWA, 'The advent of iron in Africa', in Wertime & Muhly (eds.), *The Coming of the Age of Iron,* 1980, Yale University Press

D. M. METCALF & D. R. WALKER, 'Tin as a minor constituent in two scealtas from the Shakenoak excavations', *Numismatic Chronicle,* 1976, **136,** 228–9

H. MILES, 'Barrows on the St Austell granite, Cornwall', *CA,* 1975, **14,** 5–81

H. & T. MILES, 'Excavations at Trethurgy, St Austell: interim report', *CA,* 1973, **12,** 25–9

A. MILLS, T. KING & J. WEAVER, 'Report of the gold mines in the County of Wicklow; with observations thereon, by R. Kirwan Esq.' *Trans. Dublin Soc.,* 1801, **2,** 131–48

A. L. MONGAIT, *Archaeology in the USSR,* 1961 ed., Penguin, London

A. M. T. MOORE, 'A pre-Neolithic farmers' village on the Euphrates', *Scientific American,* 1979, **241,** No. 2, 50–8

P. R. S. MOOREY, 'Luristan bronzes', *Archaeometry,* 1964, **7,** 72–80

P. R. S. MOOREY, *Catalogue of the Ancient Persian Bronzes in the Ashmolean Museum,* 1971, Oxford

P. R. S. MOOREY & F. SCHWEIZER, 'Copper and copper alloys in ancient Iraq, Syria, and Palestine: some new analyses', *Archaeometry,* 1972, **14,** 177–98

P. R. S. MOOREY & F. SCHWEIZER, 'Copper and copper alloys in ancient Turkey: some new analyses', *Archaeometry,* 1974, **16,** 112–15

C. E. MORE, 'Some observations on "ancient" mining at Phalaborwa', *J. S. African Inst. Min. & Metall.,* 1974, **74,** 227–32

C. MORRIS (ed.), *The Journey of Celia Fiennes,* 1949, London

T. A. MORRISON, *Cornwall's Central Mines: The Northern District,* 1980, Alison Hodge, Penzance

K. MUCKLEROY, C. HAZELGROVE & D. NASH, 'A pre-Roman coin from Canterbury and the ship represented on it', *PPS,* 1978, **44,** 439–44

J. D. MUHLY, *Copper and Tin: the distribution of mineral resources and the nature of the metals trade in the Bronze Age,* 1973, Archon Books, Hamden, Connecticut (reprint from *Trans. Connecticut Academy Arts & Sciences,* **43,** 155–535, with a *Supplement to Copper and Tin,* 1976, **46,** 77–136

J. D. MUHLY, 'Tin trade routes of the Bronze Age', *Scientific American,* 1973a, **61,** 404–13

J. D. MUHLY, 'New evidence for sources of and trade in Bronze Age tin', in Franklin *et al.* (eds.), 1978, 43–8

H. S. MUNROE, 'The mineral wealth of Japan', *American Inst. Min. Engineers,* 1876, **5,** 236–302 (esp. 298–9, 'Tin')

S. M. NELSON, 'Recent progress in Korean archaeology', in Wendorf and Close (eds.), *Advances in World Archaeology,* 1982, **1,** Academic Press

E. & J. NEUSTUPNY, *Czechoslovakia before the Slavs,* 1961, Thames & Hudson

R. J. NOALL (1929), 'Bussow Bronze Age village and its last inhabitants', *CA,* 1971, **10,** 29–31

H. O'BRIEN, *Phoenician Ireland,* 1833, Dublin

C. H. OLDFATHER, *Diodorus of Sicily,* 1939, Loeb Classical Library

B. H. ST J. O'NEIL, 'A Romano-British hut on St Martin's, *Scillonian Magazine,* 1949, **24,** 163–4

S. P. O'RIORDAIN, 'The halberd in Bronze Age Europe', *Archaeologia,* 1937, **86,** 195–321

P. ORLANDI & P. B. SCORTECCI, 'Minerals of the Elba pegmatites', *Mineralogical Record,* 1985, **16,** 353–63

K. S. PAINTER, *The Water Newton Early Christian Silver,* 1977, British Museum

H. PARSONS, 'The Dartmoor blowing houses', *Trans. Devon. Assoc.,* 1956, **88,** 189–96

S. R. PATTISON, 'On some post-tertiary deposits in Cornwall', *TRGSC,* 1847, **7,** 34–6

S. R. PATTISON, 'On a celt-mould found at Altarnun', *RRIC,* 1849, No. 31, 57–8

S. R. PATTISON, 'On ancient and modern tin-works in France', *JRIC,* 1867, **2,** 343–5 (taken from Mallard, 1866)

O. PAVEL, *The Birth of Greek Civilization,* 1981, London

J. C. PAYNE, 'Lapis lazuli in early Egypt', *Iraq,* 1968, **30,** 31–4

C. A. PEAL, 'Romano-British plates and dishes', *Proc. Cambridge Antiquarian Soc.,* 1967, **60,** 19–37

S. M. PEARCE & T. PADLEY, 'The Bronze Age finds from Tredavah, Penzance', *CA*, 1977, **16**, 234–7

R. A. PEEK & S. E. WARREN, in G. J. Wainwright, A. Fleming & K. Smith, 'The Shaugh Moor project', *PPS*, 1979, **45**, 1–33

W. PENALUNA, *An Historical Survey of the County of Cornwall*, 1838, Helston, 2 vols.

R. R. PENNINGTON, *Stannary Law. A History of the Mining Law of Cornwall and Devon*, 1973, David & Charles

T. C. PETER, 4th ed. of J. J. Daniell's *A Compendium of the History and Geography of Cornwall*, 1906, Truro and London

J. PHILLIPS, 'Thoughts on ancient metallurgy and mining in Brigantia and other parts of Britain, suggested by a page of Pliny's "Natural History" ', *Proc. Yorks. Phil. Soc.*, 1847, 77–92

J. A. PHILLIPS & H. LOUIS, *Treatise on Ore Deposits*, 1896, Macmillan

P. PHILLIPS, *The Prehistory of Europe*, 1981 ed., Pelican, London

I. R. PHIMSTER, 'Ancient mining near Great Zimbabwe', *J. S. African Inst. Min. Metall.*, 1974, **74**, 233–7

S. PIGGOTT, *Prehistoric India*, 1950, Penguin, London

S. PIGGOTT, *The Druids*, 1968, Thames & Hudson

S. PIGGOTT, 'Chariots in the Caucasus and in China', *Antiquity*, 1974, **48**, 16–24

S. PIGGOTT, 'A glance at Cornish tin', in V. Markotic (ed.), 1977, 141–5

S. PIGGOTT, *The Earliest Wheeled Transport from the Atlantic to the Caspian Sea*, 1983, Thames & Hudson

R. DE S. PINTO, 'Activité minière et métallurgique pendant l'âge du bronze en Portugal', *Anais de Faculdada de Ciencias do Porto*, 1933, 3–15

R. PITTIONI, 'Prehistoric copper mining in Austria: problems and facts', *University of London Inst. Archaeology*, 7th Annual Report, 1951, 16–43

R. POLWHELE, *The History of Cornwall*, 1803–8, Falmouth (reprinted by Kohler & Coombs, 1978)

R. S. POOLE, 'The Phoenicians and their trade with Britain', *JRIC*, 1865, **1**, No. 4, 1–10

J. R. POSS, *Stones of Destiny*, 1975, Houghton, Michigan Technical University

M. M. POSTAN, *The Medieval Economy and Society*, 1984 ed., Pelican, London

T. G. E. POWELL, *Prehistoric Art*, 1966, Thames & Hudson

P. PRICE, *Bells and Man*, 1983, OUP

W. PRYCE, *Mineralogia Cornubiensis*, 1778, London

F. PRYOR, *A Catalogue of the British and Irish Prehistoric Bronzes in the Royal Ontario Museum*, 1980, ROM

J. W. PYCROFT, *Arena Cornubiae*, 1856, London

H. RACKHAM (ed. *et al.*), *Pliny, Naturalis Historiae*, 1938–62, Loeb Classical Library, bilingual text

T. S. RAFFLES, 'Account of the Sunda Islands and Japan', *J. of Science*, 1817, **2**, 190–8

A. RAISTRICK, *The Hatchett Diary: A tour through the Counties of England and Scotland in 1796 visiting their mines and manufactories*, 1967, Truro

E. RANDSBORG, 'Aegean bronzes in a grave in Jutland', *Acta Archaeologia*, 1967, **38**, 1–27

P. RASHLEIGH, 'Account of antiquities discovered in Cornwall, 1774', *Archaeologia*, 1789, **9**, 187; 1794, **11**, 83

P. RASHLEIGH, *Specimens of British Minerals, selected from the cabinet of Philip Rashleigh of Menabilly*, 1797, vol. 1, 1802, vol. 2, London

T. J. REEVES, 'Gold in Ireland', *Bull. Geol. Surv. Ireland*, 1971, **1**, 73–85

C. REID, 'The island of Ictis', (read 1905), *Archaeologia*, 1906, **59**, 281–8

C. REID, *Submerged Forests*, 1913, CUP

C. REID & J. S. FLETT, *The Geology of the Land's End District*, 1907, HMSO

C. REID & J. B. SCRIVENOR, *The Geology of the Country near Newquay*, 1906, HMSO

C. REID & J. J. H. TEALL, *The Geology of Mevagissey*, 1907, HMSO

C. RENFREW, 'Wessex without Mycenae', *Annals British Sch. Archaeology Athens*, 1968, **63**, 277–85

C. RENFREW, 'The autonomy of the south-east European Copper Age', *PPS*, 1969, **35**, 12–47

C. RENFREW, 'Carbon-14 and the prehistory of Europe', *Scientific American*, 1971, **225**, No. 4, 63–72

D. M. RENNIE, 'A section through the Roman defences in Watermoor recreation ground, Cirencester', *Ant. J.*, 1957, **37**, 206–15

J. D. RIDGE, *Annotated Bibliographies of Mineral Deposits in the Western Hemisphere*, 1976, Geol. Soc. America, Memoir 313

E. V. RIEU, *Homer's The Iliad*, 1966 ed., Penguin, London

E. V. RIEU, *Homer's The Odyssey*, 1970 ed., Penguin, London

A. L. F. RIVET & C. SMITH, *The Place-Names of Roman Britain*, 1979, Batsford

W. L. ROBERTS, G. R. RAPP & J. WEBER, *Encyclopedia of Minerals*, 1974, Van Norstrand

F. R. RODD, 'Notice of a stone ladle, and of two granite troughs found at Berriow, near Trebartha, in the parish of Northill, Cornwall', *RRIC*, 1850, (32), 58

J. C. RODRÍGUEZ, 'La tecnología del bronce final en los talleres del noroeste Hispanica', *Studia Archaeologica*, 1977, **47**, 9–41

C. ROEDER, 'Prehistoric and subsequent mining at Alderley Edge, with a sketch of the archaeological features of the neighbourhood', *Trans. Lancashire & Cheshire Antiquarian Soc.*, 1901, **19**, 77–118

B. DE LA ROGERIE, 'Voyage de Mignot de Montigny en 1752', *Soc. Hist. Archaeol. Bretagne*, 1925, **6**, 225–301

H. C. ROGERS, 'Blocks of tin found in Fowey harbour', *JRIC*, 1903, **15**, 345–6

J. J. ROGERS, 'An account of the discovery of Roman fragments at Carminow, on the sea coast near Helston, in October 1860', *RRIC*, 1861, (1), 51–5

J. J. ROGERS, 'Notice of further ancient remains discovered at Carminow', *RRIC*, 1863, (45), 8–83

J. J. ROGERS, 'Saxon silver ornaments and coins found at Trewhiddle near St Austell, AD 1774', *JRIC*, 1867, **2**, 292–305

C. M. ROLKER, 'The production of tin in various parts of the world', *16th Annual Report United States Geol. Surv.*, 1894, 458–538

Y. ROMÉ, *Pénéstin d'Hier et d'Aujourd 'hui*, 1980, Lorient

R. C. A. ROTTLÄNDER, 'On the formation of amber from *Pinus* resin', *Archaeometry*, 1970, **12**, 35–51

C. T. LE ROUX, 'Informations Archéologiques (Bretagne)', *Gallia Préhistoire*, 1975, **18**, 524

J. ROWE, *Cornwall in the Age of the Industrial Revolution*, 1953, Liverpool University Press

S. ROWE, *A Perambulation of the Antient and Royal Forest of Dartmoor*, 1896, 3rd ed. revised by J. Brooking Rowe, London and Exeter

M. J. ROWLANDS, *The Organisation of Middle Bronze Age Metalworking*, 1976, British Archaeological Reports, No. 31, 2 vols.

A. L. ROWSE, *Tudor Cornwall*, 1941, Jonathan Cape

S. I. RUDENKO, *Frozen Tombs of Siberia; the Pazyryk Burials of Iron Age Horsemen*, 1970 ed., London

R. C. RUDOLPH, 'An important Dongson site in Yunnan', *Asian Perspective*, 1961, **4**, 41–54

J. RUFFLE, *Heritage of the Pharaohs: An introduction to Egyptian Archaeology*, 1977, Phaidon

S. RUNDLE, 'Cornubiana I', *JRIC*, 1892, **11**, 84–90

S. RUNDLE, 'Cornubiana III', *JRIC*, 1900, **14**, 69–84

C. W. RYAN, *A Guide to the Known minerals of Turkey*, 1957, Ankara

M. RYAN (ed.), *The Origins of Metallurgy in Atlantic Europe*, 1978, Dublin

A. H. SABET, V. CHANANENCO & V. TSOGOEV, 'Tin-tungsten and rare-metal mineralization in the central Eastern desert of Egypt', *Annals Geol. Surv. Egypt*, 1973, **3**, 75–86

L. SALWAY, *Gold and Gold Hunters*, 1978, Kestrel Books, Penguin, London

E. SANDARS, *A Beast Book for the Pocket*, 1937, OUP

N. K. SANDARS, *Bronze Age Cultures in France: the later phases from the thirteenth to the seventh century BC*, 1957, CUP

N. K. SANDARS, *The Sea Peoples: Warriors of the ancient Mediterranean 1250–1150 BC*, 1978, Thames & Hudson

A. D. SAUNDERS, 'Excavations at Castle Gotha, St Austell, Cornwall: interim report', *Proc. West Cornwall Field Club*, 1960–61, **2**, No. 5, 216–20

M. R. SAUTER, *Switzerland from the earliest times to the Roman Conquest*, 1976, Thames & Hudson

H. N. SAVORY, *Spain and Portugal: The prehistory of the Iberian peninsula*, 1968, Thames & Hudson

M. SAWADA, 'Non-destructive X-ray fluorescence analysis of ancient bronze mirrors excavated in Japan', *Ars Orientalis*, 1979, **11**, 195–213

E. SCAMUZZI, *Egyptian Art in the Egyptian Museum of Turin*, 1965, New York

H. M. SCARTH, 'Remarks on ancient chambered tumuli', *Proc. Som. Archaeol. & Nat. Hist. Soc.*, 1858, **8**, 35–62

H. SCHLIEMANN, *Troja: results of the latest researches and discoveries on the site of Homer's Troy*, 1882, London

W. H. SCHOFF, *The Periplus of the Erythraean Sea: Travel and Trade in the Indian Ocean by a Merchant of the First Century*, 1912, Longmans, Green & Co

M. SCHREIBER, 'Des substances minérales en France avant la Révolution', *Journal des Mines*, 1794, **1**, 55–92

R. D. SCHUILING, 'Tin belts around the Atlantic Ocean. Some aspects of the geochemistry of tin', *A Technical Conference on Tin*, 1967, ITC, London, **2**, 531–47

H. W. SEAGER, *Natural History in Shakespear's Time*, 1896, London

I. R. SELIMKHANOV, 'Spectral analyses of metal artifacts from archaeological monuments in the Caucasus', *PPS*, 1962, **28**, 69–79

I. R. SELIMKHANOV, 'Bronze et métaux du Caucase', in *Congrès International d'Histoire des Sciences*, 1968, **10**, (B), 87–90

I. R. SELIMKHANOV, 'Ancient tin objects of the Caucasus and the results of their analyses', in Franklin *et al.*, 1978, 53–8

A. DE SÉLINCOURT, *Herodotus, The Histories*, 1972 revised ed. by A. R. Burn, Penguin, London

A. M. B. SESTIERI, 'The metal industry of continental Italy, 13th–11th century, and its Aegean connections', *PPS*, 1973, **39**, 383–424

H. J. SEYMOUR, 'Cassiterite in the Tertiary granite from the Mourne Mountains, Co. Down', *Scientific Proc. Royal Dublin Soc.* (New Ser.), 1902, **19**, 583–4

T. SHAW, 'Spectographic analyses of the Igbo and other Nigerian bronzes', *Archaeometry*, 1965, **8**, 86–95

T. SHAW, *Nigeria: Its Archaeology and Early History*, 1978, London

C. A. SHELL, 'The early exploitation of tin deposits in south-west England', in M. Ryan (ed.), 1978, 251–63

A. SHERRATT (ed.), *The Cambridge Encyclopedia of Archaeology*, 1980, CUP

A. H. SHORTER, W. C. RAVENHILL & K. J. GREGORY, *Southwest England*, 1969, Nelson

N. SIHANOUK, *The Democratic People's Republic of Korea*, 1980, Foreign Language Publ. House, Pyongyang

I. SIMMONS & M. TOOLEY (eds.), *The Environment in British Prehistory*, Duckworth

G. SIMOENS, *The Gold and Tin in the South East of Ireland*, 1921, Dublin

L. SIMONIN, 'Sur l'ancienne exploitation des mines d'étain de la Bretagne', *Comptes Rendus Acad. Science*, 1866, 346–7

L. SIMONIN, *Mines and Miners, or Underground Life*, 1868, English ed. by H. W. Bristow, London

R. SLESSOR, 'Chinese non-ferrous metals', *Proc. Australian Inst. Min. Metall.*, 1927, **65**, 51–116

C. S. SMITH & M. T. GNUDI, *The Pirotechnia of Vannoccio*

Biringuccio, 1966, Massachusetts Inst. Technology (first issued in 1942 by the *American Inst. Min. Metall. Eng.*)

C. T. SMITH, *Historical Geography of Western Europe before 1800*, 1967, London

R. A. SMITH, *A Guide to Anglo-Saxon and Foreign Teutonic Antiquities*, 1923, British Museum

R. A. SMITH, *Guide to Early Iron Age Antiquities*, 1925, British Museum

W. W. SMYTH, 'On the mines of Wicklow and Wexford', *Records of the School of Mines*, 1853, London, **1**, 349–409

A. SNODGRASS, *Archaic Greece: the age of experiment*, 1980, Dent

J. SOUTHWOOD, 'Notes on the Nigerian tinfields', *Trans. Cornish Inst. Mining Mech. & Met. Engineers* (New Ser.), 1946, **2**, (1), 28–33

J. SOWERBY, *British Mineralogy*, 1804–6, London, (5 vols.)

K. SPINDLER & F. SCHWEINGRUBER, *Magdalenenberg VI*, 1980, Neckar Verlag, Villingen-Schwenningen

F. M. STENTON, *Anglo-Saxon England*, 1947, (2nd ed.) OUP

F. J. STEPHENS, 'Alluvial deposits in the lower portion of the Red River Valley, near Camborne', *TRGSC*, 1899, **12**, 324–35

F. J. STEPHENS, 'General notes on ancient mining in Cornwall', *RCPS* (New Ser.), 1928, **6**, 162–71

H. M. STOCKER, 'Account of some remains found in Pentuan streamwork and of the circumstances under which they were found', *PZNHAS*, 1852, **2**, 88–90

T. SULIMIRSKI, 'The Bronze Age in the USSR', *Bull. Inst. Archaeol.*, 1968, **7**, 43–83

T. SULIMIRSKI, *The Sarmatians*, 1970, Thames & Hudson

T. SULIMIRSKI, 'Late Bronze Age and Earliest Iron Age in Siberia', *Bull. Inst. Archaeol.*, 1975, **12**, 145–74

J. SWETE, 'A tour of Cornwall in 1780', *JRIC* (New Ser.), 1971, **6**, 185–219

R. SYMONS, 'Alluvium in Par Valley', *JRIC*, 1877, **5**, 382–4

R. SYMONS, *A Geographical Dictionary, or Gazetteer of the County of Cornwall*, 1884, Penzance

M. TANGYE, *Scilly 1801–1821 ... through war and peace*, 1970, Redruth

C. D. TAYLOR, 'Description of the tin stream works in Restronguet Creek near Truro', *Proc. Inst. Mechanical Eng. Birmingham*, 1873, 155–66

F. S. TAYLOR, 'The alchemical works of Stephanos of Alexandria', *Ambix*, 1937, **1**, 116–39

J. A. TAYLOR, 'Environmental changes in Wales during the Holocene period', in J. A. Taylor (ed.), *Culture and Environment in Prehistoric Wales*, 1980, British Archaeological Reports, British Series, No. 76, 101–30

J. TAYLOR, *Bronze Age Goldwork of the British Isles*, 1980, CUP

R. G. TAYLOR, *Geology of Tin Deposits*, 1979, Elsevier Scientific Publishing Co.

T. TAYLOR, 'The bronze bull of St Just', *The Sphere*, 31 Jan. 1925

M. A. TCHERINA, 'Direction des recherches archéologiques sous-marines', *Gallia Préhistoire*, 1969, **27**, 385–499

A. C. THOMAS, *Some Notes on the Folk-Lore of the Camborne Area*, 1950, Camborne

A. C. THOMAS, 'Evidence for post-Roman occupation of Chûn Castle, Cornwall', *Ant. J.*, 1956, **36**, 75–8

A. C. THOMAS, 'The character and origins of Roman Dumnonia', in A. C. Thomas (ed.), *Rural Settlement in Roman Britain*, 1966, CBA Research Report No. 7, 76–98

A. C. THOMAS, 'Roman objects from the Gwithian area', *CA*, 1972, **11**, 53–5

A. C. THOMAS, *Christianity in Roman Britain to AD 500*, 1981, Batsford

A. C. THOMAS, *A Provisional List of Imported Pottery in post-Roman Western Britain and Ireland*, 1981a, Inst. Cornish Studies Special Report No. 7

A. C. THOMAS, 'East and West: Tintagel, Mediterranean imports and the early Insular Church', in Pearce (ed.), *The Early Church in Western Britain and Ireland*, 1982, British Archaeological Reports, British Series No. 102, 17–34

A. C. THOMAS, *Exploration of a Drowned Landscape: Archaeology and History of the Isles of Scilly*, 1985, Batsford

A. P. THOMAS, 'Gurob', *Egyptology Today*, 1981, **1**, No. 5

C. THOMAS, 'Some Spanish tin deposits', *Trans. Mining Assoc. Cornwall*, 1887, **2**, 66–70

G. W. THOMAS, 'Irish mining in decline', *J. Camborne Sch. Mines*, 1983, **83**, 34–5

T. M. THOMAS, 'The mineral industries in Wales', *Proc. Geol. Assoc.*, 1972, **83**, (4), 365–83

R. L. THORP, 'Burial practices of Bronze Age China', in Wen Fong (ed.), *The Great Bronze Age of China: an Exhibition from the People's Republic of China*, 1980, Thames & Hudson

L. M. THREIPLAND, 'An excavation at St Mawgan-in-Pydar, north Cornwall', *Arch. J.*, 1965, **93**, 33–81

J. THURNAM, 'On ancient British barrows (part 2)', *Archaeologia*, 1871, **43**, 285–544

E. THURSTAN, *British and Foreign Trees and Shrubs in Cornwall*, 1930, CUP

W. TORBRÜGGE, *Prehistoric European Art*, 1968, Abrams, New York

M. TOSI & R. WARDAK, 'The Fullol hoard: a new find from Bronze Age Afghanistan', *East and West* (New Ser.), 1972, **22**, 9–17

J. M. C. TOYNBEE, *Art in Roman Britain*, 1962, London

J. M. C. TOYNBEE, *Animals in Roman Life and Art*, 1973, Thames & Hudson

W. TRAGHSLER, 'The influence of metalworking on prehistoric pottery: some observations on Iron Age pottery of the Alpine Region', and 'Precursors of polychrome painted pottery: some examples from the prehistory of Switzerland', in F. R. Matson (ed.), *Ceramics and Man*, 1966, Methuen, 140–51, 152–60

J. TROUTBECK, *A Survey of the Scilly Islands*, no date = 1796, Sherborne

D. H. TRUMP, *The Prehistory of the Mediterranean*, 1961 ed., Pelican, London

R. F. TYLECOTE, *Metallurgy in Archaeology*, 1962, Edward Arnold

R. F. TYLECOTE, 'Metallurgical examination of a socketed axe and three lumps of bronze from Gillan, St Anthony-in-Meneage', *CA*, 1967, **6**, 110–11

R. F. TYLECOTE, *A History of Metallurgy*, 1976, The Metals Society

R. F. TYLECOTE, 'Early tin ingots and tinstone from western Europe and the Mediterranean', in Franklin *et al.* (eds.), 1978, 49–52

R. F. TYLECOTE, M. S. BALMUTH & R. MASSOLI-NOVELLI, 'Copper and bronze metallurgy in Sardinia', *J. Hist. Metall. Soc.*, 1983, **17**, (2), 63–7

W. E. A. USSHER, G. BARROW, D. A. MACALISTER, with petrological notes by J. S. FLETT, *The Geology of the Country around Bodmin and St Austell*, 1909, HMSO

C. J. VEI, 'Mineral resources of China', *Economic Geology*, 1946, **41,** (2 Supplement)

P. VIGNERON, *Le Cheval dans l'antiquité Greco-Romain*, 1968, Nancy

J. VLADAR, 'Mediterranean influence in the North Carpathian Basin in the Early Bronze Age', *Preist. Alp.*, 1974, **10**, 219–36

G. A. WAINWRIGHT, 'The occurrence of tin and copper near Byblos', *J. Egyptian Archaeol.*, 1934, **20**, 29–32

W. H. WALDREN, 'A Beaker workshop area in the rock shelter of Son Matge, Mallorca', *World Archaeol.*, 1979, **11**, 43–67

T. J. WALKER, *Whale Primer: with special reference to the California gray whale*, 1962, Cabrillo Historical Assoc.

C-M WANG, 'The bronze culture of ancient Yunnan', *Peking Review*, 1960 (12 Jan.), **2,** 18–19

K. P. WANG, 'China', *Mining Annual Review*, 1971, United States Bureau of Mines, 400

R. B. WARNER, 'The Carnanton ingot', *CA*, 1967, **6**, 29–31

W. WATSON, *The Genius of China*, 1973, Times Newspapers Ltd

W. WATSON, *Ancient China*, 1974, BBC Publications

A. WAY, *Catalogue of Antiquities*, 1847, Soc. Antiq., London

A. WAY, 'Enumeration of blocks or pigs of lead and tin, relics of Roman metallurgy, discovered in Great Britain', *Arch. J.*, 1859, **16**, 22–40

T. WEAVER, 'Memoir of the geological relation of the East of Ireland', *Trans. Geol. Soc. London*, 1821, **5,** 117–304

P. S. WELLS, 'Late Hallstatt interactions with the Mediterranean: one suggestion', in V. Markotic (ed.), 1977, 189–96

T. A. WERTIME, 'The search for ancient tin: the geographic and historic boundaries', in Franklin *et al.* (eds.), 1978, 1–6

T. A. WERTIME & J. D. MUHLY (eds.), *The Coming of the Age of Iron*, 1980, Yale University Press

C. J. V. WHEATLEY, 'Aspects of metallogenesis within the Southern Caledonides of Great Britain and Ireland', *Trans. Inst. Min. Metall.*, 1971, **80**, (B), 211–23

P. WHEATLEY, *The Golden Khersonese: Studies in the Historical Geography of the Malay Peninsula before AD 1500*, 1961, University of Malaysia Press

P. WHEATLEY, 'The development of long-distance trade to and through southeast Asia', in J. A. Sabloff and C. C. Lamberg-Karlovsky (eds.), *Ancient Civilization and Trade*, 1975, University of New Mexico, 231–6

R. E. M. WHEELER, *The Indus Civilization*, 1953, CUP

R. E. M. & T. V. WHEELER, 'Report on the excavation of the prehistoric, Roman, and post-Roman site in Lydney Park, Gloucestershire', *Soc. Antiquaries Research Report*, No. 9, 1932

T. S. WHEELER, R. MADDIN & J. D. MUHLY, 'Ingots and the Bronze Age copper trade in the Mediterranean: a progress report', *Expedition*, 1975, **17**, 31–9

D. G. WHITLEY, 'On the occurrence of trees and vegetable remains in the stream tin of Cornwall', *TRGSC*, 1908, **13**, 237–56

D. G. WHITLEY, 'The Ictis of Diodorus in the light of modern theories', *TRGSC*, 1915, **15,** 55–70

D. G. WHITLEY, 'Cornish Quaternary deposits in the light of Siberian alluvial formations', *TRGSC*, 1917, **15**, 143–60

H. M. WHITLEY, 'The silting up of the creeks of Falmouth Haven', *JRIC*, 1881, **7,** 12–17

N. WHITLEY, 'Exhibition of antiquities from the Royal Institution of Cornwall', *Proc. Soc. Antiquaries London*, 1873 (2nd Ser.), **5,** 454–55

N. WHITLEY, 'An attempt to define the extent and nature of the Roman occupation of Cornwall', *JRIC*, 1875, **5,** 199–205

W. WILLETTS, *Foundations of Chinese Art*, 1965, London

D. M. WILSON, *The Anglo Saxons*, 1971 ed., Pelican, London

D. M. WILSON, *Science and Archaeology*, 1978 ed., Pelican, London

D. M. WILSON & C. E. BLUNT, 'The Trewhiddle hoard', *Archaeologia*, 1961, **98**, 75–122

J. M. WINN, 'Notice of fossil bones found in Pentewan valley and of the stream works now working therein', *RRIC*, 1839, No. 21, 45–50

A. & P. WISEMAN, *Julius Caesar in the Battle for Gaul*, 1980, Chatto & Windus

F. WÖHLER, *W. Hisinger's Versuch einer mineralogischen von Schweden*, 1826, Leipzig

R. WOLFART & N. WITTERKINDT, *Geologie von Afghanistan*, 1980, Gebröder Borntraeger, Berlin

L. K. WONG, *The Malayan Tin Industry to 1914*, 1965, University of Arizona Press, Tucson

G. B. WORGAN, *General View of the Agriculture of the County of Cornwall*, 1811, London

R. H. WORTH, 'Stray notes on Dartmoor tin workings', *Trans. Devon. Assoc.*, 1914, **46,** 284–9

R. H. WORTH, *Dartmoor*, 1953, Plymouth (reprinted 1967, David & Charles)

R. N. WORTH, 'Some old ideas about tin', *Western Chronicle of Science* (ed. J. H. Collins), 1871, **1,** 51–2

R. N. WORTH, *Historical Notes concerning the Progress of Mining Skill in Devon and Cornwall*, 1872, Falmouth

R. N. WORTH, 'Ancient mining implements of Cornwall', *Archaeol. Journal*, 1873, (1874), **31,** 53–60

R. N. WORTH, 'Some Devonshire merchants' marks', *Trans. Devon. Assoc.*, 1891, **23,** 315–17

H. E. WULFF, *The Traditional Crafts of Persia*, 1966, Massachusetts Inst. Technology

K. ZSCHOCKE & E. PREUSCHEN, 'Das urzeitliche Bergbaugebeit von Mühlbach-Bischofshofen', *Materialien zur Urgeschichte Österreichs*, 1932, **6,** Vienna

Indexes

SUBJECT INDEX

TOPOGRAPHICAL INDEX